Water and Solute Permeability of Plant Cuticles

Lukas Schreiber • Jörg Schönherr

Water and Solute Permeability of Plant Cuticles

Measurement and Data Analysis

Springer

Professor Dr. Lukas Schreiber
Ecophysiology of Plants
Institute of Cellular
and Molecular Botany (IZMB)
University of Bonn
Kirschallee 1
53115 Bonn
Germany
lukas.schreiber@uni-bonn.de

Dr. Jörg Schönherr
Rübeland 6
29308 Winsen-Bannetze
Germany
joschoba@t-online.de

ISBN 978-3-540-68944-7 e-ISBN 978-3-540-68945-4

Library of Congress Control Number: 2008933369

Cover design: WMX Design GmbH, Heidelberg, Germany

Printed on acid-free paper

springer.com

Preface

Transport properties of plant cuticles are important for different fields of modern plant sciences. Ecologists and physiologists are interested in water losses to the environment via the cuticle. Penetration of plant protecting agents and nutrients into leaves and fruits is relevant for research in agriculture and plant protection. Ecotoxicologists need to know the amounts of environmental xenobiotics which accumulate in leaves and other primary plant organs from the environment. For all of these studies suitable methods should be used, and a sound theoretical basis helps to formulate testable hypotheses and to interpret experimental data. Unnecessary experiments and experiments which yield ambiguous results can be avoided.

In this monograph, we have analysed on a molecular basis the movement of molecules across plant cuticles. Based on current knowledge of chemistry and structure of cuticles, we have characterised the aqueous and lipophilic pathways, the nature and mechanisms of mass transport and the factors controlling the rate of movement. We have focused on structure–property relationships for penetrant transport, which can explain why water and solute permeabilities of cuticles differ widely among plant species. Based on this knowledge, mechanisms of adaptation to environmental factors can be better understood, and rates of cuticular penetration can be optimised by plant physiologists and pesticide chemists.

This monograph is a mechanistic analysis of foliar penetration. We have made no attempt to review and summarise data on foliar penetration of specific solutes into leaves of specific plant species under a specific set of environmental conditions. A number of reviews can be consulted if this is of interest (Cottrell 1987; Cutler et al. 1982; Holloway et al. 1994; Kerstiens 1996a; Riederer and Müller 2006). A wealth of additional literature is cited in these books.

Once synthesised, the plant cuticle is a purely extra-cellular membrane, and metabolism or active transport which greatly affect transport across cytoplasmic membranes are not involved in cuticular penetration. For this reason, a number of books on sorption and diffusion in man-made polymeric membranes were sources of inspiration in writing this monograph. We drew heavily on the classical books by Crank (1975), Crank and Park (1968), Israelachvili (1991) and Vieth (1991).

This is not a review about foliar penetration. We aimed at writing a general textbook on sorption and diffusion in cuticles. Based on characteristic and representative examples we show (1) how problems related to water and solute transport across cuticles can experimentally be approached using suitable methods developed in the past, (2) the way in which these data can be analysed, and what we can learn from these results about structure and functioning of cuticles, and finally (3) the limitations and problems in data interpretation. At the end of each chapter, problems and solutions can be found. Some of them summarise the highlights of the text, some illustrate implications and others are intended as exercises of calculations.

The idea of analysing permeability of cuticles based on structure–property relationships was born during a stay (1967–1972) by one of us (JS) as a doctoral student in Bukovac's laboratory at Michigan State University, USA. Later, the concepts developed in the two volumes by Hartley and Graham-Bryce (1980) were of immense help to us in formulating testable hypotheses. In writing, we have relied greatly on our own work conducted at the Botany departments of the Universities of München, Bonn and Hannover, but the book could not have been written without the collaborative research in the last decades with M. Riederer (University of Würzburg), K. Lendzian (Technische Universität München), B.T. Grayson (Shell, Sittingbourne, England), P. Baur (now Bayer Crop Science) and Anke Buchholz (now Syngenta, Switzerland).

It was one of our aims to provide a better understanding of cuticular penetration, and to formulate some basic rules for predicting and optimising rates of cuticular penetration. This requires some elementary mathematics, but we have kept equations simple and calculus is not required to follow our arguments or to solve the problems. Some basic knowledge of chemistry and physics are helpful but not mandatory. We hope this book will be useful to Master and doctoral students working in different fields of plant sciences (ecology, physiology, molecular biology, ecotoxicology, plant nutrition, horticulture, pesticide science and plant protection) when faced for the first time with problems related to permeability of plant cuticles to water and solutes. Researchers at universities, applied research institutions and those in the agrochemical industry working on transport across cuticles will find numerous useful hints. This book was written as a text book and can be used for teaching, since in each chapter (1) we state the problem, (2) we describe an experimental solution, (3) we present a critical analysis of the experimental data, and (4) at the end of each chapter we add problems intended to help the student in verifying understanding of concepts and calculations.

Germany
November 2008

Lukas Schreiber
Jörg Schönherr

Acknowledgements

We gratefully acknowledge reading of preliminary chapters by Dr. Anke Buchholz, Dr. Andrea Faust, Dr. Rochus Franke, Dr. Klaus Lendzian, Dr. Jurith Montag, Dr. Kosala Ranathunge and Dr. Jiri Santrucek. Their corrections and suggestions were of immense help, and substantially improved the final version of this book.

We are thankful to Sylvia Eifinger for preparing the drawings of models and experimental equipments.

One of us (LS) is indebted to the University of Bonn and the Faculty of Mathematics and Natural Sciences for granting a sabbatical leave during the winter term 2007/2008 for writing this book.

We also acknowledge with gratitude constant support and help by Dr. Jutta Lindenborn, Dr. Christina Eckey and Dr. Dieter Czeschlik from Springer Verlag.

Finally, we thank our families for their understanding and patience during the writing and preparation of this book.

Contents

Chapter 1
Chemistry and Structure of Cuticles as Related to Water and Solute Permeability

From the very beginning of life on earth, all living organisms established protective interfaces between themselves and the aqueous or gaseous environment. In all cases these interfaces are of lipid nature. The first unicellular organisms developed cell membranes of phospholipids separating the cytoplasm from the surrounding aqueous environment. Phospholipids are major constituents of cytoplasmic membranes of contemporary organisms. Later in evolution, multicellular organism with specialised tissues and organs appeared, and the mainland was conquered successfully by plants and animals. Since the water potential of the atmosphere is always strongly negative, there is a constant loss of water from living organisms to the atmosphere. In order to survive and avoid desiccation, land-living animals and plants had to cope with this situation. With terrestrial higher plants, the evolutionary answer to this challenge was the development of a cuticle about 500 million years ago. Insects and mammals are also protected by cuticles or skins. Their cuticles have similar functions, but they differ in chemistry and structure from the plant cuticle (Andersen 1979; Rawlings 1995).

The plant cuticle is an extracellular polymer membrane which covers all primary organs such as stems, leaves, flowers and fruits. In contrast to most synthetic polymer membranes, which are mostly homogeneous in structure and composition, plant cuticles are polymer membranes characterised by a pronounced heterogeneity in both chemical composition as well as fine structure. A functional analysis of barrier properties of plant cuticles requires detailed information on chemistry and structure. It is one of our major objectives to relate chemistry and structure of cuticles to water and solute permeability. We have evaluated the literature in an attempt to find the information necessary for relating permeability of cuticles to chemistry and structure.

Using the terminology of engineering, cuticles can be classified as composite membranes. They are composed of two chemically distinct fractions, the polymer matrix membrane (MX) and soluble cuticular lipids (SCL), often called cuticular waxes. For unambiguous chemical analysis and for measuring permeability, cuticles are isolated either chemically or enzymatically (Schönherr and Riederer 1986). The method of choice is enzymatic isolation at room temperature using pectinase

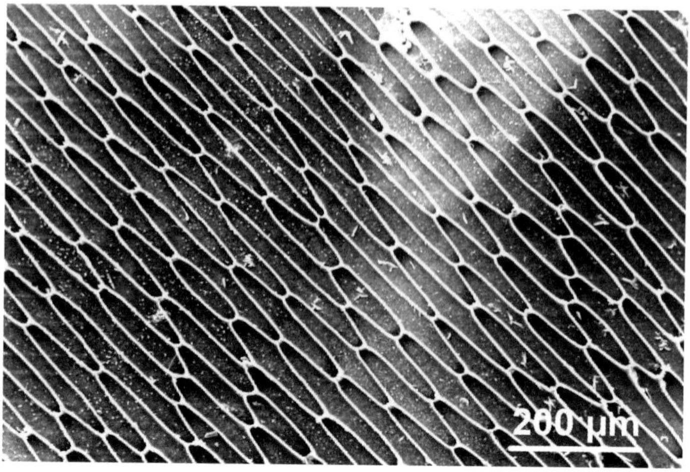

Fig. 1.1 Scanning electron micrograph of the morphological surface of a cuticle isolated with pectinase from an inner *Clivia miniata* leaf. Cuticular pegs, protruding between anticlinal cell walls, reveal the pattern of the epidermal cells

(Sect. 9.1). This avoids heat and treatment with chemicals which might cause hydrolysis or other chemical reactions. Pectinase digests the pectin layer interposed between cuticles and the cellulose wall of the epidermis. Occasionally a pectinase/cellulase mixture has been used, but the benefit of including cellulose has never been clearly demonstrated. Even when isolated using pectinase alone, the inner surfaces of the cuticular membrane look clean and cellulose residues are not detectable (Fig. 1.1).

We shall refer to isolated cuticles as cuticular membranes (CM), while the term "cuticle" is reserved to cuticles still attached to epidermis and/or organs. Cuticles cannot be isolated from leaves or fruits of all plant species. CM which can be obtained by enzymatic isolation have been preferentially used for chemical analysis, because this avoids ambiguities concerning the origin of the materials (waxes, cutin acids) obtained by extraction and depolymerisation. If enzymatic isolation of cuticles is not possible, air-dried leaves must be used. In these cases, there is a risk that some of the products obtained by solvent extraction or depolymerisation may have originated from other parts of the leaf.

1.1 Polymer Matrix

CM can be fractionated by Soxhlet extraction with a suitable solvent or solvent mixtures. The insoluble residue is the polymer matrix (MX), while the soluble lipids (waxes) can be recovered from the solvent. Chloroform or chloroform/methanol are good solvents, but many others have been used which do not quantitatively extract high molecular weight esters or paraffins, especially when used at room temperatures (Riederer and Schneider 1989).

Leaf CM {*Citrus aurantium* (bitter orange), *Hedera helix* (ivy), *Prunus laurocerasus* (cherry laurel)} preferentially used in transport experiments have an average mass of 250–400 $\mu g\,cm^{-2}$ (Schreiber and Schönherr 1996a), although CM thickness can vary between 30 nm (*Arabidopsis thaliana* (mouse-ear-cress)) and 30 μm (fruit CM of *Malus domestica* (apple)). Specific gravity of CM is around 1.1 $g\,cm^{-3}$ (Schreiber and Schönherr 1990), and using this factor the average thickness of these leaf CM can be calculated to range from about 2.3 to 3.7 μm.

If the MX is subjected to hydrolysis in 6 N HCl at 120°C, an insoluble polymer is obtained. This polymer has the consistency of chewing gum, and an elemental composition very similar to a polyester of hydroxyfatty acids (Schönherr and Bukovac 1973). It is considered to be pure cutin. The aqueous HCl supernatant contains a complex mixture of carbohydrates, amino acids and phenols, but only amino acids have been determined quantitatively (Schönherr and Bukovac 1973; Schönherr and Huber 1977). Some cuticular carbohydrates and phenolic substances have also been characterised (Marga et al. 2001; Hunt and Baker 1980). Polarised light (Sitte and Rennier 1963) and thermal expansion (Schreiber and Schönherr 1990) indicate the presence of crystalline cellulose. It is not known if polar solutes obtained by acid hydrolysis are simply trapped in the cutin as polysaccharides or polypeptides, or if they are covalently attached to cutin. Phenolic acids contained in the MX of ripe tomato fruits are released by ester hydrolysis, but it is uncertain if they were linked to cutin or to other constituents of the MX (Hunt and Baker 1980). Riederer and Schönherr (1984) have fractionated CM of leaves and fruits from various species (Table 1.1).

The mass of the CM per unit area varies widely among species between 262 $\mu g\,cm^{-2}$ (*Cucumis* (cucumber) fruit CM) and 2,173 $\mu g\,cm^{-2}$ (*Lycopersicon* (tomato) fruit CM). The wax fraction varies even more and is smallest with *Citrus* leaves (5%) and largest with *Pyrus* (pear) cv. Bartlett abaxial leaf CM (45%). The average weight fraction of the MX is 76%, with cutin and polar polymers amounting to 55% and 21% respectively. Variation among species in the fraction of polar polymers and cutin is much smaller than in mass per area of CM or in weight fraction of waxes (Table 1.1).

1.2 Cutin Composition

Cutin is insoluble in all solvents, and its composition can be analysed only following depolymerisation. MX obtained by solvent extraction of CM are usually used for these studies. This has been the standard analytical approach in the past decades for analysing the chemical composition of cutin. Cutin obtained after acid hydrolysis of polar polymers has never been analysed, and for this reason we do not know if the same cutin monomers are liberated when starting with MX or with cutin. Depolymerisation has been performed using ester cleaving reactions. In early studies, cutin was hydrolysed using methanolic or ethanolic KOH and hydroxyfatty acids were obtained by acidifying their potassium salts. After transesterification of

Table 1.1 Mass per area and composition of selected cuticular membranes (data from Riederer and Schönherr 1984)

Species	CM ($\mu g\,cm^{-2}$)	SCL (% of CM)	MX (% of CM)	CU (% of MX)	HY (% of MX)
Fruit CM					
Capsicum	1,971	10	90	61	29
Cucumis	262	20	80	55	25
Lycopersicon	2,173	7	93	69	24
Solanum	599	8	92	62	30
Leaf CM					
Citrus ab	318	5	95	73	22
Clivia ad	530	20	80	64	16
Clivia ab	466	18	82	66	16
Ficus ad	458	25	75	56	19
Ficus ab	493	37	63	52	11
Hedera ad	450	19	81	60	21
Hedera ab	430	17	83	61	22
Nerium ad	1,318	39	61	45	16
Nerium ab	1,633	37	63	46	17
Olea ad	836	29	71	50	21
Pyrus cv. Conf. ad	353	31	69	46	23
Pyrus cv. Conf. ad	324	32	68	47	21
Pyrus cv. Bartlett ad	350	38	62	43	19
Pyrus cv. Bartlett ab	421	45	55	37	18
Mean (sd)	744 (602)	24 (12)	76 (12)	55 (10)	21 (4.7)

CM, cuticular membrane; SCL, soluble cuticular lipids (waxes); CU, cutin; HY, fraction hydrolysable with HCl (polar polymers); ab, abaxial; ad, adaxial; sd, standard deviation

MX using various chemicals (boron trifluoride/methanol, methanolic HCL, KOH), methylated cutin monomers are obtained, which after silylation can be analysed by gas chromatography and mass spectrometry (Walton 1990).

This analytical approach shows that the major cutin monomers are derivatives of saturated fatty acids, predominantly in the chain length of C_{16} and C_{18}, carrying hydroxyl groups in mid-chain and end positions (Table 1.2). In addition, dicarboxylic acids having the same chain length occur in minor amounts. In some species (e.g., *Clivia miniata*, *Ficus elastica* and *Prunus laurocerasus*) C_{18}-cutin monomers with an epoxide group in the mid-chain position have been identified (Holloway et al. 1981). Primary fatty acids and alcohols with chain lengths varying between C_{16} and C_{26} are also released from the MX in minor amounts. Based on extensive studies of cutin composition, including leaves and fruits from a large number of plant species (Holloway 1982b), cutin was classified as C_{16}-, C_{18}- or a mixed-type C_{16}/C_{18}-cutin according to the dominating chain length of major cutin monomers released from the MX.

Recently, molecular biological and biochemical approaches which identify the first genes coding for the enzymes involved in cutin biosynthesis of *Arabidopsis thaliana* have been carried out successfully. In this context, cutin of *Arabidopsis thaliana* was shown to be primarily composed of one- and two-fold C_{16} and

Table 1.2 Common C_{16}- and C_{18}-monomers occurring in the polymer matrix of several plant species

Compound	Chemical Structure
C_{16}-monomers	
Palmitic acid	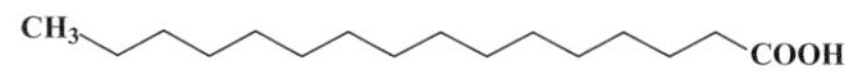
Palmitic alcohol	
16-Hydroxypalmitic acid	
1,16-Palmitic diacid	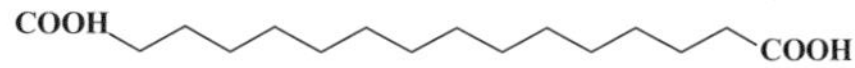
9,16-Dihydroxypalmitic acid	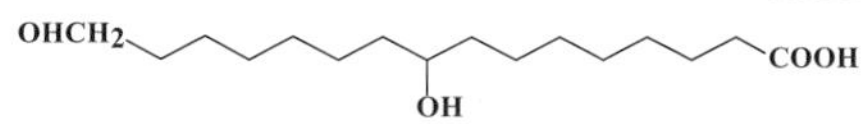
10,16-Dihydroxypalmitic acid	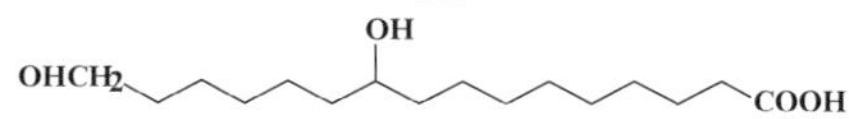
C_{18}-monomers	
Stearic acid	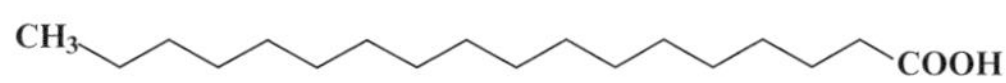
Stearic alcohol	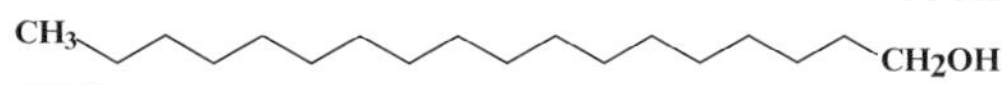
18-Hydroxystearic acid	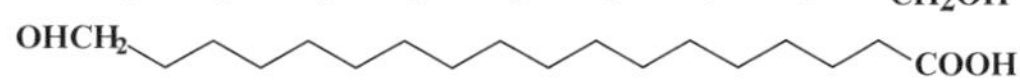
9,10,18-Trihydroxystearic acid	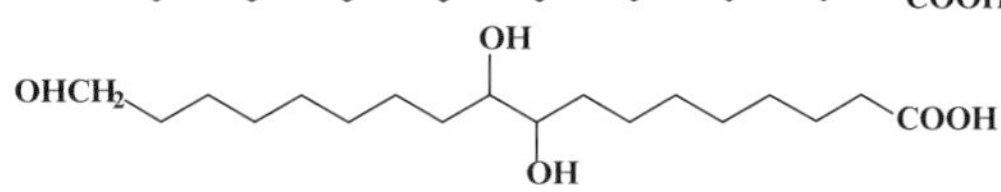
18-Hydroxy-9,10-epoxystearic acid	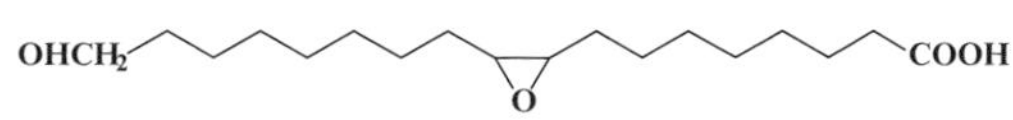

C_{18} unsaturated dicarboxylic acids (Nawrath 2006), which does not fit the picture of cutin composition derived from all previous investigations (Table 1.2). Unfortunately, barrier properties of this type of "atypical" cutin have not yet been characterised. Thus, one should be cautious before generalising cutin composition obtained by transesterification of MX of plant species from which CM can be isolated. This may represent a specific set of cutins characteristic of isolable cuticles.

We are searching for the relationship between chemistry and structure of cuticles and their permeability to water and solutes. Permeability of a membrane is related to structure, which in turn depends on chemical composition. Unfortunately, we could not find any study relating the above cutin classification to water or solute permeability. Number, type and distribution of polar functional groups in the polymer, crystallinity, and the prevalence of the glassy and rubbery states at physiological temperatures are important properties. In composite polymers, the mutual arrangement of the various polymers is also important. Simply looking at the products obtained by transesterification or acid hydrolysis reveals little about the structure and function of the polymer.

The monomer composition does not tell us much about the original composition of the polymer and the way the monomers were cross-linked and arranged in the MX. The cutin models which are based on type and predominance of cutin acids are guesswork, and are as good as the underlying assumptions (Kolattukudy 2004). A new approach is non-destructive NMR spectroscopy (Fang et al. 2001), directly mapping the intact polymer and the intermolecular cross-linking of the monomers without prior degradation of the MX. This approach allows the identification of ester linkages in cutin in vivo, and it shows that sugar moieties can be linked to cutin monomers, although the exact type of bond has not yet been identified. In spite of these numerous attempts, we must admit that we are still far away from a complete picture of the molecular architecture of cutin or the MX.

Another complication often overlooked is the fact that cuticles containing epoxy-fatty acids (Holloway et al. 1981) are only partially degraded by transesterification, and the chemical analysis of these MX is incomplete. A significant if not major fraction of the MX resists degradation, indicating the existence of bonds other than ester linkages in the cutin polymer. This "non-ester cutin" is also called cutan, whereas that fraction of the MX cross linked by ester bonds is called cutin. There is evidence that intermolecular cross-linking in cutan is mainly by ether bonds (Villena et al. 1999). This non-degradable fraction of the cuticle is still poorly characterised, because of major methodological limitations.

Clivia miniata plants are monocots, and leaves grow at their base. The age of leaf segments increases with distance from leaf base. Adaxial cuticles have been isolated from leaf strips, and the MX has been fractionated into ester cutin and cutan (Riederer and Schönherr 1988). Fractionation of total cutin into ester cutin and non-ester cutin (cutan) was possible only starting with position 3–4 cm from base. Most of the young cuticle is made up of ester cutin (Fig. 1.2), which increases rapidly with age, and in the study above its amount doubled at position 5–6 cm when epidermal

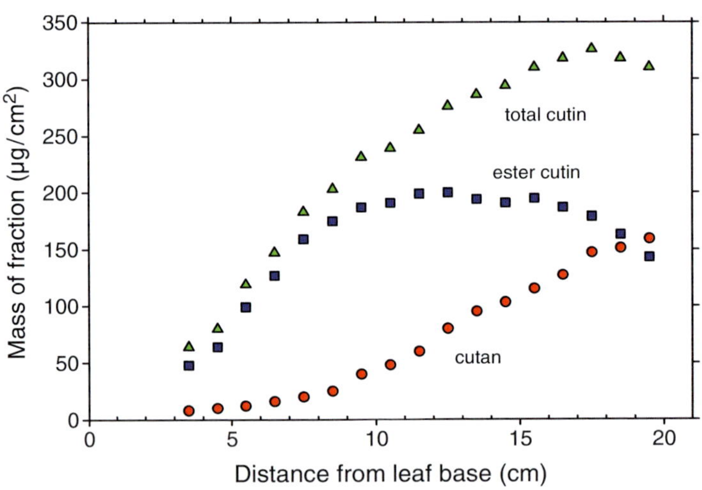

Fig. 1.2 Fractionation of total cutin of *Clivia miniata* leaves as a function of position. Total cutin was obtained by acid hydrolysis of MX leaf strips. The resulting total cutin was subjected to transesterification using BF_3–MeOH. The polymer-resisting transesterification is cutan. The amount of ester cutin was obtained by difference. (Redrawn from Riederer and Schönherr 1988)

cells had obtained their maximum area. Initially, cutan mass increased slowly up to 8–9 cm, but later its mass increased more rapidly and at 19–20 cm it was higher than the mass of ester-cutin. Ester cutin reached its maximum mass at 11–13 cm; thereafter it decreased, showing that ester cutin was converted in part to cutan.

Following transesterification the lipophilic cutin monomers are recovered with organic solvents like chloroform. Polar compounds released by transesterification are lost, since they remain in the reaction residue, which is discarded. Due to this experimental approach, it was overlooked in all previous analyses of cutin composition that glycerol, a small and highly polar organic molecule, is also released and forms an important cross-linker in the MX (Graca et al. 2002).

Polypeptides (Schönherr and Huber 1977), aromatic compounds (Hunt and Baker 1980) and carbohydrates (Wattendorff and Holloway 1980; Dominguez and Heredia 1999; Marga et al. 2001) are significant although minor constituents of the MX. The question arises as to whether they have any specific functions in the MX or if their presence is accidental. Nothing is known about the nature and the origin of the proteins. Their presence in amounts of about 1% has only been shown by amino acid analysis (Schönherr and Huber 1977) or CHN analysis (Schreiber et al. 1994) of isolated cuticles. It is not known whether proteins in the MX are structural proteins with functional stabilising properties like extensins in plant cell walls. Alternatively, it can be suggested that enzymes involved in the polymerisation of the MX (cutin esterases) were trapped during polymer formation in the MX.

More rational explanations are available for the presence in the MX of about 20–40% of carbohydrates, mainly pectin and cellulose. The outer epidermal cell wall and the cuticle on top of it must be connected to each other in some way. There is evidence that this connection can be by direct covalent links of sugar molecules to cutin molecules (Fang et al. 2001), and in addition cellulose fibrils extending into the MX network may contribute to this connection. It is obvious that a significant amount of polar functional groups on the inner physiological side of the MX is protected from enzymatic digestion by the cutin polymer. This can easily be demonstrated by testing the wettability with water of the physiological outer and inner surfaces of the MX. Contact angles on the physiological outer side are around 90°, indicating a surface chemistry of methyl and methylene groups, as should be the case with a polymer mainly composed of aliphatic monomers (Holloway 1970). Quite different from the outer side, the physiological inner side of the MX is wet by water, and droplets spread. This indicates a highly polar surface chemistry, presumably composed of hydroxyl and carboxyl groups from cellulose and pectin fibrils.

It is also evident that crystalline cellulose has a fundamental function within the amorphous cutin polymer, acting as a stabiliser which strongly affects the biomechanical properties of cuticles. Volume expansion of pure cutin, obtained by hydrolysis of carbohydrates, is much higher than expansion of the MX (Schreiber and Schönherr 1990). Unfortunately, immunological techniques with antibodies directed versus specific epitopes of cell wall carbohydrates have not yet been carried out. This type of study should allow mapping with high precision the chemical nature and the spatial arrangement of carbohydrates in cross sections of the MX, using immunogold labelling and transmission electron microscopy.

Despite the fact that the MX is a biopolymer composed of lipophilic monomers, it is also evident that significant amounts of polar functionalities (hydroxyl, carboxyl and amino groups) are present. The degree of polarity of ionisable groups depends on *p*H. Therefore, the amounts and the local distribution of these polar groups within the MX are important with respect to diffusion of polar molecules like water and ions, simply because water is sorbed to polar groups in the MX. Sorption of water, effect of *p*H on ionisation of functional groups, ion exchange capacities of cuticles and effects on transport properties are a major topic of Chap. 4. The fact that water sorbed to the MX acts as a plasticiser in the membrane is evident from investigations of the biomechanical properties of cuticles. Rheological properties like extensibility and plasticity of isolated cuticles have been shown to strongly increase upon hydration (Edelmann et al. 2005; Round et al. 2000), indicating interaction of water with polar domains within the MX.

1.3 Soluble Cuticular Lipids

1.3.1 Extraction and Classification of Waxes

The soluble fraction obtained when CM are extracted with suitable solvents is called soluble cuticular lipids (SCL) or cuticular waxes. In the following we will use the term waxes instead of SCL, since this is commonly used in literature, although this is not correct chemically because waxes in a strict sense are only wax esters as in beeswax. When cross-sections of cuticles are viewed with polarised light they are negatively birefringent due to crystalline waxes embedded in cutin (Sect. 1.4). Waxes deposited on the surface of the cuticle are called epicuticular or surface waxes. In some species, for instance barley leaves, epicuticular waxes form pronounced three-dimensional crystallites. Such glaucous leaves scatter and reflect incoming light, and they are difficult to wet. Fine structure of these epicuticular waxes (Fig. 1.3a) can be easily visualised by scanning electron microscopy (SEM). As a consequence, numerous studies have characterised and classified this wax bloom (cf. Amelunxen et al. 1967; Baker 1982; Barthlott 1990; Barthlott and Frölich 1983). Many other species lack such three-dimensional epicuticular wax crystallites, and their surfaces appear shiny or glossy. These surfaces often exhibit folding of the cuticle (Fig. 1.3b), and they are more easily wetted by water. However, absence of three-dimensional epicuticular wax crystallites does not imply that waxes are absent on these leaf surfaces. Evidence is mounting that smooth wax films occur on the cuticles of all leaves, and that wax crystallites originate from these wax films rather than from cutin. Jeffree (2006) has discussed this problem in detail.

Cuticular waxes can be obtained by extracting CM, and with some caution also by dipping leaves or by rinsing leaf surfaces with chloroform. By Soxhlet extraction of CM, total wax amounts are obtained, but it is not possible to distinguish

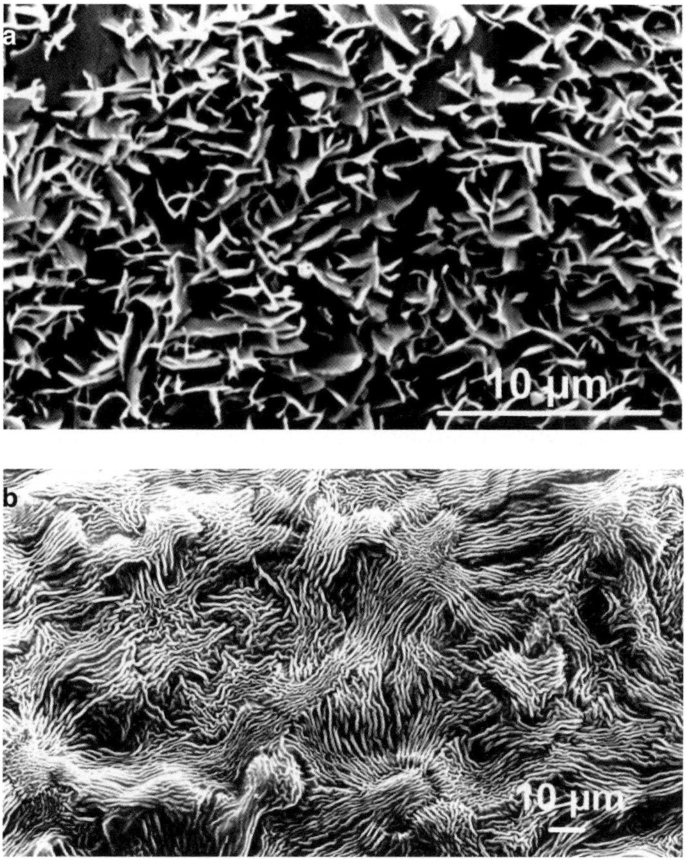

Fig. 1.3 Scanning electron micrographs of the leaf surface of (**a**) *Quercus robur* (oak) and (**b**) *Vinca major* (periwinkle). A delicate pattern of epicuticular wax crystallites is visible on the oak surface, whereas the periwinkle surface is characterised by a pronounced pattern of cuticular folding

epi- and intracuticular waxes. Dipping or rinsing leaves with chloroform at room temperature often results in partial extraction of waxes (Riederer and Schneider 1989) especially when cuticles are thick. By varying solvents and duration of extraction it has been attempted to selectively extract epicuticular waxes (Silva Fernandes et al. 1964; Holloway 1974; Baker et al. 1975; Baker and Procopiou 1975). Jetter et al. (2006) have argued that selective extraction of surface waxes with solvents is not possible, and they favour various stripping techniques (Ensikat et al. 2000; Jetter et al. 2000; Riederer and Markstädter 1996). Following stripping, cuticle surfaces appeared smooth and clean, showing that stripping removes wax bloom completely. However, there is no direct evidence that smooth continuous wax films covering the cutin are also removed. These uncertainties affect the validity of conclusions concerning amounts of surface waxes and effects of surface waxes on permeability of cuticles. We shall return to this problem in later chapters.

Total amount of waxes were obtained by Soxhlet extraction of CM from 21 species. The average mass of wax was about 100 $\mu g\,cm^{-2}$ (Schreiber and Riederer 1996a). This amount was determined gravimetrically by subtracting the mass of the MX from that of the CM. Wax coverage of the leaf CM from single species varied 40-fold between 10 (*Citrus aurantium*) and 400 $\mu g\,cm^{-2}$ (*Nerium oleander*). With *Malus domestica* fruit, wax coverage of more than 3,000 $\mu g\,cm^{-2}$ was measured.

1.3.2 Chemistry of Waxes

In most species, waxes are composed of two major substance classes: (1) linear long-chain aliphatic compounds and (2) cyclic terpenoids. Linear long-chain aliphatics can be divided into different substance classes. Compounds with chain length of C_{20} and higher most frequently belong to alkanes, primary alcohols, aldehydes, and primary fatty acids (Table 1.3). Some of the acids and alcohols found in waxes are also released by depolymerisation of cutin (see Sect. 1.1). Secondary alcohols and ketones with functional groups attached to carbon numbers between C_4 and C_{16} have also been identified. Covalent binding between primary fatty acids (C_{16}–C_{36}) and alcohols (C_{20}–C_{36}) results in long chain esters with chain lengths between C_{36} and C_{70} (Table 1.3).

Biosynthesis of these long-chain aliphatic compounds is localised in epidermal cells, and it starts from C_{16} to C_{18} fatty acids. The elongation process leading to very long chain fatty acids with the chain length between C_{20} and C_{34} (Kunst and Samuels 2003) is based on the step-by-step condensation of C_2-units to the substrate. Consequently, elongated fatty acids predominantly have even-numbered carbon chains. Oxidation leads to aldehydes and alcohols, also with even-numbered chain-length. Since esters are the condensation products of long-chain alcohols and acids, they are also characterised by even-numbered chain lengths. Alkane synthesis involves a decarbonylation step, and thus they are characterised by odd-numbered chain lengths. Secondary alcohols are synthesised from alkanes, and thus they are also odd-numbered.

Table 1.3 Most common substance classes of cuticular waxes identified by gas chromatography and mass spectrometry

Substance class	Chemical formula	Range of chain lengths	Major homologues
Acids	$CH_3–(CH_2)_n–COOH$	C_{16}–C_{32}	C_{24}, C_{26}, C_{28}
Aldehydes	$CH_3–(CH_2)_n–CHO$	C_{22}–C_{32}	C_{26}, C_{28}, C_{30}
Alcohols	$CH_3–(CH_2)_n–CH_2OH$	C_{22}–C_{32}	C_{26}, C_{28}, C_{30}
Alkanes	$CH_3–(CH_2)_n–CH_3$	C_{21}–C_{35}	C_{29}, C_{31}
Secondary alcohols	$CH_3–(CH_2)_n–CHOH–(CH_2)_n–CH_3$	C_{23}–C_{33}	C_{29}, C_{31}
Esters	$CH_3–(CH_2)_n–COO–(CH_2)_n–CH_3$	C_{36}–C_{70}	C_{40}, C_{42}, C_{44}

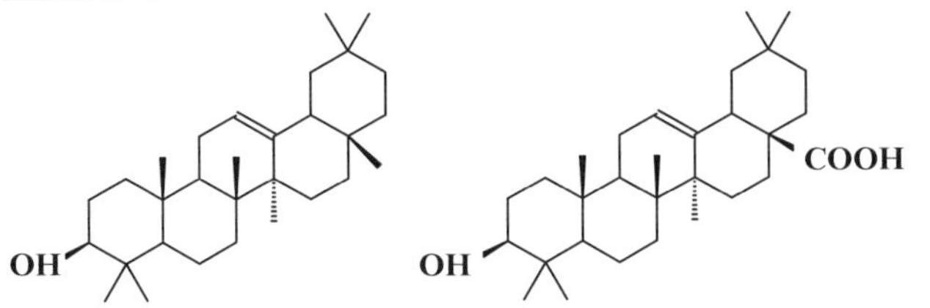

β-amyrin oleanolic acid

Fig. 1.4 Chemical structures of the triterpenoic alcohol β-amyrin and the triterpenoic acid oleanolic acid occurring in cuticular wax of various species

Triterpenoids are derived from the terpenoid metabolism (Guhling et al. 2006). Very common triterpenoids (Fig. 1.4) are pentacyclic triterpenoic alcohols (e.g., α-amyrin and β-amyrin) and acids (e.g., oleanolic acid and ursolic acid). Triterpenoids occur only in some species, whereas long-chain aliphatic compounds represent typical components of all waxes analysed so far. Occurrence of triterpenoids in larger amounts is generally limited to certain taxonomically related species. Waxes of many species of the Rosaceae, for example, are characterised by the predominance of triterpenoids amounting to 50% and more of the total wax coverage of the MX, as is the case with *Prunus laurocerasus* (Jetter et al. 2000), whereas in other species (e.g., *Hedera helix*, *Arabidospsis thaliana*) triterpenoids are present in wax extracts only in traces. Pentacyclic triterpenoids are planar molecules with very high melting points, and it is difficult to imagine how they could form homogeneous mixtures with linear long-chain aliphatic wax molecules. It is not known whether they are partially crystalline in and on the CM, as is the case with the linear long-chain aliphatic molecules.

1.3.3 Special Aspects of Wax Analysis

Analysis of the chemical composition of wax is straightforward. Intact leaves or isolated cuticular membranes are extracted with an organic solvent, e.g., chloroform, and extracted molecules are separated and quantified using capillary gas chromatography (GC). Identification is carried out via mass spectrometry (MS). Since on-column injection is used, the loss of wax compounds is minimised as long as the homologues are mobile and are not deposited on the entrance of the column. Before GC–MS became standard, wax classes were separated by thin-layer chromatography (TLC), the waxes were recovered from the TLC plates, eluted and injected in packed columns of the GC. In this procedure, recovery of the different substance classes and homologues was less constant and usually unknown (Baker 1982).

There are still unsolved problems in wax analysis. Cuticular wax samples represent fairly complex mixtures composed of different substance classes and of different chain lengths within each substance class. A representative wax sample can be composed of 50 individual compounds and more. Often, the most prominent wax compounds are identified, whereas compounds present only in traces remain unidentified. Identification of 90–95% of the compounds occurring in a specific wax sample is considered a successful analysis, and this can take a fairly long time, when time needed for identifying unknown mass spectra is included.

Quantification is normally carried out adding an internal standard (e.g., an alkane) of known amount to the wax samples. Ideally, for each individual wax compound, varying in chain length and functionalization, the best standard would of course be the identical compound. However, most wax compounds are not commercially available as standards. Furthermore, in view of the large number of wax molecules which are normally present in a typical wax sample, it would be unrealistic running a separate standard for each wax molecule, even if it was available. Therefore, in most cases an alkane, representing a linear long-chain aliphatic molecule as they are typically found in wax samples, is used as internal standard. Alkanes of even chain lengths (e.g., C_{24}) are preferred, since alkanes of uneven chain length are dominant in plant waxes.

In addition to these limitations in quantitative wax analysis, there is another problem which has rarely been addressed in the past. The total amount of cuticular wax determined gravimetrically is generally larger than wax amounts determined by GC–MS. Various reasons might contribute to this observation. Weighing does not discriminate between wax compounds and non-wax compounds, which may contribute to total weight loss. On the other hand, analysis by GC is highly specific, and permits exact identification of compounds. Wax and other compounds (e.g., sugars) can be distinguished. Esters with very high molecular weight can present formidable problems in GC due to their very long retention times (Santos et al. 2007) and the tendency to produce broad and blurred peaks. Esters with high molecular weight (>700–800) also approach the detection limit of the MS. This could lead to an underestimation of wax amounts, especially in wax samples with high amounts of esters.

Compounds with very high molecular weights have been detected in wax from *Hedera helix* leaves (Hauke and Schreiber 1998). The large difference between wax amounts determined gravimetrically and by GC–MS was mainly due to a limited resolution of the analytical approach. The difference in wax amounts depended on leaf age and season (Fig. 1.5a). When wax extracts from ivy CM were separated on a column packed with silica gel using solvents of different polarity and subjected to GS–MS, two distinct wax fractions were obtained. The long-chain aliphatics as described above were eluted from the column with the apolar solvent (diethylether), and this could be analysed directly by gas chromatography. This fraction yielded alkanes, alcohols, aldehydes, acids and esters. The eluate of the polar solvent (chloroform/methanol/water in ratios of 60:25:5) resulted in an additional wax fraction, which could not be analysed directly by gas chromatography but only after transesterification (Fig. 1.5b). Analysis of this transesterified polar wax fraction revealed that it was composed of di- and presumably trimers of ω-hydroxy fatty acids in the chain lengths C_{12}–C_{16}, which were linked by ester bonds to linear long-chain alco-

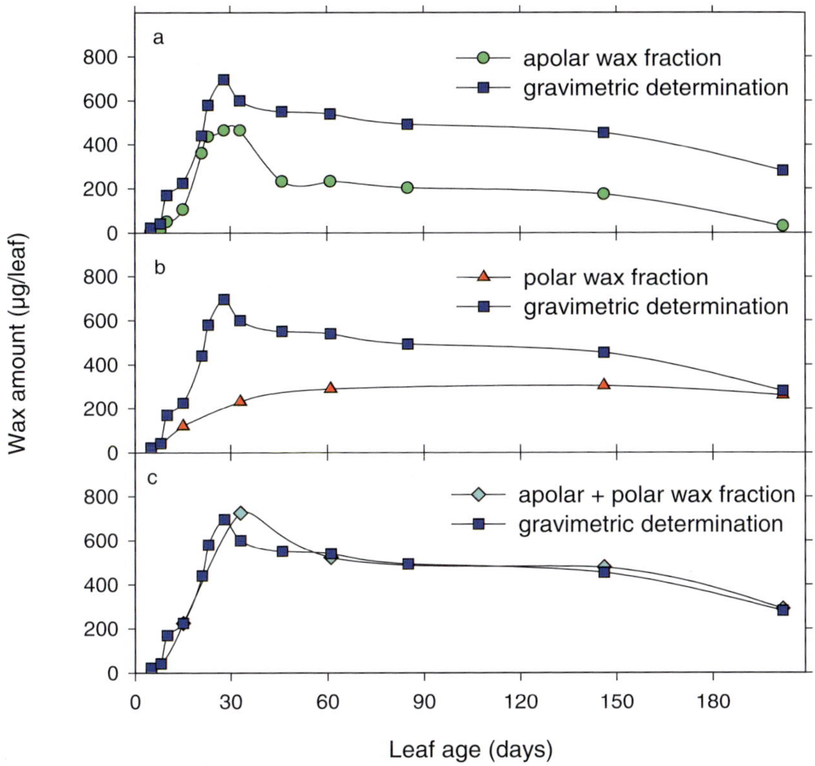

Fig. 1.5 Quantitative determination of wax in ivy (*Hedera helix*) CM over one season from bud break at the end of April to leaf senescence in November. Apolar and polar wax fractions were separated as described in the text

hols with chain lengths from C_{22} to C_{32}. Compounds of similar oligomeric structure have also been found in gymnosperms, and they are called estolides (Bianchi 1995). Molecular weights and polarity of these compounds are too high for direct GC analysis. They do not move on the column, and are overlooked if not transesterified. When apolar and polar wax fractions are combined, total amounts of wax are close to the wax amounts determined gravimetrically (Fig. 1.5c). Based on these results, it appears that gravimetric determination of wax amounts is more reliable. We don't know if the problem occurs with waxes of all species, but we suggest that wax amounts determined by GC should always be compared to those determined gravimetrically. If a significant difference is observed, fractionation of waxes by column chromatography is indicated. Deterioration of the ability of a column to separate homologues, broadening of peaks and discolouration of the column entrance could also indicate the presence of estolides in the sample which failed to move on the column. As a corollary, it should be realised that it is not a good practice to analyse waxes from leaves of unspecified age and to use only one sampling time. There simply exists no typical wax composition of leaves and fruits.

The contributions of epi- and intracuticular waxes to permeability of cuticles is not known. Such research requires that epicuticular waxes can be removed quantitatively without disturbing intracuticular waxes. As solvents very rapidly penetrate into the cuticle, the only method available today is mild stripping of waxes. Permeability would have to be determined prior to and after stripping. This has not been done so far, or at least not published. As already indicated, it is not certain that all waxes are removed by stripping, including thin continuous wax layers on the surface of cutin. In this context, significant progress has been made by a technique which first lifts off epicuticular wax from the surface before chemical analysis (Jetter et al. 2000). *Prunus laurocerasus* leaves were used. The wax of this species is dominated by triterpenoids, and it was shown that linear long-chain aliphatics were mainly deposited on the outer surface of the MX, whereas triterpenoids could rarely be lifted off. Triterpenoids were nearly exclusively found in the chloroform extracts of the stripped CM, showing that they were located within the polymer. This is good evidence that linear long-chain aliphatics and triterpenoids were spatially segregated. Effects of stripping on water permeability unfortunately have not been reported. Further evidence for a layered structure of waxes has been provided by atomic force microscopy, offering a resolution on the molecular level (Koch et al. 2004). After epicuticular waxes were lifted off using an epoxide glue, the surface of the CM appeared smooth. Within 80 min, a new film of 3–5 nm thickness was regenerated on the surface of the cuticle. The chemical identity of these regenerated films and the effects of stripping with epoxy glue were not investigated. These two approaches (Jetter et al. 2000; Koch et al. 2004) immediately raise a series of important questions: (1) how do these observations relate to barrier properties of CM? (2) do mainly linear long-chain aliphatics contribute to the formation of the transport barrier, or are triterpenoids also important? (3) where is the waxy cuticular transport barrier located? and (4) how is the waxy barrier structured on the molecular level? These are some of the questions which we will address in the following chapters.

Extraction of wax has been shown to increase water permeability of CM 50- to 1,000-fold (see Chap. 4). Anyone trying to relate water and solute permeability to wax amounts, location and composition should have reliable data. Permeability and wax composition should be studied using identical or at least comparable samples. This requires reliable and reproducible methods of wax analysis. To our knowledge, only three studies comparing water permeability of *Citrus* leaf wax with wax composition using the same population of isolated cuticles have been published (Haas and Schönherr 1979; Geyer and Schönherr 1990; Riederer and Schneider 1990). Results and conclusions are presented in Sect. 4.6.

1.4 Fine Structure of Cuticles

In his recent review, Jeffree (2006) has summarised the literature on fine structure of cuticles and epicuticular waxes, and there is no need to repeat this. Instead, we shall evaluate the literature for hints concerning a correlation between fine structure and permeability. From the 372 studies reviewed, only two explicitly dealt with diffusion. Wattendorff and Holloway (1984) used potassium permanganate as tracer. Schmidt et al. (1981) attempted to find a correlation between water permeability and fine structure of *Clivia* CM at different stages of development. All other workers rationalised their work by alluding to the barrier function of cuticles, but they simply used standard procedures to generate pictures, while permeability was not estimated. Nevertheless, some useful terminology and information about structure–permeability relationships may be obtained from some of these studies.

Extracting waxes increases permeability by 1–3 orders of magnitude (Chap. 4 and 6). This shows that cuticular waxes play a decisive role in water and solute permeability, and both localisation and structure of waxes are important in understanding structure–property relationships. The presence of polar paths in lipophilic cuticles is another topic of importance, because it is a prerequisite for penetration of hydrated ionic solutes but not necessarily of water (Schönherr 2006).

1.4.1 Nomenclature

We adopt the definitions and nomenclature of Jeffree (2006), which is also used by most of the workers in the field. The cuticle is a polymeric membrane located on the epidermal wall of primary organs. It has a layered structure. The outermost layer is called cuticle proper (CP), and the layer underneath is the cuticular layer (CL). In many species, an external cuticular layer (ECL) located under the CP and an internal cuticular layer (ICL) facing the epidermal wall can be distinguished. Soluble cuticular lipids or waxes occur as epicuticular waxes and as embedded or intracuticular waxes. CP, CL and waxes constitute the cuticle (CM), which in some species can be isolated enzymatically. Due to its layered structure, the cuticle is a heterogeneous membrane. We distinguish transversal heterogeneity which is apparent in cross-sections, and lateral heterogeneity which arises due to the presence of trichomes and stomata.

1.4.2 Transversal Heterogeneity

1.4.2.1 Light Microscopy

In the light microscope, cross-sections of cuticles appear homogeneous. When *Clivia* cuticles are stained with Sudan III, transversal heterogeneity is not visible (cf. Fig. 2.6). Sudan III is a non-ionic lipophilic dye, but it does not stain solid paraffin or carnauba wax (Sitte and Rennier 1963). The dye is sorbed in polymeric cutin, but wax is not stained. Prominent cuticular pegs extend deep between the anticlinal walls of the epidermal cells. The thick epidermal wall is stained at *p*H 4 with toluidine blue, which is an anionic dye that binds to carboxyl groups of pectins in the cell wall. If cuticles are stained with toluidine blue at *p*H9, the cutin also binds toluidine blue due to the presence of carboxyl groups (unpublished results). Cation-exchange capacity of cuticles will be dealt with later (Sect. 4.3). Resolution of the light microscope is of the order of 0.5 μm, and this prevents the study of transversal heterogeneity of very thin cuticles.

With polarised light, location and orientation of crystalline waxes have been studied. All CM investigated with polarised light exhibited birefringence or double refraction. Birefringence is evidence for the presence of crystalline structures. In cross-sections of cuticles, waxes give negative and cellulose gives a positive birefringence (Meyer 1938; Roelofsen 1952; Sitte and Rennier 1963). There are a few examples of positive birefringence due to waxes in cuticles (Sitte and Rennier 1963). Extracting or melting waxes eliminates wax birefringence. On cooling, birefringence reappears, showing re-crystallisation from the melt.

Extracted CM exhibit form double refraction, indicating the presence of lamellar voids. Form birefringence disappears when the polymer matrix is imbibed with solvents having the same refractive index as cutin, which is 1.5. In periclinal positions the lamellar voids are oriented parallel to the surface of the cuticle, and in vivo they are filled with wax platelets in which the long axes of the paraffinic chains are oriented perpendicular to the surface of the CM. If viewed from the top, the waxes of *Clivia* cuticles appear isotropic, but near the anticlinal walls lamellae bend down towards the anticlinal walls, and in these positions waxes appear birefringent when viewed from the top (Meyer 1938). This shows an oblique orientation of the wax molecules in anticlinal pegs.

The presence of cellulose in cuticles has been a subject of controversy among microscopists, because it does not exhibit the typical histochemical reactions when embedded in cutin. Crystalline cellulose exhibits positive birefringence, and the outer epidermal walls are always positive birefringent (Fig. 1.6). The CL of most cross-sections of cuticles did not show positive birefringence. Extracting or melting waxes to eliminate possible interference of negative birefringence of waxes did not change this picture. It appears that the CL of most plant cuticles contains little if any crystalline cellulose. Most researchers agree that the CP is free of crystalline cellulose. Polarised light does not detect amorphous polysaccharides, but with the transmission electron microscope (TEM) polysaccharides can be demonstrated in the CM. They are amorphous, since they are not birefringent (see below).

Using *Ficus elastica* leaves, Sitte and Rennier (1963) observed form double refraction in the outer regions of the CM very early, when it was only about 2.5 μm thick and the leaf was still unfurling. Incorporation of wax into these preformed voids occurred early, but reached its maximum only when the leaf was fully expanded. At the same time, the thickness of the CM increased by interposition of cutin between the CL and the epidermal wall, and the layered structure shown in Fig. 1.6 was formed. The mature leaf had a lamellated CP of 1–2 μm, the ICL measured about 2.5 μm and the ECL was about 3 μm thick.

The formation of voids in cutin prior to deposition of crystalline intracuticular waxes is an astounding phenomenon, yet a necessity because crystallisation of wax in a dense polymer network is improbable. Cutin monomers are likely to

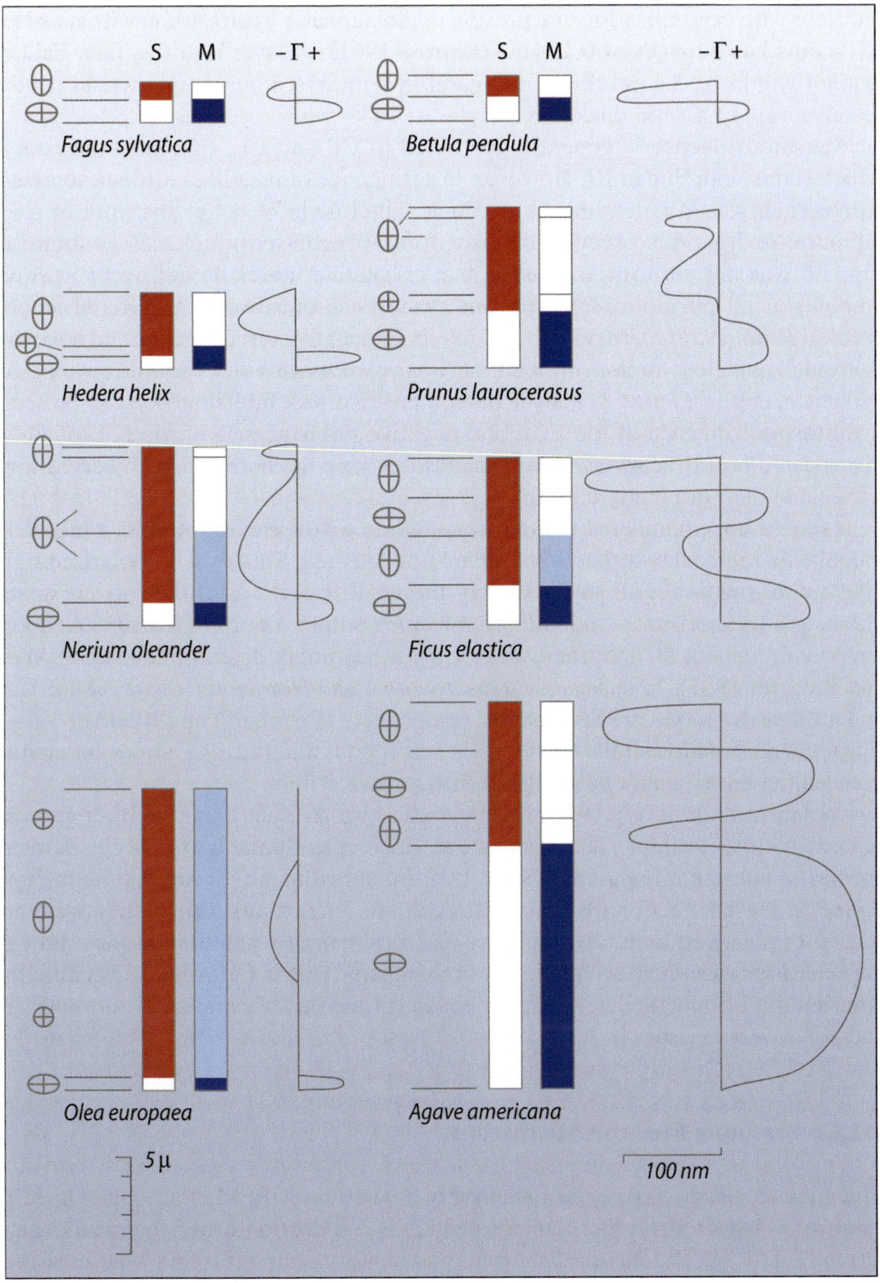

Fig. 1.6 Staining of cross-sections of cuticles from selected species and birefringence. The cuticle stained red with Sudan III and the cellulose wall stained blue with methylene blue. The signs of birefringence and of intensity (Γ) are shown on the *left* and the right side of each species respectively. (Redrawn from Sitte and Rennier 1963)

interfere with crystallisation and prevent it. An aliphatic hydrocarbon with n carbon (C) atoms has a length of 0.154 nC (Barrow 1961). A layer of a C_{20} fatty acid or alcohol would be 3.1 nm thick, and paraffin with 31 carbon atoms would need a lamellar void of 4.8 nm thickness.

Wax birefringence is generally restricted to CP and CL, that is to the cuticle which stains with Sudan III. However, birefringence of the CP is difficult to assess with certainty, as it is less than 1 μm thick and close to or below the limit of resolution of the light microscope. Intensity of birefringence of thick CM as shown in Fig. 1.6 was not uniform, indicating that crystalline waxes do not occur in equal amounts at all positions. Intensity and occurrence of anisotropy differed among species. *Prunus laurocerasus* had two layers of negative birefringence and a narrow isotropic zone close to the cell wall. *Olea europaea* exhibited very little wax birefringence, and in *Ficus* CL a layer having positive wax birefringence can be seen. Positive birefringence of the ECL and negative birefringence of the ICL of *Ficus* disappeared on extraction, which indicates that they are both caused by crystalline waxes, but their orientation differs.

It should be remembered that only crystalline waxes are anisotropic, while individual wax molecules sorbed in cutin are not detected. Studies with polarised light give no information if all waxes are crystalline or if portions of the wax are amorphous and are sorbed as individual molecules within amorphous cutin. At room temperature, about 80% of the wax of *Citrus aurantium* is amorphous (Reynhardt and Riederer 1991). In leaves of *Fagus sylvatica* and *Hordeum vulgare*, about 72% and 48% of the waxes are amorphous, respectively (Reynhardt and Riederer 1994). Thus, a large fraction of the total wax is amorphous and must be somewhere on or in the cutin, but it cannot be localised with polarised light.

There is no hint in Fig. 1.6 that epicuticular waxes contributed to birefringence of cuticle cross-sections. This is amazing, since epicuticular waxes occur in many species in substantial amounts (Sect. 1.3). Epicuticular waxes are definitely crystalline, at least a fraction of them (Jeffree 2006; Jetter at al. 2006). They were not seen with polarised light, and this may be a problem of resolution, or they may go unnoticed because their orientation is not uniform. This is unfortunate, because the contribution of epicuticular waxes to barrier properties of cuticles is an important and controversial issue.

1.4.2.2 Scanning Electron Microscopy

Epicuticular waxes have been extensively studied with SEM. Each plant species exhibits a characteristic fine structure which is a reflection of its chemical constitution (Fig. 1.3). The most recent review is that by Jeffree (2006), who classifies surface waxes as granules, filaments, plates, tubes, rods and background wax films. Jeffree (2006) remarks that the range of types of epicuticular wax structures recognised very nearly consumes the available words in the English language. The SEM does not tell us if epicuticular waxes are amorphous or crystalline. If the wax bloom is sufficiently thick to scatter light, the leaf and fruit surfaces appear matt, bluish and glaucous. Bright, green and glossy leaves have a smooth layer of epicuticular wax that reflects light effectively. Thickness and structure of this wax layer cannot be investigated with the SEM, because wax-free cutin and a smooth wax layer look very similar.

Microcrystalline wax blooms have two obvious functions. Light reflection can reduce heat damage to leaves, and it renders their surfaces difficult to wet. This prevents leaching of solutes from the apoplast during rain. The function of epicuticular waxes as a barrier to solutes is a matter of debate and conjecture, because it has not been investigated or at least not published. Foliar application of chemicals requires that leaves are wet, and this is realised by adding surfactants. This assures that aqueous spray droplets are retained, but this is only an indirect effect, and it is not known whether permeability of cuticles is affected by the wax bloom. If epicuticular waxes occur as a continuous wax layer on top of the CP, this would have a substantial effect on water and solute permeability (Chap. 4). In glossy leaves which have little microcrystalline wax bloom, wax crusts can be seen with the SEM. Is there such a continuous wax film under the wax bloom? Haas and Rentschler (1984) painted the adaxial leaf surface of blackberry leaves (*Rubus fruticosus*) with cellulose nitrate dissolved in amyl acetate (6% w/v). After evaporation of the solvent, it was possible to strip off the cellulose nitrate film. The surface of the cuticle looked perfectly smooth after stripping, and the wax bloom was entrapped in the film. It is possible that a smooth wax layer remained on the cuticle after stripping, because it did not adhere to polar cellulose nitrate. Transpiration of leaves before and after stripping was not measured, but surface wax entrapped in the cellulose nitrate and total wax obtained by washing of adaxial leaf surfaces with chloroform were analysed. Total wax amounted to $14.4\,\mu g\,cm^{-2}$, most of which was located on the surface ($12.9\,\mu g\,cm^{-2}$). Epicuticular wax contained mainly alcohol acetates (36%) n-alcohols (30%) and n-alkyl esters (25%), while major components of intracuticular waxes were fatty acids (20%), alcohols (44%) and alcohol acetates (28%). Triterpenoid acids were detected only in the intracuticular wax.

Jetter et al. (2006) have criticised this approach, because partial extraction of intracuticular waxes by amyl alcohol cannot be precluded. Jetter et al. (2000) used cryo-adhesive sampling to obtain epicuticular wax from adaxial surfaces of *Prunus laurocerasus* leaves. Total cuticular wax was $28\,\mu g\,cm^{-2}$, and epicuticular wax amounted to $13\,\mu g\,cm^{-2}$. The epicuticular wax consisted exclusively of aliphatic constituents, while intracuticular wax contained large amounts of triterpenoids and small amounts of aliphatics. The effect of removal of epicuticular waxes on water permeability was not investigated, and it is not clear whether cryo-adhesive sampling also removes the background wax film completely.

A recent study using the atomic force microscope (AFM) and vital leaves of *Euphorbia latyrus, Galanthus nivalis* and *Ipheion uniflorum* revealed that progressively the surface of the cuticle is covered completely with monomolecular and then bimolecular layers (Koch et al. 2004). The surface of the cuticle was first cleaned from surface wax with an epoxy resin glue. Parallel to the formation of wax layers, rod-like crystals arose that grew at their tips. This is a fascinating study, because it demonstrates within 80 min the presence of highly ordered mono- and bilayers which developed rapidly. In contrast to intracuticular wax, where interference by the cutin polymer with crystallisation is probable, this problem does not arise when wax layers form on the surface of the cuticle proper. Stripping of surface wax with epoxy glue in the study by Koch et al. (2004) did not appear to damage the cuticle or the leaf, and the method seems to be suitable to study effects of surface wax on permeability. Barrier properties of lipid monolayers are addressed in Sect. 4.6.

1.4.2.3 Transmission Electron Microscopy

Before cuticle cross-sections can be viewed with the TEM, leaf tissue is fixed with glutaraldehyde, stained en bloc with OsO_4 or $KMnO_4$, dehydrated with ethanol or acetone, infiltrated with epoxy resins (Epon-Aradite) and then polymerised at 60°C. From the block, ultra-thin sections are cut, which are usually contrasted with aqueous uranyl acetate and lead citrate. It is not very likely that cuticular waxes survive these procedures without change in structure or dislocation. The resin monomers may dissolve waxes, and when the resin is cured at 60°C most waxes will become fluid and redistribute. Epicuticular waxes or remnants of them are rarely seen in the TEM, indicating that they have disappeared. Rather strangely, solvent properties and melting behaviour of waxes in Epon-Araldite seem not to have been investigated so far, and localisation of waxes in thin sections should not be attempted. Hence, fine structure seen with the TEM is that of cutin, cutan, polysaccharides and polypeptides.

Unstained cuticles appear in the TEM similar to the embedding media, both of them being polyester polymers composed mainly of carbon, hydrogen and oxygen. Staining is necessary to obtain micrographs with fine structure. The stains contain atoms with higher masses, and they absorb electrons much more effectively. Stains either bind selectively to functional groups or they react with them. In either case, they must diffuse into the tissue when staining is en bloc or in the embedding medium during section staining.

Glutaraldehyde is used routinely to preserve cytological details. It is not known if it affects fine structure of cuticles. OsO_4 is a non-ionic and lipophilic stain. At 25°C solubility in water is small (7.24 g/100 g water), while in carbon tetrachloride 375 g/100 g can be dissolved. OsO_4 is used as a buffered aqueous solution. It is an oxidising agent, and converts olefins to glycols. It also oxidises peroxides (Budavary 1989). Cutin acids with double bonds do not occur frequently in cutin monomers (Holloway 1982b), but in cutin from *Clivia* leaves, 18% of the identified cutin acids was 18-hydroxy-9-octadecenoic acid (Riederer and Schönherr 1988). In cuticles lacking unsaturated cutin acids, the contrast after en bloc staining with OsO_4 is probably caused by sorption in cutin, while in *Clivia* cuticle oxidation of double bonds may contribute to fine structure. We could not find any study dealing with changes in chemistry of cutin acids following treatment with OsO_4.

$KMnO_4$ is a salt; it has a high water solubility and is used as aqueous solution. It is a very strong oxidising agent which breaks double bounds and converts alcoholic

OH-groups to carboxyl groups. It oxidises cellulose, and glass or asbestos filters must be used for filtration of $KMnO_4$ solutions. During oxidation MnO_2 is formed, which is not water-soluble and precipitates (Falbe and Regitz 1995). The contrast after en bloc staining with $KMnO_4$ most likely arises due to insoluble MnO_2 precipitates which form at positions where double bonds and hydroxyl groups were present. Epoxy fatty acids which occur in large amounts in *Clivia* cuticles are likely converted to vicinal alcohols, which subsequently may be oxidised to carboxyl groups. To our knowledge, chemical changes in cutin and cutin acids following treatment with $KMnO_4$ have not been studied.

En bloc staining requires penetration of reagents into the tissue and in the cuticle. Entrance into the CM occurs both from the cell wall and the outer surface of the cuticle. Penetration is not interfered by the epoxy resin as it is not yet present, but crystalline waxes are expected to slow penetration of ionic $KMnO_4$.

Section staining with aqueous ionic compounds is hampered by the epoxy resin but the resin itself remains electron lucent, showing that it does not contain reactive functional groups which could react with uranyl acetate, lead citrate or the acidic solutions of iodide and silver proteinate used for localising epoxide groups in cuticles (Holloway et al. 1981). Ions are hydrated and do not penetrate into hydrocarbon liquids or solid waxes (Schönherr 2006). Epoxy resins are expected to be insurmountable barriers to ions. Chemical reaction with epoxy groups is probably limited to those groups exposed on the surface of the thin section.

Holloway (1982a) and Jeffree (2006) have classified cuticles based on most prominent fine structural details. Type 1 cuticles have a polylaminated outer region (CP) which is sharply delineated against a mainly reticulate inner region (CL). The cuticle of *Clivia* is a typical example. In Type 2 cuticles, the outer region is faintly laminate, which gradually merges with the inner mainly reticulate region. Leaf cuticles from *Hedera helix* and *Ficus elastica* and the onion bulb scale cuticle belong to this class. In Type 3 cuticles the outer region is amorphous, no lamellae are visible and the inner region is reticulate (*Citrus limon* (lemon), *Citrus aurantium, Malus* sp., *Prunus laurocerasus, Prunus persica* (peach), *Pyrus communis*). Type 4 cuticles are all reticulate (tomato and pepper fruit cuticles and leaf cuticles of *Vicia faba* (broad bean), *Citrus sinensis* (orange) are typical examples). There are two more types (all lamellate, or all amorphous) but we have not studied their permeability in detail.

There is no obvious correlation between the above classification and permeability, except that tomato and fruit cuticles are much more permeable than leaf cuticles of types 1–4. This is not too surprising, since waxes determine water and solute permeability and they are not visible in TEM. There is no hint how lamellae of the CP affect permeability. Polar polymers most likely contribute to the fibrillar reticulum, but its three-dimensional structure has not been studied by serial sectioning. In most cuticles the density of the fibrillar network decreases towards the cuticle proper, and they become more faint. Studies with *Clivia* cuticles demonstrate that appearance and fine structure of transverse sections change during development and aging of cuticles (cf. Fig. 1.8).

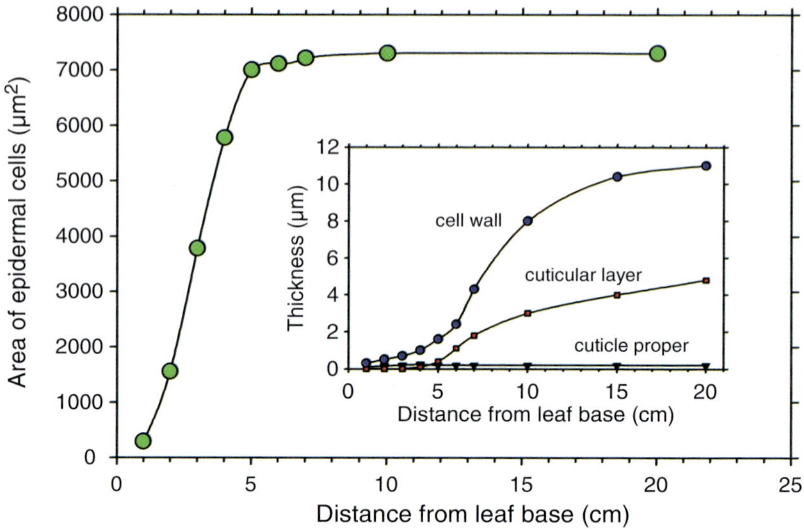

Fig. 1.7 Development of epidermal cells and adaxial epidermis of *Clivia miniata* leaves. (Redrawn from Riederer and Schönherr 1988)

The adaxial cuticle of *Clivia miniata* leaves has a laminated CP, and is ideally suited to study cuticle development. *Clivia* is a monocot, and leaves grow at their base such that cuticle age increases in direction to the leaf tip. Size of epidermal cells, thickness of cuticles and cell wall, cutin composition, cutin biosynthesis and fine structure have been investigated as a function of position, that is of age (Mérida et al. 1981; Schmidt and Schönherr 1982; Lendzian and Schönherr 1983; Riederer and Schönherr 1988).

Between position 1 cm and 5 cm, the projected area of epidermal cells increased about ninefold from 800 μm^2 to 7,000 μm^2. Afterwards, cell area no longer changed. At the same positions, cell length increased from 50 to 250 µm (Riederer and Schönherr 1988); that is, epidermis cells increased both in length and width up to position 5 cm (Fig. 1.7).

The CP was synthesised first and its thickness increased up to 3 cm from leaf base. Fine structure also changed. Maximum thickness of CP was about 200–250 nm, and it increased no further between 3 and 20 cm (Figs. 1.7 and 1.8 inset). Starting at position 3 cm the cuticular layer developed, and it increased in thickness up to position 20 cm from leaf base. Fine structure of the CL (Fig. 1.8) and chemistry (Riederer and Schönherr 1988) changed significantly.

Lamellation of the CP is best seen at 3 cm from base. At higher positions (>4 cm) the central part of the CM has little contrast, probably because OsO_4 does not penetrate because the CP is incrusted with waxes. The cuticular layer starts to develop at position 3 cm and increases in thickness at higher (older) positions. The CL is initially reticulate, but in old and mature positions fine structure has disappeared. It is

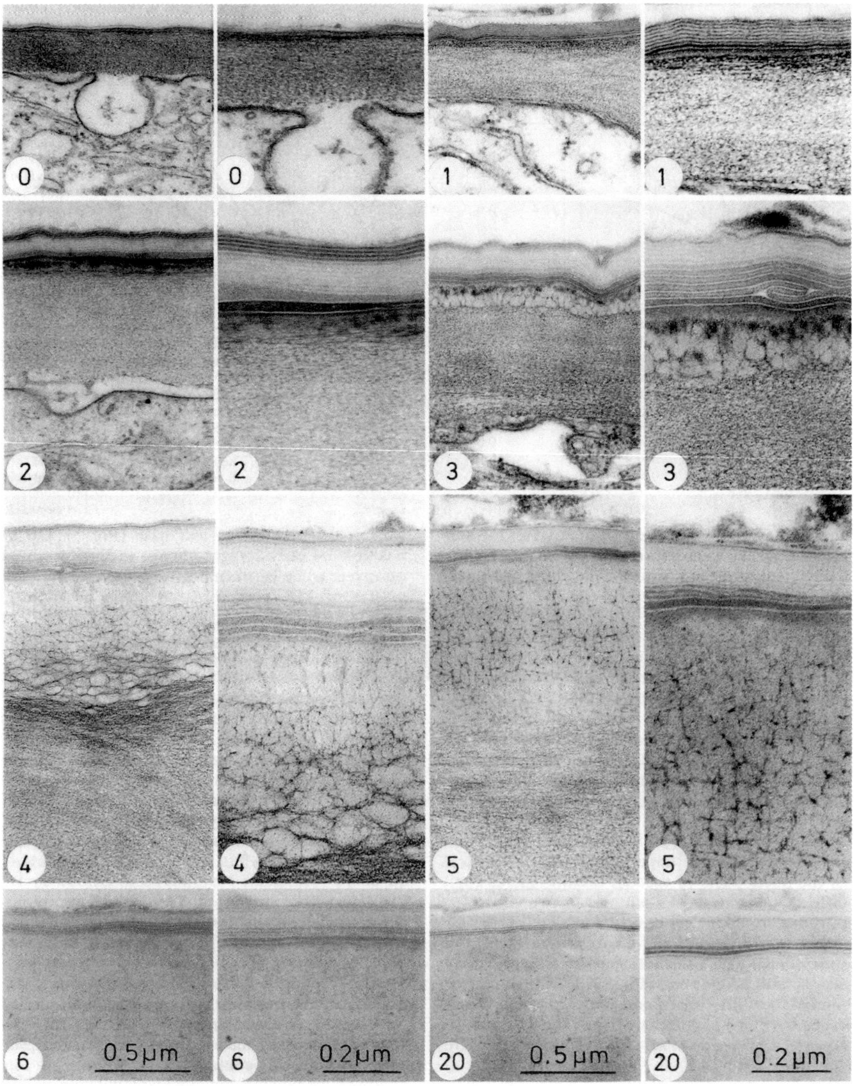

Fig. 1.8 Transmission electron micrographs of transverse sections of adaxial cuticles from *Clivia miniata* at different stages of development. Numbers at the lower *left corners* refer to distance from leaf base. Fixed en bloc with OsO_4, and sections were stained with uranyl acetate and lead citrate. (Taken from Riederer and Schönherr 1988)

not known if incrustation with waxes prevents penetration of OsO_4 or if polar reactive functional groups have disappeared or are covered up. At position 20 cm cutan occurs in very large amounts (Fig. 1.2), and during transformation of ester cutin to cutan epoxy groups and double bonds are most likely consumed. It is believed that the dark fibrillar network (5 cm) marks the location of polar polymers embedded in cutin, but it is not known why it is no longer visible in old cuticles.

The chemical nature of the CP is a matter of debate (Jeffree 2006). Some workers believe that electron-lucent lamellae are waxes, while the more dense lamellae are made of cutin. This is pure speculation, because it has not been established that typical ester cutin is present in the CP at all. The mass of the CP is very small relative to the total mass of the CM, and constituents that occur only in traces might be lost during processing. After transesterification of cutin from positions 2–3 cm where the CP contributes most of the mass of the CM (Fig. 1.8), only nine n-fatty acid (C_{12}, C_{14}, C_{16} and C_{18}) homologues have been identified which amounted to 52% of the total mass of cutin. The most frequent cutin acid (11%) was 9,10-epoxy-18-hydroxyoctadecanoic acid (Riederer and Schönherr 1988). It is difficult to envision a polymer composed of 50% of simple fatty acids.

The CP of *Clivia* cuticle survived extraction of CM with chloroform (Fig. 1.9) and exhibited heavy contrast. Electron-lucent lamellae were preserved, and this is unlikely to happen if they were made of waxes. The CL was differentiated into an external (ECL) and internal (ICL) layer, with large differences in contrast. Since the specimen was extracted, penetration by $KMnO_4$ was not hindered by waxes, and failure of the ECL to develop contrast indicates that reactive groups had been eliminated during cutan formation.

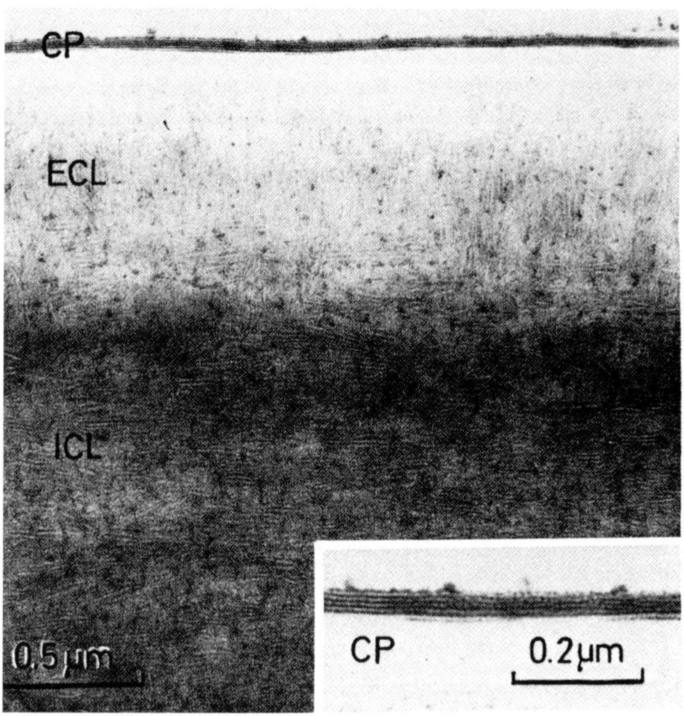

Fig. 1.9 Transverse section of a polymer matrix membrane obtained from the adaxial surface of a young *Clivia miniata* leaf. The MX was stained with $KMnO_4$ prior to embedding, and sections were stained with uranyl acetate and lead citrate. (Taken from Schmidt and Schönherr 1982)

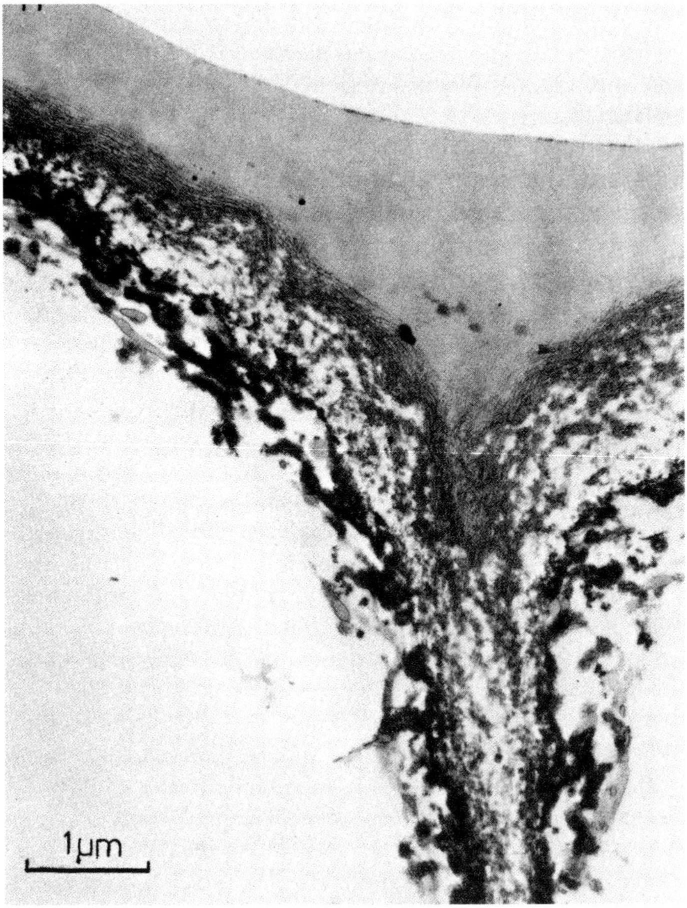

Fig. 1.10 Transverse sections of *Clivia* polymer matrix treated with BF_3–MeOH. Sections were stained with lead citrate. (Taken from Schmidt and Schönherr 1982)

The presence of cutan in the ECL is clearly seen in a specimen extracted with chloroform and depolymerised with BF_3–MeOH (Fig. 1.10). This treatment eliminated ester cutin and left cutan and the polar polymers behind. These polar polymers strongly reacted with lead citrate applied as section stain. Cutan did not exhibit any fine structure, and it is not known if this is due to failure of uranyl acetate to penetrate cutan and/or the embedding medium.

Periclinal penetration of $KMnO_4$ during en block staining was considerably faster in electron dense lamellae than in electron-lucent lamellae of *Agave* and *Clivia* cuticles (Wattendorff and Holloway 1984), but the contribution of the lamellated CP to water or solute permeability of CM has not been studied and is not known. Schmidt et al. (1981) studied water permeability of isolated *Clivia* CM and MX. Water permeability of young and mature CM was the same, but extracting waxes increased permeance by factors of 438 and 216 in young and mature CM respectively. This demonstrates that water permeance of MX decreased during leaf development, when MX mass increased from 0.4 to 0.7 mg cm^{-2}. The lower permeance of MX from mature leaves indicates that cutan has a lower permeability than cutin (Figs. 1.8, 1.9 and 1.10).

1.4.3 Cuticle Synthesis

Lendzian and Schönherr (1983) studied cutin synthesis in adaxial cuticles of *Clivia miniata* leaves. Aqueous solutions of ^{3}H-hexadecanoic acid buffered at *p*H5 were applied as 10 μl droplets to the surface of detached leaves. The leaves were incubated in the dark at 100% humidity for 24 h. After incubation, the leaves were exhaustively extracted with methanol/chloroform (1:1) in a Soxhlet apparatus to remove ^{3}H-hexadecanoic acid. After drying, an autoradiograph was made (Fig. 1.11). It shows black spots at the positions of the droplets. Their intensity decreased from the base to the tip of the leaves, except that little blackening was obtained at the leaf base. Greatest intensity can be seen at the position 1–5 cm from leaf base, and the droplets are not circular but oblong. At higher positions most spots are circular, but some are irregular because not all droplets spread evenly and were segments of spheres.

The ^{3}H-radioactivity was obviously immobilised in the cuticle, and was called ^{3}H-cutin. When depolymerised with BF_3–MeOH, the radio-TLC showed at least four peaks which were not identified, but the methyl ester of hexadecanoic acid was not detected as depolymerisation product. The most likely sequence of events is as follows: ^{3}H-hexadecanoic acid administered to the cuticle surface penetrated into epidermal cells and was hydroxylated. These cutin acids somehow reached the cuticle and were attached to cutin, probably catalyzed by cutinases.

Maximum rates of ^{3}H-cutin synthesis coincide with regions of maximum expansion of epidermal cells (Fig. 1.7). At a hexadecanoic acid concentration of 0.023 g l^{-1} as donor, the authors calculated maximum rates of ^{3}H-cutin synthesis of around 0.3 μg cm^{-2} h^{-1}. The ^{3}H-label was attached to carbon atoms number 9 and

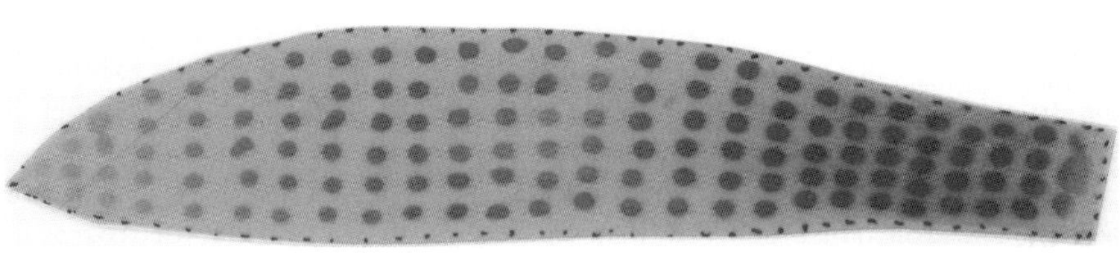

Fig. 1.11 Autoradiograph of the adaxial surface of young *Clivia miniata* leaf of 17 cm length. Droplets containing ^{3}H-hexadecanoic acid were placed on the cuticle, and incubated for 24 h in the dark at 100% humidity. Before applying the X-ray film, the leaf was extracted exhaustively with chloroform/methanol to remove all soluble radioactivity. (Taken from Lendzian and Schönherr 1983)

10. One of them is eliminated during hydroxylation in synthesis of cutin acids of the C_{16}-family. Thus, the specific radioactivity of the cutin acids was only half as high as that of ^{3}H-hexadecanoic acid, and the maximum rate of ^{3}H-cutin synthesis should be twice as high, that is 0.6 μg cm^{-2} h^{-1}. At this rate, 14.4 μg cm^{-2} of ^{3}H-cutin could be synthesised in 24 h. Between positions 3 and 5, the ester cutin fraction increased by about 50 μg (Fig. 1.2). We do not know the growth rate of leaves and epidermis cells. If they need a day for 1 cm, about one third of the cutin synthesised would belong to the C_{16}-family.

About 2% of the weight of the CM were C_{16}-cutin acids at position 2–3 cm (Riederer and Schönherr 1988). At 5–6 cm, C_{16}-cutin acids amounted to about 25%. This shows that C_{16}-cutin acids occur in *Clivia* CM, but it is also clear that ester cutin is made up primarily by C_{18}-cutin acids. Some of the ^{3}H-hexadecanoic acid molecules may have been elongated and turned into C_{18}-cutin acids. Unfortunately, the radioactive cutin acids were not identified chemically.

These studies demonstrate that *Clivia* cuticles are very dynamic structures that greatly change during development. The CP appears first and is maintained, but thickness of the CL and its chemical composition undergo major changes. Cutin synthesis occurred even at the tip of the leaf. After cell expansion is complete, non-ester cutin occurs in large amounts because ester cutin is converted into non-ester cutin (Fig. 1.10). There are no comparable studies of epidermis and cuticle development in other plant species.

1.4.4 Lateral Heterogeneity

The cuticle over ordinary epidermal cells (pavement cells) covers the surfaces of stems, leaves, flowers and fruits. Many mechanistic studies into permeability of cuticles were carried out using isolated cuticular membranes which lack trichomes and stomata. Much less is known about the involvement of specialised epidermal cells in water and solute permeation. However, trichomes and stomata occur on most leaves and stems (Glover and Martin 2000: Bird and Gray 2003), and it is well-established that permeability of cuticles over these special structures differs from that over pavement cells (Strugger 1939; Bauer 1953; Franke 1960; 1967; Meier-Maercker 1979). The cuticle over guard cells and trichomes is often traversed by aqueous pores, and they play a decisive role in foliar penetration of ionic solutes (Schönherr 2006). This aspect of foliar penetration is treated comprehensively in Chap. 5.

Problems

1. What is the average thickness of a cuticle, and what are the upper and lower values of very thick and very thin cuticles?

2. In textbooks it is usually stated that cutin is a polymer composed of hydroxylated fatty acids cross linked via ester bonds. Is this statement correct, partially correct or wrong?
3. What is the average amount of cuticular wax?
4. What is the chemical composition of cuticular waxes?
5. What are the major problems often occurring in wax analysis?
6. Is there an easy way to separate epicuticular waxes from intracuticular waxes?
7. Before looking at cuticles with the TEM, they are often treated with OsO_4 and $KMnO_4$. What is the reason for this treatment and how do these chemicals react with cuticles?
8. Why are cuticles heterogeneous membranes? What is the difference between transversal and lateral heterogeneity?

Solutions

1. Most cuticles are about 2–3 μm thick (*Citrus aurantium*, *Hedera helix*, *Prunus laurocerasus*); however, depending on the species thickness of cuticles can vary between 30 nm (leaf cuticle of *Arabidopsis*) and 30 μm (fruit cuticle of apple).
2. This statement is partially correct. Plant cuticles are composed only to a certain degree of hydroxylated fatty acids which are cross-linked via ester bonds. Many cuticles contain cutan, which is less well characterised and cross-linked by other bonds (ether bonds and probably direct C–C-bonds) than ester bonds. Therefore, it cannot be degraded by transesterification reactions. Furthermore, there are polar compounds (carbohydrates, proteins and phenols) forming a small but important fraction of the cuticle mass.
3. On average, the cuticle contains about $100\,\mu g\,cm^{-2}$. However, depending on the species, wax coverage can vary tremendously between 10 (leaf cuticle of *Citrus aurantium*) and 400 (leaf cuticle of *Nerium oleander* (oleander)) to $3{,}000\,\mu g\,cm^{-2}$ (fruit cuticle of *Malus domestica*).
4. Cuticular waxes are in most cases composed of linear long-chain aliphatic compounds and cyclic terpenoids. Linear long-chain aliphatics are composed of different substance classes (e.g., acids, aldehydes, alcohols, alkanes, secondary alcohols and esters) with chain length ranging from C_{20} to C_{36}. Esters composed of primary fatty acids (C_{16}–C_{36}) and alcohols (C_{20}–C_{36}) have chain lengths between C_{36} and C_{70}.
5. A major problem often encountered in wax analysis is the fact that gravimetrically determined amounts are often higher than wax amounts determined by gas chromatography. The reasons for this observation are variable and not fully understood. Gravimetric determination of wax amounts could lead to an overestimation, whereas determination by GC could lead to an underestimation of wax amounts.
6. Different approaches (e.g., varying extraction times and mechanical removal of waxes) have been suggested, but most of these procedures can be questioned.

Even if there is no epicuticular wax film visible with the SEM, it is not clear whether there is still a thin mono- or bimolecular layer of wax on the outer surface of the cuticle.

7. Treating cuticles with chemicals before looking at them with the TEM is necessary to increase their contrast. OsO_4 is lipophilic and thus is sorbed to lipophilic cutin domains. In addition, it oxidises double bonds. $KMnO_4$ is a strong oxidising agent breaking double bounds and converting alcoholic OH-groups to carboxyl groups. In cuticles, OH-groups of carbohydrates are probably oxidised by $KMnO_4$. However, no systematic studies concerning how OsO_4 and $KMnO_4$ exactly react with cuticles have been carried out.
8. Cuticles are heterogeneous membranes because they are composed of cutin, cuticular waxes and polar polymers. The cutin polymer itself is composed of a lipophilic and polar fraction. Cuticles are characterised by transversal heterogeneity. A thin outer layer called cuticle proper (CP) of unknown composition is succeeded by a much thicker inner layer called cuticular layer (CL) composed of cutin and polar polymers. Lateral heterogeneity arises because, besides normal epidermal cells, stomata and trichomes occur in most leaf epidermises.

Chapter 2
Quantitative Description of Mass Transfer

Plant cuticles are thin membranes. Thickness typically ranges from 1 to 15 μm. The inner surface of the cuticle faces the apoplastic fluid, which is an aqueous solution of mineral ions and small organic molecules. Most of the time, the morphological outer surface of the cuticle of terrestrial plants is in contact with air having humidity ranging from 20% to 90% or higher. During fog or rain, the outer surface of the cuticle can be wet by water.

Water and solutes can cross the cuticle in both directions. Normally humidity is below 100%, and water flows towards the outer surface where it evaporates. This is called cuticular transpiration. During rain and fog, with humidity close to 100% and wet leaf surfaces, the flow in the opposite direction can also happen. In addition, leaching of solutes from the apoplast to the leaf surface occurs (Tuckey 1970). Cuticles are permeable to many solutes such as nutrients, growth regulators, insecticides, fungicides and environmental chemicals. The importance of these transport processes for survival of plants, plant production and environmental pollution is obvious. Hence, plant scientists have studied them extensively for more than five decades. In spite of these efforts, mass transport across cuticles is still not well-understood. Rates of cuticular penetration differ greatly among plants species and solutes, but the reasons are still obscure.

Once synthesised, the cuticle represents a purely physical system. It does not actively interact with water and solutes. Penetration is a physical process. For this reason the term "cuticular uptake" is inappropriate, as it insinuates active participation in mass transfer by plants, cuticles or parts of them. Unfortunately, "foliar uptake" or "cuticular uptake" have often been used in the literature.

In the majority of cases, water and solutes cross cuticles by diffusion, which is based on random molecular motions over small molecular distances. Quantitative description of diffusion involves a mathematic model based on fundamental physical properties. This kind of approach is not very popular with many biologists, who when confronted to Fick's law of diffusion are tempted to change the subject. However, Fick, one of the pioneers in diffusion, was a biologist, more precisely a physiologist, who among other things worked on astigmatism of the eye, functioning of muscles and thermal functioning of the human body (quoted from Cussler 1984).

L. Schreiber and J. Schönherr, *Water and Solute Permeability of Plant Cuticles.*

In this chapter, we consider mass transfer across homogeneous membranes. We define variables and transport parameters, and introduce quantitative transport models. Cuticles are not homogeneous, but our analysis of chemistry–structure–permeability relationships is based on these concepts and calculations.

2.1 Models for Analysing Mass Transfer

The flow of molecules (F) across an interface, a membrane or a stagnant layer is expressed as amount (M) per time. Mol per second or mass per second are frequently used. The choice depends on the specific situation. If the flow is divided by the area (A) across which transport takes place, the flux (J) is obtained with units of $\mathrm{mol\,m^{-2}\,s^{-1}}$ or $\mathrm{kg\,m^{-2}\,s^{-1}}$. For a net flow to occur, there must be a driving force. This may be a difference of concentration (C in $\mathrm{mol\,m^{-3}}$ or $\mathrm{kg\,m^{-3}}$) or a difference in pressure (MPa).

When solvent or solutes are labelled with stable or radioactive isotopes, a flux can be measured even in the absence of a concentration difference. In this case, the driving force is the difference in concentration of isotope. This is a very attractive alternative, because the flux and the difference in concentration are very easy to measure, and there is no need to establish a concentration or a pressure difference across the membrane. Random motion of a molecule marked by stable (^{18}O) or radioactive isotopes (^{3}H, ^{14}C) in an environment of identical but unlabelled molecules is called self-diffusion.

The choice of the model depends on what we want to find out and which information is available. This can be explained on the basis of an idealised experimental setup. A typical transport apparatus is shown in Fig. 2.1. The membrane may be a synthetic polymer, a cuticle or a stagnant layer of water. Those who find it difficult to visualise a stagnant water film separating stirred solutions can think of a gel of 10% gelatine or agar-agar. Transport across a stagnant water film is the simplest case, as there is a water continuum between the two solutions and the membrane. The solute is always surrounded by water. If the membrane is made up of a dense non-porous polymer such as polyethylene, the solute must leave the water phase and enter the polymer, which is not aqueous.

There are basically two ways to study mass transport:

(1) The difference in concentration remains practically constant. This type of experimental setup is called "steady state".
(2) The concentrations of the compartments change with time. This affects the mathematics but not the choice of model. We shall explain these models and their characteristics on the basis of simple experiments.

We want to analyse permeability of a gelatine membrane to urea. Urea is a non-volatile solid, and it must be dissolved in water. The membrane is inserted between the compartments which are open for taking samples (Fig. 2.1). This also assures equal pressure on both sides of the membrane.

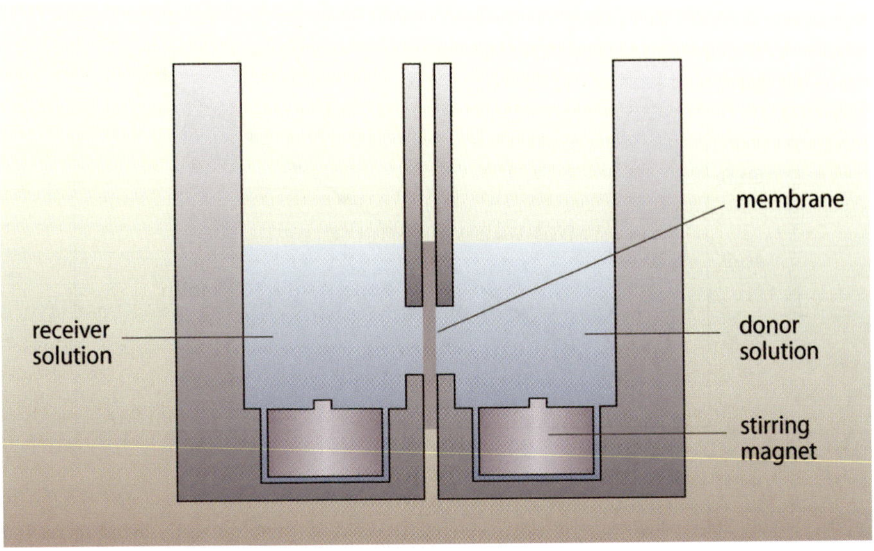

Fig. 2.1 Schematic drawing of a typical transport apparatus. Donor and receiver solutions are separated by a membrane

Water is added to both compartments. Solutions are stirred for mixing. The apparatus is maintained at constant temperature, and at time zero we add to one of the compartments a small amount of urea. This compartment is called donor. The other compartment is the receiver from which samples are withdrawn periodically. Alternatively, the total volume of the receiver is withdrawn for chemical analysis and an equal amount of water is returned to the receiver. Urea concentration in the receiver is measured by a suitable method, and the total amount of urea that penetrated is calculated. Sampling intervals are so short that the concentration in the donor practically remains constant, while at the same time sufficient urea must penetrate into the receiver to allow chemical analysis. These data can be analysed using three different models.

2.1.1 Model 1

We first plot the amount of urea (mol) that penetrates against time (h) and observe that after some time the amount (M) increases linearly with time (t) (Fig. 2.2). The slope of the linear portion of the plot is the flow (M/t), and the intersection with the x-axis is the extrapolated hold-up time (t_e). It takes some time before the first molecules penetrate the membrane and appear in the receiver solution. Some additional time passes before the transport becomes steady. The sum of both is the hold-up time. To make the result more general we divide the flow by the area (A) of the membrane exposed to urea ($10\,cm^2$) and we obtain the flux (J), which amounts

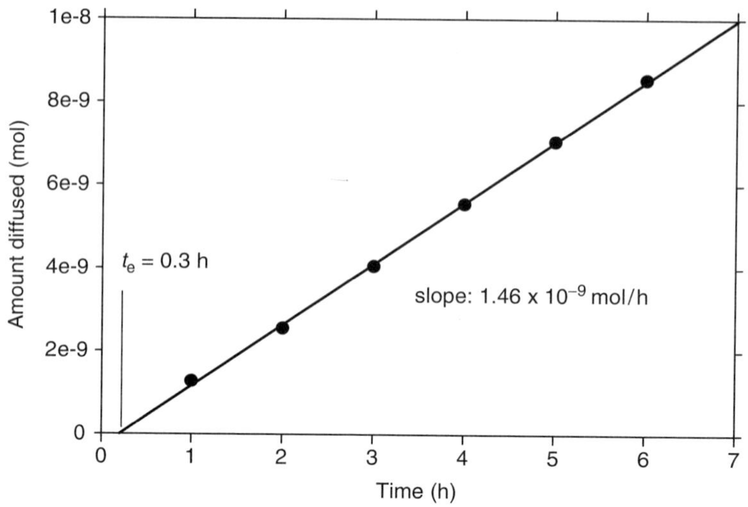

Fig. 2.2 Steady state penetration of urea showing the hold-up time (t_e) and a linear increase in amount diffused per time. Donor concentration was $1 \times 10^{-3}\,\mathrm{mol\,m^{-3}}$

to $1.46 \times 10^{-10}\,\mathrm{mol\,cm^{-2}\,h^{-1}}$, or in SI units $4.06 \times 10^{-10}\,\mathrm{mol\,m^{-2}\,s^{-1}}$

$$J = \frac{M}{t} \times \frac{1}{A} = \frac{F}{A}. \tag{2.1}$$

Increasing the membrane area by a factor of 2 increases the flow of urea by the same factor. J is the normalised flow also called flow density and is independent of membrane area. J is useful to compare results of experiments with different membrane areas.

Next we want to find out what happens when we vary concentration differences. In our experiment, the urea concentration in the receiver remained practically zero and the donor concentration was constant. Now we conduct a number of experiments using the same membrane but different donor concentrations, determine the urea flux and plot it against the concentration of the donor using SI units (Fig. 2.3).

In this particular case we obtain a linear plot (but this must not always be the case), which tells us that the flux is proportional to concentration in the donor or more precisely to the difference in urea concentration between donor and receiver. The slope of the plot is the coefficient of proportionality (P)

$$J = P\left(C_{\mathrm{donor}} - C_{\mathrm{receiver}}\right). \tag{2.2}$$

This mass transfer coefficient is often called permeability coefficient (P), and in our example it has the dimension of a velocity ($4.16 \times 10^{-7}\,\mathrm{m\,s^{-1}}$). If flux measurements are conducted at only one concentration, P can still be calculated from data given in Fig. 2.2, but in this case we would not know if P is constant or if it depends on concentration. In this model, the thickness of the membrane need not be known.

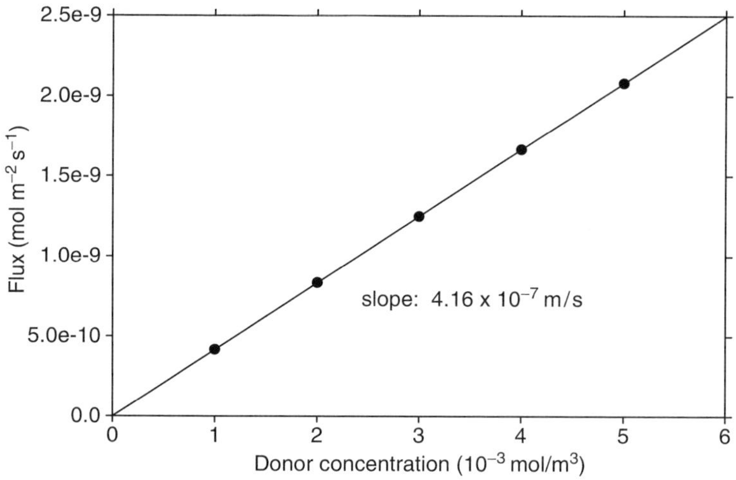

Fig. 2.3 The effect of donor concentration on steady state flux of urea

Hence, model 1 is suitable in all cases when membrane thickness is not known or difficult to estimate accurately.

2.1.2 Model 2

We next want to find out how the flux varies when we change membrane thickness (ℓ). With biological membranes thickness cannot be manipulated, but for the present purpose we use a gelatine membrane which can be prepared at different thicknesses by casting 10% hot aqueous gelatine between two glass plates separated by spacers. After cooling to room temperature, stable membranes are obtained that can be used in experiments. As above, we determine the flux of urea at a given urea concentration of the donor, but we use membranes of different thicknesses (ℓ). When we plot J vs ℓ, we find that the flux is inversely related to membrane thickness. The flux is reduced by one half when the membrane thickness is doubled (Fig. 2.4). At a given donor concentration, the product of flux and membrane thickness is constant

$$J\ell = D\left(C_{\text{donor}} - C_{\text{receiver}}\right). \qquad (2.3)$$

Equation (2.3) is a form of Fick's law of diffusion, and the new proportionality coefficient is the well-known diffusion coefficient (D) having the dimension $\mathrm{m^2\,s^{-1}}$.

D may not be constant, and it often depends on concentration. However, this is easy to find out by conducting experiments using a membrane with constant thickness but different donor concentrations. In our example, the plot J vs $1/\ell$ is linear and has the slope of $1.25 \times 10^{-12}\,\mathrm{mol\,m^{-1}\,s^{-1}}$ (Fig. 2.4 inset). D is obtained by dividing this slope by the concentration difference between donor and receiver (here $1 \times 10^{-3}\,\mathrm{mol\,m^{-3}}$). In our example, this leads to a D of $1.25 \times 10^{-9}\,\mathrm{m^2\,s^{-1}}$.

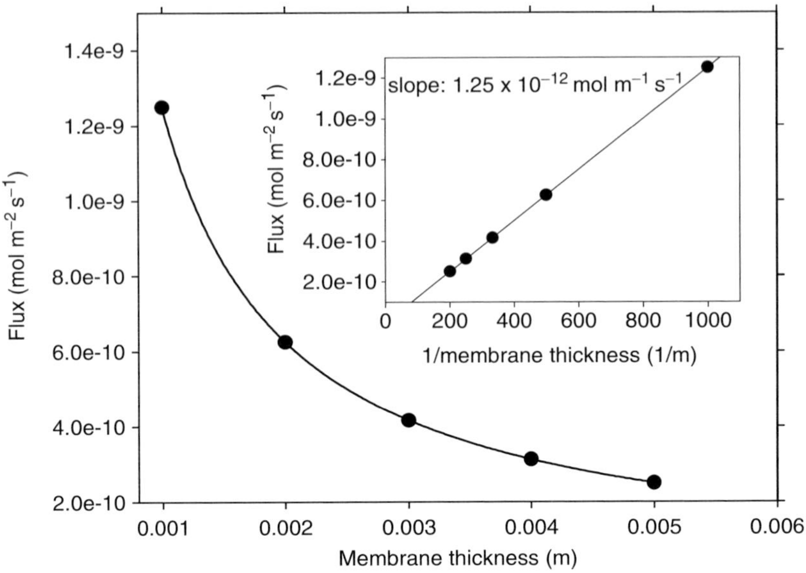

Fig. 2.4 The effect of membrane thickness on steady state flux of urea and calculation of the diffusion coefficient (*inset*). Donor concentration was $1 \times 10^{-3}\,\mathrm{mol\,m^{-3}}$

Multiplying P (2.2) by the membrane thickness in meters yields the permeability coefficient of a membrane having 1 m thickness ($\mathscr{P}$) which has the dimension $\mathrm{m^2\,s^{-1}}$ and is numerically equal to D since the membrane is aqueous

$$P\ell = D = \mathscr{P}. \tag{2.4}$$

Using the data of Fig. 2.3, which were generated using a membrane having 3 mm thickness, we obtain $\mathscr{P} = 1.25 \times 10^{-9}\,\mathrm{m^2\,s^{-1}}$. $\mathscr{P}$ is very useful for comparing permeability of homogeneous membranes to various solutes and to water.

The diffusion coefficient may also be calculated from the extrapolated hold-up time (t_e) and the square of the membrane thickness (ℓ^2):

$$D = \frac{\ell^2}{6t_e}. \tag{2.5}$$

The derivation of (2.5) can be found in Crank (1975). Using the hold-up time given in Fig. 2.2 and a membrane thickness of 3 mm, we obtain $(3 \times 10^{-3}\,\mathrm{m^2})/(6 \times 1{,}200\,\mathrm{s}) = 1.25 \times 10^{-9}\,\mathrm{m^2\,s^{-1}}$. Since we worked with an aqueous membrane, $\mathscr{P} = D$.

2.1.3 Model 3

Mass transfer across a membrane may also be analysed in analogy to a first order chemical reaction. In this case, the change in urea concentration in the receiver with time ($\mathrm{mol\,m^{-3}\,s^{-1}}$) is taken to be proportional to donor concentration:

$$\frac{\Delta C_{\mathrm{receiver}}}{\Delta t} = kC_{\mathrm{donor}}. \tag{2.6}$$

This proportionality coefficient (k) is a rate constant having the dimension $\mathrm{s^{-1}}$. Membrane area and thickness do not enter into calculation. Example calculations will be presented below.

All of these three models have been used in studies of permeability of cuticles. The choice among the models depends on the particular situation. P, D or k are parameters which contain information about properties of cuticles and solutes and their interactions. They are essential when we want to find out why permeability of cuticles from different genotypes differs, and why it depends on the type of solute and on environmental factors. These models and equations are simple and straightforward to use. In spite of this, most researchers in the past measured penetration or "uptake" during only a single time interval. From these data P, D or k can not be calculated, and we are left with huge amounts of uncorrelated data.

2.1.4 Conductance and Resistance

Most students are familiar with Ohm's law, which is analogous to models 1 and 2 except that it deals with electrons rather than molecules (mass transfer). If the ends of a piece of wire (a conductor of electricity) are connected to a battery, a current will flow. Current is the number of electrons that flow per unit cross-section of the wire under a difference of electrical potential (volt):

$$\text{current} = \text{conductance} \times \text{potential difference}. \tag{2.7}$$

If the length of the conductor is also considered, the factor of proportionality is called conductivity:

$$\text{current} \times \text{lenght} = \text{conductivity} \times \text{potential difference}. \tag{2.8}$$

These equations are of the same type as (2.2) and (2.3), which state that mass flux (2.2) or mass flux times membrane thickness (2.3) are proportional to concentration difference. This analogy may ease the apprehension some biologists experience when dealing with Fick's law.

The reciprocal of conductance is the electrical resistance, and 1/conductivity is called resistivity. This convention can also be used with mass conduction. For a perfect analogy, we rename the permeability coefficient (P) from (2.2) and call it permeance (P) and its reciprocal value (mass) resistance ($\mathscr{R}$). The reciprocal value of $\mathscr{P}$ (permeability coefficient) from (2.4) then becomes resistivity. This terminology was suggested by Hartley and Graham-Bryce (1980), and we shall be using it throughout this book.

To further illustrate the use and the usefulness of the above models and calculations, we will add a few examples related to problems in the plant sciences. Initially, steady state transport is considered. Later we deal with situations when concentrations change with time.

2.2 Steady State Diffusion Across a Stagnant Water Film

Thin water films occur often in plants. For instance, after spray application, sessile droplets may form on the cuticle, or the droplets may merge into a film if an effective wetting agent was added. The epidermal cell wall is made of cellulose, pectins and water. The water in the outer epidermal wall is stagnant most of the time. As a first approximation, we treat the cell wall as pure aqueous phase. We look at the fate of the foliar nutrient urea that was applied to the leaves, penetrated into the cuticle and just arrived at the cuticle/cell wall interface. Urea molecules cross this water film by diffusion and once they arrive at the plasmalemma they penetrate it and enter the cytoplasm and are metabolised. This maintains the urea concentration at the plasmalemma/cell wall interface at practically zero, while urea is constantly delivered from the cuticle. The situation is depicted in Fig. 2.5.

At the cuticle/cell wall interface, urea is constantly delivered, and this provides a constant urea concentration to the water in contact with the cuticle. This concentration we call C_{donor}. Urea arriving at the plasmalemma instantly disappears in the cell, and C_{receiver} is zero. This establishes a linear concentration gradient across the water film of the epidermal cell, which is maintained as long as the cuticle serves as a constant source (steady state). Since $C_{\text{donor}} - C_{\text{receiver}}$ is constant, the flux across the water film is constant (steady), and according to (2.3) we can analyse this situation using a slightly modified form of (2.3)

$$J = \frac{D}{\ell}\left(C_{\text{donor}} - C_{\text{receiver}}\right). \tag{2.9}$$

D for urea in water at 25°C can be looked up in tables (Cussler 1984). At a concentration of $10^{-3}\,\text{mol}\,\text{l}^{-1}$ it amounts to $1.38 \times 10^{-9}\,\text{m}^2\,\text{s}^{-1}$. Let's assume a thickness of the epidermal wall of 10 μm for calculating $D/\ell = 1.38 \times 10^{-4}\,\text{m}\,\text{s}^{-1}$, and with $C_{\text{don}} = 10^{-3}\,\text{mol}\,\text{l}^{-1}$ the steady state flux of urea is $1.38 \times 10^{-7}\,\text{mol}\,\text{m}^{-2}\,\text{s}^{-1}$. According to (2.2) and (2.3) D/ℓ equals P, which means that the permeance of the epidermal wall is $1.38 \times 10^{-4}\,\text{m}\,\text{s}^{-1}$. This is a very interesting and useful result, because we can use it to prove that solutes applied to the outer surface of the cuticle will not accumulate in the cell wall. Diffusion coefficients in cuticles of solutes are always smaller than $10^{-15}\,\text{m}^2\,\text{s}^{-1}$ (Chap. 6), and with a cuticle thickness of 5 μm we obtain a permeance of $2 \times 10^{-10}\,\text{m}\,\text{s}^{-1}$. This is more than six orders of magnitude

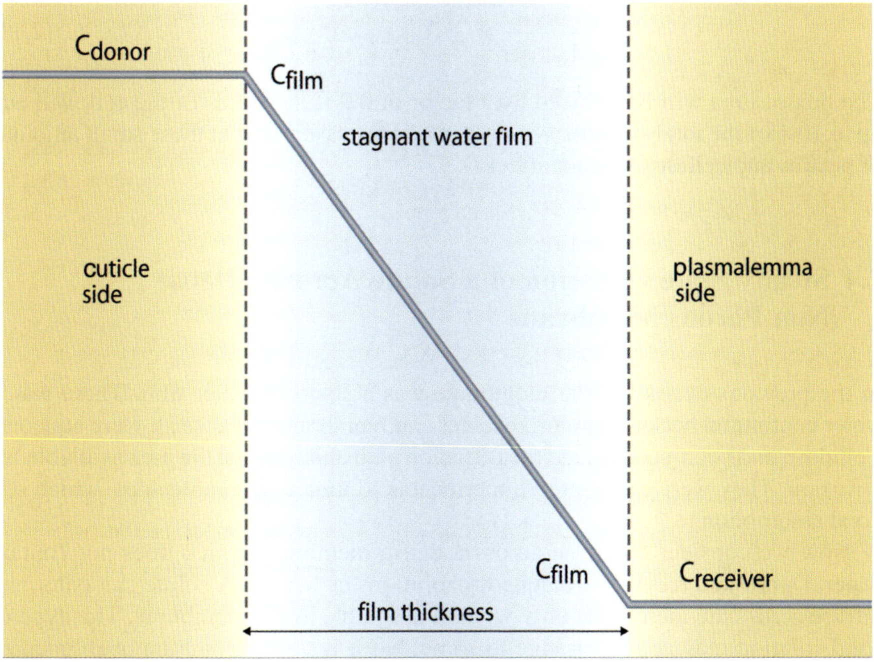

Fig. 2.5 Schematic drawing (not to scale) of a water film having the thickness ℓ and separating plasmalemma and cuticle

smaller than permeance of cell walls, and accumulation of solutes in the cell wall is a highly unlikely event. This can be shown to be true in a more formal way by considering the cuticle and the cell wall as two resistances in series.

2.3 Steady State Diffusion Across a Stagnant Water Film Obstructed by Cellulose and Pectin

Cellulose and pectin occupy some of the volume of the cell wall, and this can be expected to slow diffusion of solutes. The question is how much, and this can again be estimated without having to conduct the experiments. The volume fraction of water is

$$\text{volume fraction of water} = \frac{\text{volume of water}}{\text{volume of cell wall}} \tag{2.10}$$

and in the present example we work with a volume fraction of water of 0.9, which means that 90% of the cell wall is made up of water. Volume is $A \times \ell$, and with the thickness of the cell walls being constant the area fraction of water is equal to the volume fraction of water. In other words, 90% of the cell wall area is occupied by water, and we can modify (2.9) by including the area fraction:

$$J = \frac{A_{\text{water}}}{A_{\text{cell wall}}} \times \frac{D}{\ell} \left(C_{\text{donor}} - C_{\text{receiver}} \right). \tag{2.11}$$

The flux of urea will be reduced by a factor of 0.9 if the solids of the cell wall add up to 10% of the total volume. We have implicitly assumed that these small amounts of pectins and cellulose do not affect D.

2.4 Steady State Diffusion of a Solute Across a Dense Non-Porous Membrane

In the previous examples, the membrane was a stagnant water film. There was a water continuum between donor, receiver and membrane because all were aqueous. Cellulose and pectin obstructed the diffusion path and reduced the area available for diffusion. They were considered impermeable to urea solute molecules, which is a good assumption.

Now we consider diffusion across a dense membrane which does not contain water. Donor and receiver are aqueous solutions as before. A solute can enter and diffuse across the membrane only when it is soluble in the membrane. This type of solid solution may appear strange to some, but it is a very common phenomenon. Stained wax candles, fibres and plastics are examples. Histological sections are usually stained to better visualise different organs and tissues. The cuticle depicted in Fig. 2.6 stained orange using the lipophilic stain Sudan III. The stain dissolved in cutin, and this is a good example of a solid solution.

Cuticles are lipid membranes, composed mainly of cutin and waxes. Solutes can be classified as hydrophilic or lipophilic. Solutes having high water solubility such as amino acids, sugars or inorganic salts are hydrophilic. Solutes which better dissolve in lipid solvents (i.e., ether, hexane, chloroform, octanol, olive oil) are lipophilic. When working with cuticles it is useful to classify solutes according to their solubility in cuticles and in water. The ratio of the two solubilities is the partition coefficient K, which can be determined easily. A piece of cuticle is equilibrated in an aqueous solution of the solute. When equilibrium is established, the concentrations in the cuticle and in water are determined. As it is not easy to determine the volume of the piece of cuticle precisely, molal concentrations ($\text{mol}\,\text{kg}^{-1}$)

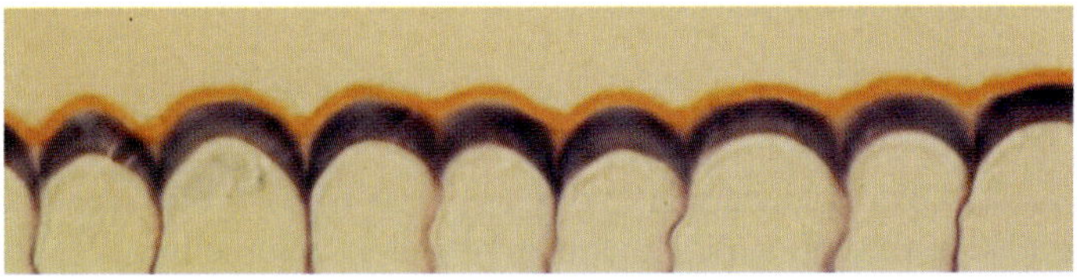

Fig. 2.6 Photomicrograph of a thin cross-section of an adaxial *Clivia* leaf epidermis. The lipophilic cuticle is stained orange with Sudan III, and the acidic cell wall is blue after staining with Toluidine blue at *p*H 4.0

rather than molar concentrations ($\mathrm{mol\,l^{-1}}$) are used

$$K = \frac{\text{concentration in cuticle } (\mathrm{mol\,kg^{-1}})}{\text{concentration in water } (\mathrm{mol\,kg^{-1}})}. \tag{2.12}$$

K is dimensionless, and depending on type of solute it can vary by more than 10 orders of magnitude. If $K = 1$ the solute is equally soluble in the cuticle and in water. If $K < 1$ we consider the solute as hydrophilic. The solute is lipophilic when $K > 1$.

2,4-Dichlorophenoxyacetic acid (2,4-D) is an important substance in research and agriculture. It is an artificial auxin (growth regulator), and can be used as a selective herbicide. It is a weak acid having a $p\mathrm{K_a}$ of 2.73 and a molecular weight of 221 $\mathrm{g\,mol^{-1}}$. The cuticle/water partition coefficient K of non-ionised, protonated 2,4-D varies among plant species, and for pepper fruit cuticle it is about 600 (Sect. 6.1).

In an experiment to determine K, a piece of isolated pepper fruit cuticle is dropped into an aqueous solution of 2,4-D. Since it is a weak acid, the solution must be buffered to maintain a constant pH and known concentrations of ionised and non-ionised 2,4-D molecules. We decide to use a buffer having pH 2.73 which provides us with 50% ionised and non-ionised molecules each [cf. (6.5a)]. The dissociated form is negatively charged, and it is not lipid-soluble because the ionised carboxyl group is surrounded by water molecules. Only the non-ionised 2,4-D molecules dissolve in cuticles and can diffuse in it (Riederer and Schönherr 1984). Initially the cuticle is free of 2,4-D, but as time progresses 2,4-D accumulates in the cuticle until equilibrium is established. At equilibrium, the concentration of non-ionised species in the cuticle is 600 times higher than in the surrounding solution. This appears to be a transport from lower to higher concentration. How can this enigma be explained? Equilibrium means that the chemical potentials in the cuticle and the buffer are equal, not the concentrations. We shall explain this by considering steady state diffusion of a solute across a non-porous homogeneous membrane inserted between two aqueous solutions.

Differential solubility in solutions and in the membrane results in a discontinuity in the concentration–distance profile (Fig. 2.7). If $K > 1$, the concentration in the membrane at the solution–membrane interface is $K \times C_{\mathrm{donor}}$. This is higher than the concentration of the donor, and the concentration gradient across the membrane (which is linear since we are in the steady state) is steeper by the factor K than that between the solutions (Fig. 2.7a).

When the chemical potential is plotted no discontinuity is seen, and the gradient in the membrane is no longer linear, because chemical potential varies with the logarithm of concentration (2.13). The chemical potential (μ) of a neutral species (j) is (Nobel 1983)

$$\mu_j = \mu_j^* + RT \cdot \ln a_j + \bar{V}_j p + m_j g h. \tag{2.13}$$

For diffusion across cuticles this equation can be simplified, as not all of the terms make significant contributions. The gravitational term $m_j g h$ accounts for the work required to raise an object of mass m_j per mole to a vertical height h, where g is the

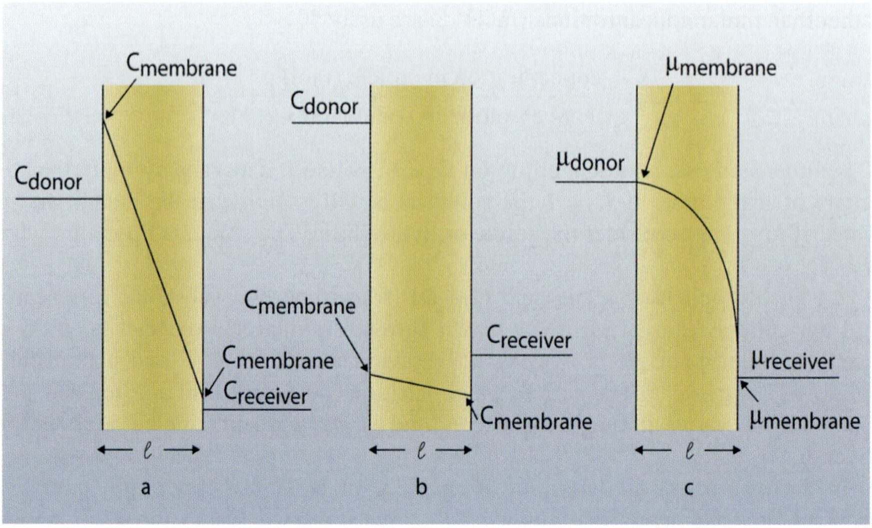

Fig. 2.7 Concentration profiles across a lipophilic membrane. **(a)** Linear profile obtained with a lipophilic solute having a partition coefficient $K > 1$. The concentration in the membrane at the solution/membrane interface is higher by the factor K than concentration of donor and receiver respectively. **(b)** Linear profile with a solute having $K < 1$. **(c)** Profile of the chemical potential (μ) across a membrane. The profile is not linear, and there are no discontinuities at the solution/membrane interface on either side

gravitational acceleration. This term can be omitted since Δh is extremely small in diffusion across cuticles. The effect of pressure (p) on chemical potential can also be neglected, even though a significant pressure gradient exists across cuticles due to the turgor pressure of the leaf cells. However, at 20°C the pressure term in (2.13) amounts to only 200 J mol^{-1} if the partial molar volume of a molecule ($\bar{V}_j$) is 0.2 L and the pressure is 10 bar. By comparison, the term RT is 2.44 kJ mol^{-1} at the same temperature.

The effect of solute activity (a_j) on chemical potential is contained in the second term on the right side of the equation. The activity of a species in a solution or a membrane is usually not known precisely and the concentration (C_j) is used instead, assuming the activity coefficient to be 1.0. Multiplying ln a_j with RT (where R is the gas constant in J mol^{-1} K^{-1} and T the temperature in Kelvin) gives the term RT ln a_j with the units of energy per mole of solute. With these simplifications, the chemical potential of a non-electrolyte (or a non-ionised weak acid) is

$$\mu_j = \mu_j^* + RT \ln C_j. \tag{2.14}$$

Chemical potential, like electrical potential, has no absolute zero. It is a relative quantity, and it must be expressed relative to an arbitrary level. Electrical potential is measured against ground, and in the equation for the chemical potential an unknown additive constant, the reference potential μ_j^* is added. The absolute value for the chemical potential is therefore never known, but this has little consequence as long as chemical potential difference for the same solvent is considered. In calculating the difference in chemical potential between two locations (i.e., across a membrane with water on both sides), the reference potential cancels.

The situation is different when we consider chemical potentials in two adjacent immiscible phases; for instance, water and octanol or water and cuticles. The value of the chemical potential for a solute in the standard state (μ_j^*) depends on the solvent, and in calculating the difference in chemical potential between two different and immiscible phases the two reference potentials do not cancel. For example, if a lipophilic solute is added to a vessel containing water and pieces of cuticle and the vessel is shaken vigorously, the concentrations in the two phases at equilibrium will not be the same. If the solute is more soluble in the cuticle than in water, the activity in the cuticle (a_{cuticle}) will be higher than in water (a_{water}). Since we are in equilibrium, the chemical potential of the solute in water and cuticle is the same ($\mu_{\mathrm{cuticle}} = \mu_{\mathrm{water}}$). Hence, the standard chemical potentials must differ, such that $\mu^*_{\mathrm{cuticle}} < \mu^*_{\mathrm{water}}$.

2.4.1 The Experiment

We want to study diffusion of 2,4-D across an isolated pepper fruit cuticle. This cuticle ($\ell = 10\,\mu\mathrm{m}$, $A = 1\,\mathrm{cm}^2$) is inserted into an apparatus similar to that shown in Fig. 2.1. To donor and receiver chambers, 100 ml citric acid buffer of pH2.73 are added. At this pH, 50% of the 2,4-D molecules are ionised, the other 50% are non-ionised. Donor and receiver solutions are stirred to assure mixing, and once the apparatus has obtained the desired temperature of 25°C 2,4-D is added to the donor solution at a total concentration of $2 \times 10^{-3}\,\mathrm{mol\,l^{-1}}$; hence, the concentration of non-ionised 2,4-D molecules is $1 \times 10^{-3}\,\mathrm{mol\,l^{-1}}$. The receiver solution is withdrawn quantitatively every hour and replaced by fresh buffer. The receiver solution is assayed for 2,4-D.

Results are shown in Fig. 2.8. The amount that diffused into the receiver increased linearly with time. The extrapolated hold-up time was 0.8 h or 2,880 s. The slope of the plot is $1 \times 10^{-8}\,\mathrm{mol\,cm^{-2}\,h^{-1}}$, which amounts to $2.78 \times 10^{-8}\,\mathrm{mol\,m^{-2}\,s^{-1}}$. The linear plot suggests that diffusion was in the steady state. We can check this by comparing the amount diffused in 5 h ($\sim 5 \times 10^{-8}$ mol) with the amount of 2,4-D in the donor, which is 1×10^{-4} mol. Indeed, only 0.05% of the amount in the donor diffused into the receiver, and the concentration of the donor solution practically remained constant. According to (2.2), the permeance can be calculated

$$P = \frac{J}{C_{\mathrm{donor}} - C_{\mathrm{receiver}}} = \frac{2.78 \times 10^{-8}\,\mathrm{mol\,m^{-2}\,s^{-1}}}{1\,\mathrm{mol\,m^{-3}}} = 2.78 \times 10^{-8}\,\mathrm{m\,s^{-1}}. \quad (2.15)$$

In this calculation we used the concentration of non-ionised 2,4-D in the aqueous donor, even though we know (Fig. 2.7a) that the concentration difference across the cuticles was in fact 600 times steeper than that between donor and receiver. Hence

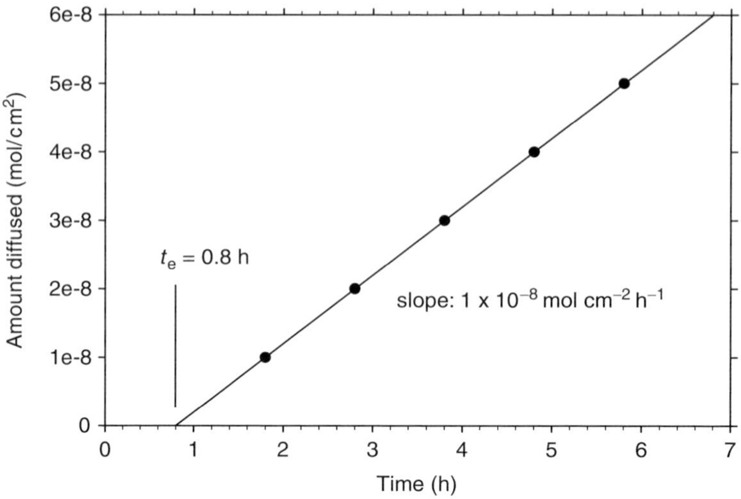

Fig. 2.8 Steady state diffusion of 2,4-D across a pepper fruit CM. Membrane area was $1\,\text{cm}^2$ and donor concentration was $1\,\text{mol}\,\text{m}^{-3}$

the real driving force was much greater than assumed in our calculation. This could be accounted for by including the partition coefficient K in (2.2):

$$J = P\left(KC_{\text{donor}} - KC_{\text{receiver}}\right) = PK\left(C_{\text{donor}} - C_{\text{receiver}}\right). \tag{2.16}$$

If we decide to use model 2 for analysing the data, (2.3) assumes the form

$$J = \frac{DK}{\ell}\left(C_{\text{donor}} - C_{\text{receiver}}\right), \tag{2.17}$$

which shows that permeance P is a mixed parameter which includes the diffusion and partition coefficient and membrane thickness:

$$P = \frac{DK}{\ell}. \tag{2.18}$$

It is a general convention not to include K as part of the driving force, because it is not always known, while concentrations are known or can be measured. If P and K are measured and ℓ is known, D can be calculated as $P\ell/K$. For pepper fruit cuticles we get $4.63 \times 10^{-16}\,\text{m}^2\,\text{s}^{-1}$. Alternatively, we can use (2.5) and calculate D from the hold-up time, and we obtain $5.79 \times 10^{-15}\,\text{m}^2\,\text{s}^{-1}$, which is 12.5 times larger. The reason for this discrepancy is the heterogeneity of cuticles. In deriving the above equations, it was assumed that membranes are homogeneous in chemistry and structure. Heterogeneity of cuticles is a major obstacle in analysing permeability of cuticles in terms of structure–property relationships (Chaps. 4–6).

2.5 Diffusion Across a Membrane with Changing Concentrations

In the previous examples, the volumes (V) of the receiver and donor solutions were not considered. However, V_{donor} and $V_{receiver}$ must not be equal, and in fact they are often adjusted to extend the duration of the steady state. In deriving an equation allowing C_{donor} and $C_{receiver}$ to vary, the volumes of the compartments are included explicitly (Hartley and Graham-Bryce 1980).

The total amount (M) in the system is assumed constant, and the amount in the membrane is taken to be negligible. Hence, the solute is either in the donor or in the receiver:

$$M_{total} = M_{donor} + M_{receiver} = C_{donor}V_{donor} + C_{receiver}V_{receiver}. \tag{2.19}$$

Initially, when $t = 0$ the donor concentration is C_0 and $C_{receiver} = 0$. The system is in equilibrium when concentrations in receiver and donor are equal (C_∞):

$$C_\infty = \frac{C_0 V_{donor}}{V_{donor} + V_{receiver}}. \tag{2.20}$$

As long as $C_{donor} \gg C_{receiver}$, we can write for the mean flow rate (F)

$$F = \frac{V_{donor}\left(C_{donor,1} - C_{donor,2}\right)}{(t_2 - t_1)} = \frac{V_{donor}\Delta C_{donor}}{\Delta t}, \tag{2.21}$$

where numerical suffixes denote times (t). For experiments allowing major change of ΔC to occur, we must use the differential form of (2.21)

$$V_{receiver}\frac{\mathrm{d}C_{receiver}}{\mathrm{d}t} = -V_{donor}\frac{\mathrm{d}C_{donor}}{\mathrm{d}t} = F = PA\left(C_{donor} - C_{receiver}\right). \tag{2.22}$$

The integral solution can be expressed as

$$-PA\left(\frac{1}{V_{donor}} + \frac{1}{V_{receiver}}\right)t = \ln\frac{C_{donor} - C_{receiver}}{C_0}. \tag{2.23}$$

Equation (2.23) assumes more convenient forms under the following situations. If C_{donor} is held constant by having $V_{donor} \gg V_{receiver}$ or by other means, V_{donor} becomes unimportant and we can write

$$\frac{-PAt}{V_{receiver}} = \ln\frac{C_0 - C_{receiver}}{C_0}. \tag{2.24}$$

If $C_{receiver}$ is held zero or nearly so by having $V_{receiver} \gg V_{donor}$, (2.23) becomes

$$\frac{-PAt}{V_{donor}} = \ln\frac{C_{donor}}{C_0}. \tag{2.25}$$

When working with weak electrolytes, C_{receiver} may be maintained at zero by using a buffer as receiver in which the solute is fully ionised. Lipophilic solutes may be scavenged and trapped in phospholipid vesicles, which maintains the concentration of the aqueous phase of the receiver at practically zero. Rearranging (2.25) leads to

$$\frac{\ln(C_{\text{donor}}/C_0)}{t} = \frac{-PA}{V_{\text{donor}}} = -k. \tag{2.26}$$

When $\ln(C_{\text{donor}}/C_0)$ is plotted vs time, a straight line is obtained with slope $-k$. Equations (2.4)–(2.26) treat mass transfer as a first-order process (model 3), and k is the first-order rate constant. Permeance (P) can be calculated when k, A and V_{donor} are known. Equation (2.26) states that C_{donor}/C_0 decays exponentially with time

$$\frac{C_{\text{donor}}}{C_0} = e^{-kt}, \tag{2.27}$$

and once k has been determined experimentally, the donor concentration can be calculated for any time interval. The time required for the donor concentration to decrease to one half is

$$\ln 0.5 = -kt \tag{2.28}$$

and the half time ($t_{1/2}$) is $0.693/k$. A simple experiment will help to demonstrate the use of the above equations.

2.5.1 The Experiment

Again, we want to study diffusion of 2,4-D across an isolated pepper fruit CM. It is inserted into our transport apparatus (Fig. 2.1). As donor solution, a buffer (1 ml) having *p*H 2.73 is used, and as receiver (10 ml) we use borax buffer with *p*H 9.2. At this *p*H, 2,4-D is fully ionised, and the concentration of the non-ionised species is always zero. The membrane area exposed to solutions is $1\,\text{cm}^2$. The experiment is started by adding 2,4-D to the donor solution at a total concentration of $2\,\text{mol}\,\text{m}^{-3}$. At 24-h intervals the receiver solution is quantitatively withdrawn, replaced by fresh buffer and analysed for 2,4-D. The solute fraction (M_t/M_0) of 2,4-D which penetrated into the receiver and was sampled from it is calculated and plotted against time (Fig. 2.9a). It increases in a non-linear fashion (circles), while the fraction lost from the donor decreases with time (squares). It took nearly 3 days for half of the 2,4-D to penetrate into the receiver. Plotting $\ln(C_{\text{donor}}/C_0)$ vs time results in a straight line with a slope of $-0.25\,\text{day}^{-1}$ (Fig. 2.9b). This is evidence that our mass transport of 2,4-D was in fact a first-order process. Next we calculate $t_{1/2}$ as $\ln 0.5/-0.25\,\text{day}^{-1}$, which is 2.77 days. Permeance is obtained using (2.26) and SI units

$$P = \frac{kV_{\text{donor}}}{A} = \frac{(2.89\times10^{-6}\,\text{s}^{-1})\,(1\times10^{-6}\,\text{m}^3)}{1\times10^{-4}\,\text{m}^2} = 2.89\times10^{-8}\,\text{m}\,\text{s}^{-1}. \tag{2.29}$$

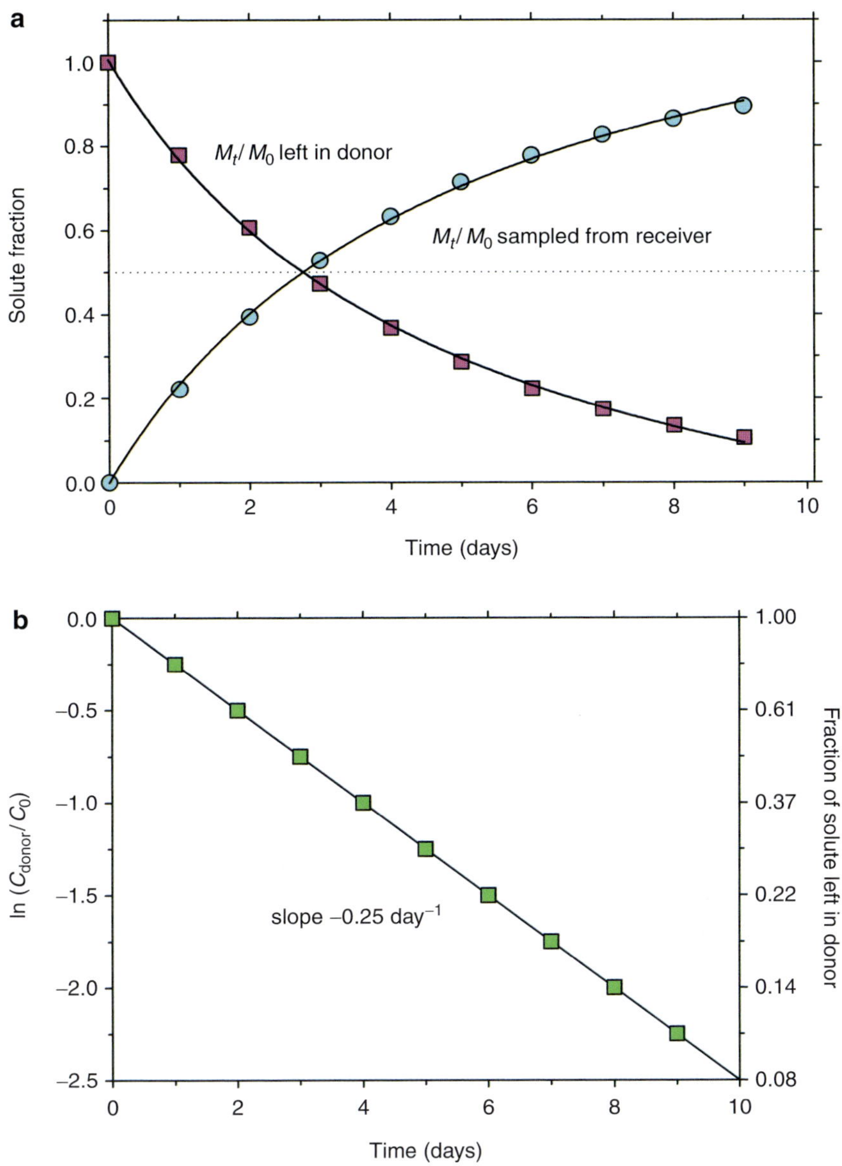

Fig. 2.9 Non-steady penetration of 2,4-D across a pepper fruit CM. The fraction of 2,4-D left in the donor decreases with time (*pink squares*), and the fraction of 2,4-D which was sampled from the receiver (*cyan dots*) increases (**a**). Both fractions add up to 1.0 at all times. The *dotted line* marks the point at which M_t/M_0 is 0.5. Plotting ln (C_{donor}/C_0) vs time results in a linear plot (**b**). The slope represents the rate constant k

An estimate of P may also be obtained from the raw data in Fig. 2.9a. During the first day C_{donor} decreased from C_0 ($1\,\mathrm{mol\,m^{-3}}$) by a factor of 0.221. These data can be used to calculate the flow of 2,4-D from (2.21)

$$F = \frac{V_{\mathrm{donor}}\Delta C_{\mathrm{donor}}}{\Delta t} = \frac{\left(1\times 10^{-6}\,\mathrm{m^3}\right)\left(0.221\,\mathrm{mol\,m^{-3}}\right)}{86{,}400\,\mathrm{s}} = 2.56\times 10^{-12}\,\mathrm{mol\,s^{-1}}. \tag{2.30}$$

This figure must be divided by membrane area and driving force. Using the initial concentration of the donor of $1\,\mathrm{mol\,m^{-3}}$, we obtain

$$P = \frac{F}{AC_{\mathrm{donor}}} = \frac{2.56\times 10^{-12}\,\mathrm{mol\,s^{-1}}}{\left(1\times 10^{-4}\,\mathrm{m^2}\right)\left(1\,\mathrm{mol\,m^{-3}}\right)} = 2.56\times 10^{-8}\,\mathrm{m\,s^{-1}}, \tag{2.31}$$

which is a little smaller than that calculated using (2.29). The donor concentration decreased during the first day from $1\,\mathrm{mol\,m^{-3}}$ to $0.779\,\mathrm{mol\,m^{-3}}$, and this decrease in driving force with time is responsible for the non-linearity seen in Fig. 2.9a. If we want to calculate P assuming steady state conditions, we could use the mean donor concentration (2.21) during the first day, which is the concentration after 12 h of $0.89\,\mathrm{mol\,m^{-3}}$. With this driving force we obtain a P of $2.88\times 10^{-7}\,\mathrm{m\,s^{-1}}$, and this is identical to that calculated from the first-order rate constant (2.29). This calculation nicely shows that the error in P is not really large when we assume constant driving force, while in fact C_{donor} decreased significantly. However, it is essential to use the average donor concentration and not the initial one.

2.6 Determination of the Diffusion Coefficient from Sorption or Desorption Kinetics

So far we have introduced two ways to estimate diffusion coefficients. D can be obtained from steady state rates of diffusion (2.3) or from the extrapolated hold-up time (2.5). D can also be determined from rates of sorption or desorption data. Crank (1975) gives a set of equations for various geometries of solids sorbers (sphere, cylinder, cube or thin sheets). In all cases, fractional sorption or desorption (M_t/M_0) is measured as a function of time, with M_t being the amount sorbed at time t and M_0 being the amount sorbed at infinite time. Thus, M_t/M_0 varies between 0 and 1. The medium may be a gas, a vapour or a liquid.

We restrict our attention to two equations for planar membranes. In deriving the equations it was assumed that both surfaces are accessible to the vapour or sorbate, and that equilibration between membrane surfaces and surrounding medium is instantaneous. If D does not depend on concentration of sorbate in the membrane, it is calculated from the square of the thickness of membrane (ℓ^2) and time ($t_{1/2}$) needed for half maximum sorption ($M_t/M_0 = 0.5$):

$$\left(\frac{t}{\ell^2}\right)_{1/2} = -\frac{1}{\pi^2 D}\ln\left[\frac{\pi^2}{16} - \frac{1}{9}\left(\frac{\pi^2}{16}\right)^9\right]. \tag{2.32}$$

The quantity in square brackets contains only the constant π, and can be evaluated. After rearranging we have

$$D = \frac{0.049}{(t/\ell^2)_{1/2}}. \quad (2.33)$$

This equation is applicable to all temperatures, but D should not depend on concentration. If the latter restriction is abandoned, the appropriate equation is

$$\frac{M_t}{M_0} = \frac{4}{\sqrt{\pi}}\sqrt{\frac{Dt}{\ell^2}} \quad (2.34)$$

and D can be obtained from the slope of a plot M_t/M_0 vs $\sqrt{t}$. This equation can be used to calculate D from sorption or desorption kinetics.

2.6.1 The Experiment

Sorption of water vapour in a thin sheet (1.83×10^{-4} m) of polymethacrylate (PMA) was studied at a vapour pressure (p) of 2,288 Pa and a temperature of 35°C. At this T, saturation vapour pressure (p_0) is 5,623 Pa, hence the partial pressure (p/p_0) was 0.41 and relative humidity 41%. Water content of the polymer at 35°C and $p/p_0 = 1$ was 9.9 mg water per g polymer, which amounts to about 1% by weight. This is a low water content, and the polymer did not measurably swell; that is, the thickness of the sheet did not increase during sorption. In PMA all carboxyl groups are esterified with methyl groups, and the oxygen atoms of the ester group are the sole dipoles in the polymer. Data were taken from the work of Kishimoto et al. (1960), and are shown in Fig. 2.10.

When M_t/M_0 was plotted against time, maximum curves were obtained for both sorption and desorption, and the plateau was reached after about 1,200 s (Fig. 2.10a). It took about 175 s to reach half maximum sorption, which is marked by a dotted line. Plotting data as M_t/M_0 vs the square root of time $\sqrt{t}$. resulted in a straight line up to about $M_t/M_0 = 0.7$ (Fig. 2.10b).

D can be calculated from the slope of the linear portion of the plot using (2.34), which after rearrangement results in

$$D = \frac{\pi(\text{slope})^2\ell^2}{16}. \quad (2.35)$$

With a slope of 0.03896 ($\sqrt{}$s) we obtain a diffusion coefficient of $9.98 \times 10^{-12}\,\text{m}^2\,\text{s}^{-1}$. From the raw data shown in Fig. 2.10a, we can calculate D using (2.33) and a half time of 165 s, and we obtain $9.95 \times 10^{-12}\,\text{m}^2\,\text{s}^{-1}$. These two results are very similar, and for most purposes (2.33) will suffice. However, (2.34) should be used since it uses more information. It also demonstrates if the data can be linearised as stated by theory. Finally, it is difficult to know a priori how long it will take to reach $M_t/M_0 \geq 0.5$. Hence, it is better to calculate D from the initial slope and (2.35).

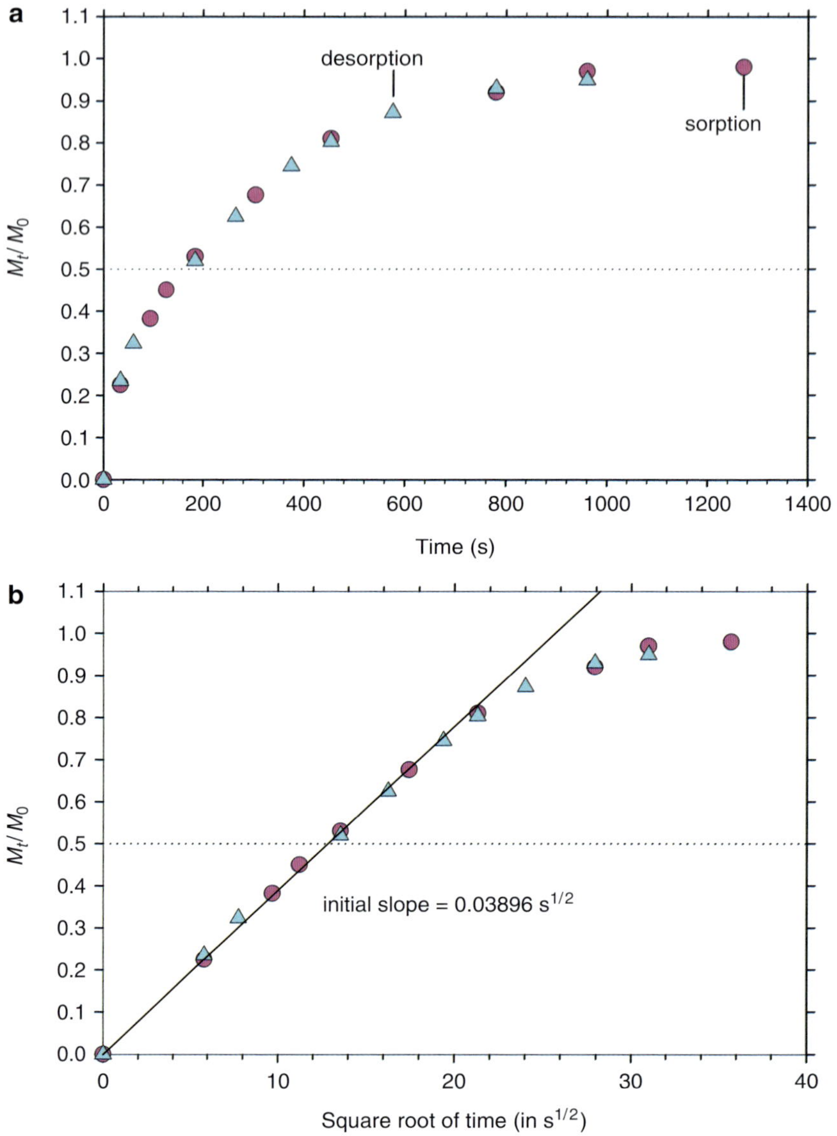

Fig. 2.10 Water vapour sorption (*pink circles*) and desorption (*cyan triangles*) at 35°C in a polymethacrylate membrane having a thickness of 183 μm plotted vs time (**a**) or vs the square root of time (**b**). (Redrawn from data of Kishimoto et al. 1960)

If diffusion coefficients are identical when calculated from sorption and desorption experiments, this is good evidence that D does not depend on concentration. This was the case in our example, since sorption and desorption plots were superimposed up to $M_t/M_0 = 1$ (Fig. 2.10a). This is not too surprising, since water content was only 1%, even at the highest vapour pressure.

2.7 Summary

In this chapter we have presented three basic transport models, and we have shown how permeance (P), diffusion coefficients (D) and first-order rate constants (k) can be calculated from measurements of mass flux and membrane thickness. We have pointed out that it is not a good practice to compare data on mass transfer across cuticles from different species and various solutes when data are given as percent penetrated during a single arbitrary time interval. Such data can not be related to properties of cuticles or solutes.

Most problems can be handled by analogy to the above models and equations. In the chapters to follow, some additional equations are presented for analysing flux data. If new problems should arise which have not been treated here, the readers will find assistance in the books by Crank (1975), Crank and Park (1968), Cussler (1984), Hartley and Graham-Bryce (1980) and Vieth (1991).

Problems

1. What are the numerical values of the resistance in cell wall ($P = 1.38 \times 10^{-4}\,\mathrm{m\,s^{-1}}$) and cuticle ($P = 1.46 \times 10^{-7}\,\mathrm{m\,s^{-1}}$), and what is the total resistance?
2. What are the steady state fluxes of urea and sucrose across a cell wall with $\ell = 1\,\mu\mathrm{m}$ when the concentration difference between donor and receiver is $1 \times 10^{-3}\,\mathrm{mol\,l^{-1}}$? The diffusion coefficient of sucrose in water is $5.23 \times 10^{-10}\,\mathrm{m^2\,s^{-1}}$.
3. In calculations (2.15), we assumed that 2,4-D was either in the donor or in the receiver solutions. Since K is 600, some of the 2,4-D must have been in the cuticle. How much was it, and did this omission significantly affect permeance? Use a specific weight of the cuticle of $1{,}000\,\mathrm{kg\,m^{-3}}$.
4. If the experiment shown in Fig. 2.9 had been terminated after 2 or 3 days, which permeance would have resulted using the steady state assumption, and how much would it differ from the true value?
5. Diffusion of lipophilic solutes in cuticular waxes is a very slow process, and diffusion coefficients in the range of 10^{-18} to $10^{-21}\,\mathrm{m^2\,s^{-1}}$ have been measured. How long would it take to reach $M_t/M_0 = 0.5$ if the wax layer is $2\,\mu\mathrm{m}$ thick?

Solutions

1. The resistance of the cell wall is $7{,}246\,\mathrm{s\,m^{-1}}$, and the resistance of the cuticle is $6.849 \times 10^6\,\mathrm{s\,m^{-1}}$. The sum of the two resistances is $6.857 \times 10^6\,\mathrm{s\,m^{-1}}$, and this is not significantly different from the resistance of the cuticle. Hence, the resistance of the cell wall is negligible.

2. We use (2.9) and obtain the steady state fluxes of urea and sucrose across the cell wall as 1.38×10^{-3} and $5.23 \times 10^{-4}\,\mathrm{mol\,m^{-2}\,s^{-1}}$ respectively.
3. We solve this problem in four steps. (1) The amount of non-ionised 2,4-D in the donor is the product of mass (0.1 l = 0.1 kg) and concentration of donor ($1 \times 10^{-3}\,\mathrm{mol\,l^{-1}}$), which is 1×10^{-4} mol. (2) The concentration of 2,4-D in the cuticle can be calculated using (2.12). The concentration of 2,4-D in the cuticle is the product of the donor concentration ($1 \times 10^{-3}\,\mathrm{mol\,kg^{-1}}$) and partition coefficient (600), which is $0.6\,\mathrm{mol\,kg^{-1}}$. (3) The area of cuticle exposed to the donor was $1\,\mathrm{cm^2}$ and the thickness was 10 μm, which results in a volume of cuticle of $1 \times 10^{-9}\,\mathrm{m^3}$. With a specific weight of cuticle of $1{,}000\,\mathrm{kg\,m^{-3}}$, the mass of the cuticle exposed to the donor is 1×10^{-6} kg. The amount of 2,4-D in the cuticle is the product of mass of cuticle times 2,4-D concentration in cuticle, which amounts to 6×10^{-7} mol. (4) According to Fig. 2.7 the solute concentration in the cuticle during steady state is only half the maximum concentration. Hence, in the steady state 1×10^{-4} mol 2,4-D are in the donor, while only 3×10^{-7} mol are sorbed in the cuticle. This is a negligible amount, and we can safely use the donor concentration of $1 \times 10^{-3}\,\mathrm{mol\,l^{-1}}$ in calculating permeance. However, it should be clear from the calculation that sorption in the cuticle would not have been negligible if the donor volume had been only 1 ml or less.
4. From Fig. 2.9a we read the fractions of 2,4-D that would have been collected from the receiver, had we taken only one sample after the second (0.39) or the third day (0.53) respectively. Using (2.30) and these figures, we calculate the flow (F) in 2 days or 3 days as 2.25×10^{-11} and $2.04 \times 10^{-11}\,\mathrm{mol\,s^{-1}}$ respectively. Before we can calculate P using (2.31), we must decide which donor concentration to use. We are ignorant, and use the initial concentration of $1\,\mathrm{mol\,m^{-3}}$ and obtain a P of $2.25 \times 10^{-7}\,\mathrm{m\,s^{-1}}$ if we terminated the experiment after 2 days, and $2.04 \times 10^{-7}\,\mathrm{m\,s^{-1}}$ if the experiment lasted 3 days. The true P calculated from the rate constant k is $2.89 \times 10^{-7}\,\mathrm{m\,s^{-1}}$; hence, we underestimated P by factors of 1.28 and 1.42 respectively. This may not seem a lot, but we must remember that the 2,4-D concentration of the receiver was zero throughout, because we used a pH of 9.2 and all 2,4-D molecules reaching the receiver were ionised. Without this trick, ΔC across the membrane would have been much smaller and the error in P much larger, the magnitude depending on the volume of the receiver.
5. We use (2.34) to solve this problem, and obtain 55 h and 2,273 days respectively.

Chapter 3
Permeance, Diffusion and Partition Coefficients: Units and Their Conversion

Since 1960, units of measurement have been based on the SI (Systéme International) system, and we shall be using it in this book. Basic units in this system are (among others) metre (m), kilogram (kg), mol, seconds (s), Pascal (Pa) and Kelvin (K). This has not always been the case, and before 1960 the CGS system (cm, g, s) was common in the natural sciences. In engineering, a great variety of non-metric units (i.e., inch, pound, fluid ounce, and horse-power) are still in use, particularly in the Anglo-Saxon countries.

In Chap. 2, mass transfer across membranes was characterised by permeance, diffusion coefficients and first-order rate constants. We shall use these parameters to explain why permeability of cuticles from different plant species to various solutes differs greatly. Permeability depends on chemistry and structure of membranes, and we shall relate chemistry and structure of cuticles to their permeability. To this end, we have incorporated studies by physical scientists, engineers and biologists on permeability of natural and synthetic membranes published during the last 7 decades. This necessitates converting older units into SI units.

3.1 Units of Permeability

Permeability or permeance is defined as flux per unit driving force. Both can be based on volume or mass. A common driving force is mass (mol, kg) per volume (m^3 of air, vapour, gas or liquid). Depending on reference state (liquid or vapour), they differ numerically by orders of magnitude, while dimensions can be identical. This is not always clear by looking at the dimensions.

During the 1950s and 1960s, permeability of synthetic membranes to permanent gases (O_2, N_2) and vapours was studied extensively. To biologists, water permeability is of particular interest. The units used in these studies are closely related to the experimental approaches. Membranes were placed between donor and receiver, and both compartments were evacuated. The experiment was started by establishing a constant pressure in the donor compartment. Penetration of gas or vapour caused

L. Schreiber and J. Schönherr, *Water and Solute Permeability of Plant Cuticles.*

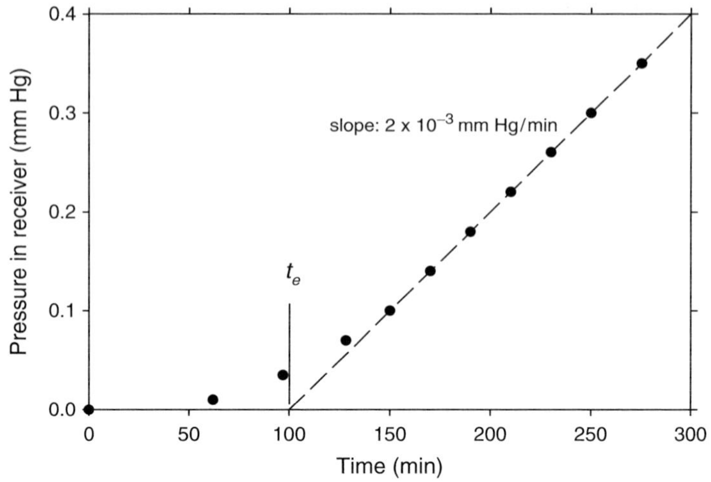

Fig. 3.1 Water vapour transport across an ethyl cellulose (EC) membrane at 25°C. The pressure increase in the receiver compartment was plotted vs time. (Redrawn from Yasuda and Stannett 1962)

an increase in pressure of the receiver, and this was used to calculate the flux. The steady state flux was obtained by keeping the pressure difference between donor and receiver practically constant (steady state). Gas or vapour pressures were measured in cm mercury (cmHg). Vapour pressure greatly depends on temperature, which necessitates rigorous temperature control. In this book, we shall deal with permeability of membranes to water and water vapour. Penetration of permanent gases (O_2, N_2, CO_2) has been reviewed by Lendzian and Kerstiens (1991). While the following conversions also apply to permanent gases, we shall simply use the term vapour. A typical example is steady state diffusion of water vapour at 25°C across an ethyl cellulose (EC) membrane, as shown in Fig. 3.1.

After some time the flux becomes steady, and from these data the extrapolated hold-up time (t_e) and the steady state flux (amount per unit area and time) can be calculated. Volumes of gases or vapours greatly depend on temperature, and fluxes were expressed as volume at standard temperature (273.15 K) and pressure (101,325 Pa), abbreviated as STP. For calculating permeability (P_{Hg}), the volume of gas or vapour at STP (J_v) was multiplied by membrane thickness (ℓ in cm) and divided by the pressure in the donor (p_{donor} in cmHg)

$$P_{Hg} = \frac{J_v \ell}{p_{donor}} = \frac{cm^3(STP)cm^{-2}s^{-1}cm}{cmHg}. \tag{3.1}$$

In words, this permeability coefficient has the dimension cubic centimetres of vapour at STP passing per second under a gradient of 1 cmHg per centimetre membrane thickness and square centimetre of membrane area. Before converting this rather awkward unit to the SI system, we shall present an example taken from Yasuda and Stannett (1962).

3.1.1 Example

Diffusion of water vapour across a polyethylene terephthalate (PET) membrane of 0.1 cm thickness was measured at 25°C. The vapour pressure in the donor chamber was 2.375 cmHg, which is the saturation vapour pressure of water at 25°C. J_v was $4.2 \times 10^{-7}\,\text{cm}^3$ vapour (STP) per cm^2 and s. From these data P_{Hg} can be calculated using (3.1):

$$P_{\text{Hg}} = \frac{4.2 \times 10^{-7}\text{cm}^3(\text{STP}) \times 0.1\,\text{cm}}{\text{cm}^2\text{s} \times 2.375\,\text{cmHg}} = 1.77 \times 10^{-8}\frac{\text{cm}^3(\text{STP})\,\text{cm}}{\text{cm}^2\,\text{s}\,\text{cmHg}}. \tag{3.2}$$

One might be tempted to simplify this dimension of P_{Hg} and write $\text{cm}^2\,\text{s}^{-1}$, because in writing P_{Hg} driving force was 1 cmHg. This was never done, and we shall stick to the extended unit found in the literature, which does not use the suffix Hg to characterise this type of permeability coefficient.

In biology, it is not customary to use fluxes based on vapour volume. Instead, mass fluxes (mol or kg) are used. Hence, we must convert the volume flux of gases or vapours (J_v) into mass fluxes (J). The volume of water vapour at STP can be converted to gram water by considering the vapour as an ideal gas, and this was always assumed when calculating J_v from experimental flux data expressed in pressure units (Fig. 3.1). The volume (V) of 1 mol of an ideal gas at STP can be calculated from the ideal gas law:

$$V = \frac{n \times R \times T}{p} = \frac{1\,\text{mol} \times 8.3143\,\text{m}^3\,\text{Pa} \times 273.15\,\text{K}}{\text{mol}\,\text{K} \times 101{,}325\,\text{Pa}} = 0.022414\,\text{m}^3, \tag{3.3}$$

where R is the gas constant. One mol of water has a mass of 18 g, and the density of water vapour (δ_{wv}) at STP is

$$\delta_{\text{wv}} = \frac{\text{mass}}{\text{volume}} = \frac{18\,\text{g}}{22.414 \times 10^3\,\text{cm}^3} = 8.03 \times 10^{-4}\,\text{g}\,\text{cm}^{-3} \tag{3.4}$$

and the mass flux of water (J_w) can be calculated

$$\begin{aligned} J_w &= J_v \times \delta_{\text{wv}} = \left[4.2 \times 10^{-7}\,\text{cm}^3\ (\text{STP})\ \text{cm}^{-2}\,\text{s}^{-1}\right]\left(8.03 \times 10^{-4}\,\text{g}\,\text{cm}^{-3}\right) \\ &= 3.37 \times 10^{-10}\,\text{g}\,\text{cm}^{-2}\,\text{s}^{-1}. \end{aligned} \tag{3.5}$$

Before permeance or permeability can be calculated in SI units, the driving force must be converted. Yasuda and Stannett (1962) and others used cmHg, and conducted various experiments with different pressures in the donor. Results are shown in Fig. 3.2. The volume flux was proportional to pressure in cmHg or to partial pressure.

In the SI system the driving force is often mass per volume, which is $\text{mol}\,\text{m}^{-3}$ or $\text{kg}\,\text{m}^{-3}$. This is practically always the case with solutes dissolved in water. With water vapour, partial pressure or activity of water can be used as an alternative. At this point we decide to use partial pressure as driving force, but we shall deal with

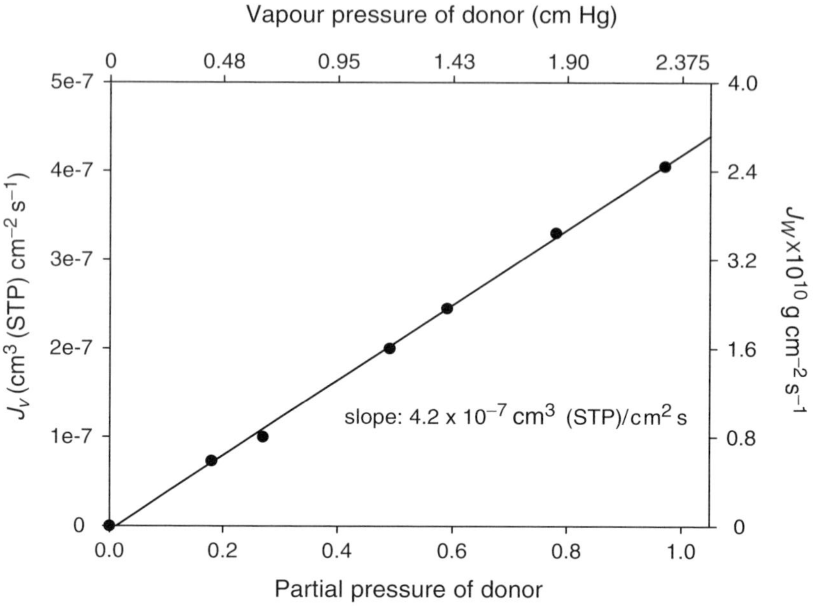

Fig. 3.2 Volume flux of water vapour across a polyethylene terephthalate (PET) membrane [J_v (STP) plotted vs partial pressure of donor (p/p_0)]. Membrane thickness was 1 mm and temperature 25°C. (Redrawn from Yasuda and Stannett 1962)

other driving forces in more detail later. Since water activity of the vapour is equal to partial or fractional vapour pressure (p/p_0), that is $a_{wv} = p/p_0$, we can convert driving force in cmHg to partial pressure simply by dividing actual vapour pressure by saturation vapour pressure. With water at 25°C, saturation vapour pressure is 2.375 cmHg.

Based on the CGS system of the original literature, permeability $(\mathscr{P}_w)$ of a 1 cm thick membrane is

$$\mathscr{P}_w = \frac{J_w \times \ell}{a_w} = \frac{3.37 \times 10^{-10}\,\mathrm{g}\,(0.1\,\mathrm{cm})}{\mathrm{cm}^2\,\mathrm{s} \times 1} = 3.37 \times 10^{-11}\,\mathrm{g\,cm}^{-1}\,\mathrm{s}^{-1} \tag{3.6}$$

and since 1 g of water at 25°C has a volume of 1 cm^3, $\mathscr{P}_w$ is $3.37 \times 10^{-11}\,\mathrm{cm}^2\,\mathrm{s}^{-1}$ or $3.37 \times 10^{-15}\,\mathrm{m}^2\,\mathrm{s}^{-1}$. In this calculation the water flux enters as mass (g) and is converted to volume of liquid water. The driving force is water activity (or partial vapour pressure). With pure water as donor, $\mathscr{P}_w$ can be calculated by dividing $(J_w \times \ell)$ by the water concentration (C_w), which at 25°C is 1 g cm^{-3} or 1,000 kg m^{-3}. C_w depends much less on temperature than C_{wv}. For a solvent, $a = \gamma N$ where γ is the activity coefficient and N is the mol fraction. Considering water as pure ideal solvent, both γ and N are 1.0, and we have $C_w = a_w$. Hence, $\mathscr{P}_w$ is numerically identical when driving forces are either C_w, a_{wv} or p/p_0.

In a closed system, equilibrium between vapour and liquid water is established. By definition pure water has an activity of 1.0, and water activity is numerically identical for vapour and liquid water as long as temperature is the same ($a_w = a_{wv}$). Hence, permeability coefficients are numerically identical when driving force is expressed as water activity of vapour or liquid. However, ecologists prefer to use the water vapour concentration (C_{wv}) over leaves as driving force of transpiration. At 100% humidity ($p/p_0 = a_{wv} = 1$) and at 25°C, 1 m^3 of air contains 23.05 g water in the vapour phase (Nobel 1983), and the permeability coefficient now becomes

$$\begin{aligned}\mathscr{P}_{wv} &= \frac{J_w\,(0.1\,\text{cm})}{C_{wv}} = \frac{3.37\times10^{-10}\,\text{g}\,(0.1\text{cm})}{\text{cm}^2\,\text{s}(23.05\times10^{-6}\,\text{g}\,\text{cm}^{-3})}\\ &= 1.46\times10^{-6}\,\text{cm}^2\,\text{s}^{-1} = 1.46\times10^{-10}\,\text{m}^2\,\text{s}^{-1}.\end{aligned} \tag{3.7}$$

$\mathscr{P}_w$ and $\mathscr{P}_{wv}$ have identical dimensions, but they differ by a factor equal to the ratio of density of liquid water and water vapour concentration. At 25°C we have

$$\frac{\delta_w}{C_{wv}} = \frac{1{,}000\,\text{kg}\,\text{m}^{-3}}{23.05\times10^{-3}\,\text{kg}\,\text{m}^{-3}} = 43{,}384. \tag{3.8}$$

This has caused considerable confusion among workers in different fields, and for this reason it is absolutely necessary to specify which type of driving force was used in calculating permeance or permeability coefficients for diffusion of water.

Using the above derivations and definitions, the various permeability coefficients can be easily converted. The numbers given as examples refer to a temperature of 25°C

$$\begin{aligned}\frac{\mathscr{P}_{wv}}{P_{Hg}} &= \frac{\delta_{wv}(\text{STP})\times\text{cmHg}_{\text{saturation}}}{C_{wv}}\\ &= \frac{(8.03\times10^{-4}\,\text{g}\,\text{cm}^{-3})\,(2.375\,\text{cmHg})}{23.05\times10^{-6}\,\text{g}\,\text{cm}^{-3}} = 82.7.\end{aligned} \tag{3.9}$$

Note that only δ_{wv}(STP) is constant, while the saturation vapour pressure (cmHg$_{\text{saturation}}$) and concentration of water vapour in air (C_{wv}) vary with temperature and must be looked up in tables (i.e., Nobel 1983)

$$\begin{aligned}\frac{\mathscr{P}_w}{P_{Hg}} &= \delta_{wv}(\text{STP})\times\text{cmHg}_{\text{saturation}}\\ &= (8.03\times10^{-4}\,\text{g}\,\text{cm}^{-3})\,(2.375\,\text{cmHg}) = 1.91\times10^{-3}.\end{aligned} \tag{3.10}$$

P_{Hg} has the dimension cubic centimetres of vapour at STP passing per second under a gradient of 1 cmHg per centimetre thickness and square centimetre of membrane area. Multiplication of P_{Hg} by 82.7 results in the dimension of cm^2 s^{-1} for $\mathscr{P}_{wv}$. The same is true when $\mathscr{P}_w$ is calculated from P_{Hg} by multiplying it by 1.91×10^{-3}. SI units (m^2 s^{-1}) of $\mathscr{P}_{wv}$ and $\mathscr{P}_w$ are obtained by multiplication with 10^{-4}. It follows from (3.9) and (3.10) that

$$\frac{\mathscr{P}_{wv}}{\mathscr{P}_w} = \frac{\delta_w}{C_{wv}} = \frac{1\,\text{g}\,\text{cm}^{-3}}{23.05\times10^{-6}\,\text{g}\,\text{cm}^{-3}} = 43{,}384. \tag{3.11}$$

3.2 Diffusion Coefficients

No problems arise with D when calculated from the extrapolated hold-up time (t_e) using (2.5) or from sorption data (2.33) or (2.34). The dimension is always $m^2\,s^{-1}$ or $cm^2\,s^{-1}$.

3.3 Partition Coefficients

In Chap. 2, we defined the partition coefficient as ratio of molal concentration in the membrane and in the surrounding solution (2.12). In the polymer literature dealing with gas or vapour permeability across homogeneous membranes, another variable is used which is related to the partition coefficient. The vapour sorption coefficient ($\mathscr{S}$) can be calculated from P_{Hg} and the diffusion coefficient

$$\mathscr{S} = \frac{P_{Hg}}{D}. \tag{3.12}$$

For the PETP membrane shown in Fig. 3.2, P_{Hg} was $1.77 \times 10^{-8}\,cm^3$ (STP) cm per $cm^2\,s\,cmHg$, and D amounted to $3.94 \times 10^{-9}\,cm^2\,s^{-1}$ (Yasuda and Stannett 1962). With these data we obtain

$$\begin{aligned}\mathscr{S} &= \frac{P_{Hg}}{D} = \frac{1.77 \times 10^{-8}\,cm^3\,(STP)\,cm}{cm^2\,s\,cmHg\,(3.94 \times 10^{-9}\,cm^2\,s^{-1})} \\ &= 4.49\frac{cm^3\ vapour\ (STP)}{(cm^3\ polymer)\,cmHg}.\end{aligned} \tag{3.13}$$

This is the amount of water vapour sorbed in PETP at a vapour pressure of 1 cmHg or at a partial pressure of 0.421. When the sorption isotherm is linear, the amount sorbed is proportional to vapour pressure. Thus, sorption at 100% humidity (2.375 cmHg) is 10.66 cm^3 (STP) water vapour per cm^3 polymer. The partition coefficient K_{Hg} is

$$K_{Hg} = \frac{\mathscr{S}}{p/p_0} = \frac{4.49\,cm^3\,(STP)}{(cm^3\ polymer)0.421} = 10.66\frac{cm^3\,(STP)}{cm^3 polymer}. \tag{3.14}$$

Multiplying K_{Hg} by the density of water vapour at STP results in a new partition coefficient K_w which is on the basis mass per volume:

$$\begin{aligned}K_w &= K_{Hg} \times \delta_{wv}(STP) \\ &= \left(10.66\frac{cm^3\ vapor\,(STP)}{cm^3\ polymer}\right)\left(8.03 \times 10^{-4}\frac{g\,water}{cm^3\ vapour}\right) \\ &= 8.56 \times 10^{-3}\frac{g\,water}{cm^3\ polymer}.\end{aligned} \tag{3.15}$$

The density of PETP at 25°C is $1.39\,\mathrm{g\,cm^{-3}}$, and dividing the above figure by density of the polymer we finally arrive at a partition coefficient of 6.16×10^{-3} on a mass basis ($\mathrm{g\,g^{-1}}$ or $\mathrm{kg\,kg^{-1}}$). This lengthy calculation can be shortened by using permeability coefficients other than P_{Hg}. With $\mathscr{P}_{\mathrm{w}}$, which can be obtained from P_{Hg} using (3.10), we obtain K_{w} directly:

$$K_{\mathrm{w}} = \frac{\mathscr{P}_{\mathrm{w}}}{D} = \frac{3.37 \times 10^{-11}\,\mathrm{cm^2\,s^{-1}}}{3.94 \times 10^{-9}\,\mathrm{cm^2\,s^{-1}}} = 8.56 \times 10^{-3}\,\frac{\mathrm{cm^3\ water}}{\mathrm{cm^3\ polymer}} \tag{3.16}$$

and using $\mathscr{P}_{\mathrm{wv}}$ we obtain

$$K_{\mathrm{wv}} = \frac{\mathscr{P}_{\mathrm{wv}}}{D} = \frac{1.46 \times 10^{-6}\,\mathrm{cm^2\,s^{-1}}}{3.94 \times 10^{-9}\,\mathrm{cm^2\,s^{-1}}} = 370.6\,\frac{\mathrm{cm^3\ vapor}}{\mathrm{cm^3\ polymer}}. \tag{3.17}$$

In calculating the ratios $\mathscr{P}/D$, we must use the same units. Above, we used the CGS system. If SI units are used for $\mathscr{P}$ and D the numerical values are the same, and the dimensions of the partition coefficients would be $\mathrm{m^3}$ vapour per $\mathrm{m^3}$ polymer.

The concentration of water in the polymer can be obtained by multiplying the partition coefficient by the appropriate driving force, which is the driving force used in calculating the permeability coefficient ($\mathscr{P}$). For instance:

$$\begin{aligned} C_{\mathrm{w}}^{\mathrm{polymer}} &= K_{\mathrm{w}} a_{\mathrm{w}} = (8.56 \times 10^{-3})\,(1.0) \\ &= 8.56 \times 10^{-3}\,\mathrm{cm^3\ water/cm^3\ polymer}. \end{aligned} \tag{3.18}$$

Alternatively, we may use the concentration of water instead of a_{w} and obtain

$$\begin{aligned} C_{\mathrm{w}}^{\mathrm{polymer}} &= K_{\mathrm{w}} C_{\mathrm{w}} = (8.56 \times 10^{-3})\,(1.0\,\mathrm{g\,cm^{-3}}) \\ &= 8.56 \times 10^{-3}\,\mathrm{g\ water/cm^3\ polymer}. \end{aligned} \tag{3.19}$$

The results are numerically identical only at 4°C when density of water is $1\,\mathrm{g\,cm^{-3}}$. At higher and lower temperatures water activity is always 1.0, but concentration of water is lower or higher. However, at physiological temperatures density of liquid water varies little with temperature, and is practically $1\,\mathrm{g\,cm^{-3}}$. With K_{wv}, the concentration of water vapour (C_{wv}) must be used

$$\begin{aligned} C_{\mathrm{w}}^{\mathrm{polymer}} &= K_{\mathrm{wv}} \times C_{\mathrm{wv}} \\ &= 370.6\,(23.05 \times 10^{-6}\,\mathrm{g\,cm^{-3}}) = 8.54 \times 10^{-3}\,\mathrm{g\,cm^{-3}}. \end{aligned} \tag{3.20}$$

All of these partition coefficients can be found in the literature, but often without the super- and subscripts used here. One should be sure to understand how they were derived, since all are dimensionless. This also applies to K_{Hg} when (STP) is omitted. We have added the same suffixes to the partition coefficients which were used in calculating permeability coefficients, to make it clear how they are defined and calculated. This was not always the case in the literature.

This complexity observed with water in polymers usually does not exist with solutes. As long as the same molal dimensions are used for polymer and solution, they cancel in calculating partition coefficients.

Problems

1. At 25°C, the concentration of water vapour in air at 100% humidity is $23.05\,\mathrm{g\,m^{-3}}$, while in vacuum it is $803\,\mathrm{g\,m^{-3}}$ (3.4) when $p/p_0 = 1$. What is the reason for this difference?
2. The units of pressure have changed. It is useful to remember that 1 Torr = $133.32\,\mathrm{Pa} \approx 1\,\mathrm{mmHg}$; 1 atm = 760 Torr, 1 bar $\approx$ 750 Torr. Normal pressure is defined as 760 mmHg = 101,325 Pa = 1.01325 bar. What would be the steady state slope in Fig. 3.1 in $\mathrm{Pa\,s^{-1}}$?
3. If the EC-membrane (Fig. 3.1) had a thickness of 1 mm, what would be the diffusion coefficient?
4. Using the data given in Fig. 3.1, calculate the steady state flux (J_v) of water vapour in $\mathrm{cm^3(STP)\,cm^{-2}\,s^{-1}}$ by using the equation

$$J_v = \frac{\Delta p_{\mathrm{receiver}}}{\Delta t \times A} \times \frac{273}{273+25} \times \frac{V_{\mathrm{receiver}}}{760\,\mathrm{mmHg}}. \tag{3.21}$$

   The volume of the receiver was $100\,\mathrm{cm^3}$, A was $1\,\mathrm{cm^2}$ and temperature was 25°C.
5. With the result obtained in problem 4, calculate P_{Hg} of the membrane. Membrane thickness was 1 cm, and the pressure in the donor was 2.375 cmHg.
6. Assuming a linear sorption isotherm, what is the equilibrium concentration ($\mathrm{g\,g^{-1}}$) of water in EC (density $1.13\,\mathrm{g\,cm^{-3}}$) at 25°C at 50% humidity?

Solutions

1. In air at normal pressure, the total pressure is the sum of the partial pressures of O_2, N_2 and water vapour. In the absence of O_2 and N_2, much more water vapour is soluble per unit volume.
2. We obtain $0.267\,\mathrm{Pa\,min^{-1}}$ or $4.44 \times 10^{-3}\,\mathrm{Pa\,s^{-1}}$
3. We use $D = \ell^2/6t_e$ and obtain $D = 2.78 \times 10^{-11}\,\mathrm{m^2\,s^{-1}}$ or $2.78 \times 10^{-7}\,\mathrm{cm^2\,s^{-1}}$.
4. The steady state flux is $4.02 \times 10^{-6}\,\mathrm{cm^3\,(STP)\,cm^{-2}\,s^{-1}}$.
5. Using (3.1) we obtain $P_{\mathrm{Hg}} = 1.96 \times 10^{-6}\,\mathrm{cm^3\,(STP)\,cm^{-2}\,s^{-1}}$ per cmHg.
6. We solve this problem in four steps. We already calculated P_{Hg} and D and (i) we now calculate $\mathscr{P}_w$ using (3.10) and obtain $3.74 \times 10^{-9}\,\mathrm{cm^2\,s^{-1}}$. (ii) Dividing this number by D (3.16) we obtain $K_w = 1.35 \times 10^{-2}\,\mathrm{cm^3}$ water per $\mathrm{cm^3}$ polymer. (iii) From (3.18) with $p/p_0 = a_w = 0.5$, we calculate the water content at 50% humidity of $6.73 \times 10^{-3}\,\mathrm{cm^3}$ water per $\mathrm{cm^3}$ polymer. (iv) Dividing this figure by $1.13\,\mathrm{g\,cm^{-3}}$ and assuming a density of liquid water of $1\,\mathrm{g\,cm^{-3}}$, we obtain the water content of 5.95 g water per g polymer.

Chapter 4
Water Permeability

All terrestrial organisms have a problem in common; they must minimise water loss to the atmosphere and prevent desiccation. The driving force of transpiration at 25°C and 50% humidity is −95 MPa, and at lower humidity it is even greater. Water supply is often short. Higher plants, insects and mammals use similar strategies to save water. They have generated membranes of very low water permeability at their interface with the dry air surrounding them most of the time. Synthetic polymers used for membranes, tubing, containers and other packaging materials also have low permeability to gases, water and other solvents to protect goods. Before we turn to permeability of cuticles and to strategies of plants to built effective barriers for protection against adverse influences from the environment, we will briefly compare water permeability of synthetic membranes with permeance of plant polymer matrix membranes. Synthetic polymer membranes have been studied extensively during the last decades, and structure–permeability relationships have been established. What can we learn from homogeneous synthetic membranes to better understand permeability of heterogeneous plant cuticles?

4.1 Water Permeability of Synthetic Polymer Membranes and Polymer Matrix Membranes: A Comparison of Barrier Properties

Plant cuticles are polymeric membranes. The polymer matrix (MX) is composed of lipophilic cutin and hydrophilic polar polymers (Sect. 1.1). Cutin is a polyester composed of hydroxyfatty acids, and depending on the number of hydroxyl groups and extent of cross linking it contains 20–25% oxygen. The polymer matrix of tomato fruits is composed of carbon (67%), oxygen (23%), hydrogen (8.2%) and nitrogen (0.65). After acid hydrolysis, which eliminates polar polymers (polysaccharides, polypeptides and phenolic compounds), no nitrogen was found and elemental composition was 71% C, 20% O and 9.4% H (Schönherr and Bukovac 1973). This is very similar to the composition calculated for a linear polyester of C_{16}

dihydroxyfatty acids. Composition of ivy leaf cuticles is 65% C, 25% O, 9.3% H and 0.8% N (Schreiber et al. 1994). Polar functional groups are permanent dipoles which are involved in hydrogen bonding and they affect sorption of water and water permeability of a polymer (Schönherr 2006). Electron microscopy suggests that cutin and polar polymers do not form a homogeneously mixed phase (copolymers). In TEM polar polymers are seen as a network of anastomosing fibrils embedded in cutin (Sect. 1.4.2).

In this chapter, we analyse contributions of polar polymers, cutin and waxes to water permeability, and present evidence showing that cutin and polar polymers form two independent parallel pathways for transport of water and highly water soluble solutes. The effect of waxes on water permeability and their deposition in cutin and on the surface of cuticles is another important topic.

When comparing effectiveness of cuticles as water barriers with man-made polymers, it is useful to treat water permeability of the polymer matrix and cuticular membranes separately. There is a large number of synthetic polymers, and we selected some which resemble cuticles structurally and chemically. Cellulose acetate (CA), polyvinyl acetate (PVA), polyethyl methacrylate (PEMA) and polyethylene terephthalate (PET) are polyesters. Nylon is a polyamide, and ethyl cellulose (EC) is a cellulose ether. Depending on degree of substitution they contain 30–50% oxygen by weight. Polyethylene (PE) and polypropylene (PP) are polymers lacking functional groups. Polymers can be partially crystalline (PE, PP, PET), and depending on temperature they occur in the glassy or the rubbery state. In the glassy state, the polymer chains are stiffer and the polymer is more brittle than in the rubbery state (Park 1968). All selected polymers are homogeneous, and diffusion coefficients have been determined from sorption/desorption or from time lag (see Chap. 2). With a few exceptions (Rust and Herrero 1969) permeability to water vapour has been given as P_{Hg}. This can be converted to P_{wv} and $\mathscr{P}_{\mathrm{wv}}$ as explained in Chap. 3. With $\mathscr{P}_{\mathrm{wv}}$ and D known, the partition coefficient and water content of membranes can be calculated using (3.17) and (3.20) respectively.

Polymer matrix membranes are obtained by extracting waxes from cuticular membranes. These MX membranes had a thickness of about 3 µm, and P_{w} was determined gravimetrically using the cup method (Sect. 9.7) with water inside the chambers and humidity outside being practically zero (Schönherr and Lendzian 1981).

Water permeability of selected MX-membranes and synthetic polymer membranes is listed in Table 4.1. Some polar synthetic polymers swell, and sorption isotherms may not be linear when partial pressure is >0.5. In these cases, values for P_{wv}, D and K_{wv} at $p/p_0 < 0.5$ were selected. Among the synthetic polymers, PP is the one with the lowest and EC has the highest $\mathscr{P}_{\mathrm{wv}}$. They differ by a factor of 1,000, while their diffusion coefficients differ only by a factor of 39. PVA was the polymer with the lowest diffusion coefficient, and D for the other polymers ranged from 10^{-11} to $10^{-13}\,\mathrm{m}^2\,\mathrm{s}^{-1}$. $\mathscr{P}_{\mathrm{wv}}$ and D are not identical, because sorption of water vapour differs among the polymers (2.4) and (2.17). Partition coefficients (K_{wv}) were calculated as $\mathscr{P}/D$, which is valid when polymers are homogeneous. Water concentration of membranes in equilibrium with 100% humidity (C_{w}) was calculated by multiplying K_{wv} by $23.05 \times 10^{-3}\,\mathrm{kg\,m^{-3}}$, the water vapour concentration of air at 100% humidity and 25°C (3.20). Concentration of water in synthetic polymers in equilibrium with 100% humidity (or $p/p_0 = 1$) ranged from 0.65 to $168\,\mathrm{kg\,m^{-3}}$.

P_{wv} was obtained by multiplying P_w by 43,384 (3.11). Permeances of the selected MX membranes ranged from 1.2×10^{-3} to $2.5 \times 10^{-4}\,\mathrm{m\,s^{-1}}$. In order to compare synthetic polymers and MX membranes, both types of membranes must have the same thickness. Permeances for the synthetic membranes were calculated as $\mathcal{P}/\ell$ for a thickness of 3 µm, which is similar to thicknesses of the MX membranes in Table 4.1. As these synthetic polymers are homogeneous, this is perfectly legitimate. Permeances of MX membranes are similar to those calculated for the polar polymers EC, CA, PMA, PEMA and Nylon. Permeances of PP, PE, PVA and PET membranes are considerably lower.

Comparing diffusion coefficients meets with some difficulties (Table 4.1). Becker et al. (1986) determined D values for *Ficus* and *Citrus* MX using the hold-up time method (2.5), while Chamel et al. (1991) estimated D from sorption isotherms (2.33), and their D are considerably higher than those of Becker et al. (1986). Chamel et al. (1991) also determined water vapour sorption in CM and MX gravimetrically. Sorption in MX and CM was similar or identical because most sorption sites (dipoles) are contributed by cutin and polar polymers, and access of water to

Table 4.1 Water permeability at 25°C of selected synthetic polymer membranes and plant polymer matrix membranes isolated from astomatous leaf surfaces

Polymer	$\mathcal{P}_{wv}$ ($\mathrm{m^2\,s^{-1}}$)	D ($\mathrm{m^2\,s^{-1}}$)	K_{wv}	C_w ($\mathrm{kg\,m^{-3}}$)	P_{wv} ($\mathrm{m\,s^{-1}}$) ($\ell = 3\,\mu m$)	J_{wv} ($\mathrm{g\,m^{-2}\,h^{-1}}$)
PP[a]	1.9×10^{-11}	4.9×10^{-13}	39	0.90	6.3×10^{-6}	0.52
PE[a]	3.3×10^{-11}	1.2×10^{-12}	28	0.65	1.1×10^{-5}	0.91
PVA[b]	3.7×10^{-11}	5.1×10^{-15}	7,269	168	1.2×10^{-5}	1.00
PET[a]	1.3×10^{-10}	2.7×10^{-13}	484	11.2	4.3×10^{-5}	3.57
Nylon[b]	3.3×10^{-10}	1.2×10^{-13}	2,750	63	1.1×10^{-4}	9.13
PEMA[c]	2.9×10^{-9}	1.1×10^{-11}	264	6.1	9.7×10^{-4}	80.5
CA[b]	9.3×10^{-9}	3.1×10^{-12}	3,000	69	3.1×10^{-3}	257.3
EC[d]	1.9×10^{-8}	1.9×10^{-11}	1,000	23	6.3×10^{-3}	522.8
Ficus MX[e]	1.4×10^{-9}	1.8×10^{-14}	7.8×10^{4}	1.8×10^{3}	2.5×10^{-4}	20.7
Ficus MX[f]	–	4.1×10^{-13}	2,000	34	-	–
Citrus MX[e]	5.2×10^{-9}	6.0×10^{-15}	8.7×10^{5}	2.0×10^{4}	1.8×10^{-3}	149.3
Citrus MX[f]	–	2.6×10^{-14}	2,828	41	-	–
Pyrus MX[g]	–	–			4.7×10^{-3}	390.0
Hedera MX[g]	–	–	–	–	1.2×10^{-3}	99.6

[a]Rust and Herrero (1969)
[b]Hauser and McLaren (1948)
[c]Stannett and Williams (1965)
[d]Wellons and Stannett (1966)
[e]Becker et al. (1986)
[f]Chamel et al. (1991)
[g]Schönherr and Lendzian (1981)

these polar functions was apparently not reduced by waxes. They reported 34 and $41\,\mathrm{kg\,m^{-3}}$ for *Ficus* and *Citrus* MX. This is the range of C_w observed with EC, Nylon and CA.

A further complication arises when we calculate K_{wv} as $\mathscr{P}/D(=P_{wv}\ell/D)$. These partition coefficients are larger by orders of magnitude than those obtained from sorption experiments (Table 4.1), and as a consequence water concentrations (C_w) are much higher than those determined gravimetrically. Precision of the sorption experiments is very good, and no assumptions are needed to calculate water concentration in MX. Hence, these values are reliable, and values calculated as $\mathscr{P}/D$ must be in error. Becker et al. (1986) calculated D from the hold-up time (t_e) and membrane thickness ℓ ($D = \ell^2/6t_e$). Membrane thickness also enters in calculating $\mathscr{P}$ ($= P_{wv}\ell$). Combining the above equations we obtain $K_{wv} = P_{wv} \times 6t_e/\ell$. Since water concentration in MX obtained using the $\mathscr{P}/D$ ratio is larger by factors of 53 (*Ficus*) and 488 (*Citrus*) respectively, it appears that the real diffusion paths are much longer than thickness of the MX. This tortuosity (τ) of the MX is astounding and difficult to explain. Later (Sect. 4.5) we shall present evidence that in MX membranes water flows in two parallel pathways — that is, in polar polymers and in the cutin polymer. Diffusion coefficients calculated from hold-up times are some averages for the two pathways, and no structural information can be extracted from them.

This large difference in C_w depending on method of determination is excellent evidence that MX membranes are not homogeneous. It is not possible to characterise water transport in MX using unique values for $\mathscr{P}$, D or K_{wv}, as is possible with synthetic polymer membranes. The constituents of the MX (polar polymers and cutin) form separate phases, and each phase has its own diffusion and partition coefficient. Both coefficients vary with position, and the values derived from sorption experiments and hold-up times are some type of average, which is not very useful for analysing water permeability of cutin and polar polymers in MX.

In homogeneous membranes, effectiveness of water barriers is characterised by $\mathscr{P}$ and D. Since MX membranes are not homogeneous, this comparison is not meaningful. Hence, we compare efficacy of membranes based on maximum water fluxes (J_{wv}). We calculated the maximum water fluxes across $3\,\mu\mathrm{m}$-thick membranes (Table 4.1). Maximum flux occurs when driving force is maximum, that is, when humidity on the donor side is 100% and on the receiver side 0%. The maximum flux of water vapour (J_{wv}) was obtained by multiplying P_{wv} by the vapour concentration of $23.05\,\mathrm{g\,m^{-3}}$, which at 25°C amounts to 100% humidity.

Maximum moisture flux across PP and PE is small, as would be expected for packaging materials. Commercial household foils or bags are 3–10 times thicker, and fluxes are only a third or a tenth of those shown in Table 4.1. With the more polar polymers, J_{wv} ranged from 80.5 to $522.8\,\mathrm{g\,m^{-2}\,h^{-1}}$, and with the MX-membranes water loss amounts to 20.7–$390\,\mathrm{g\,m^{-2}\,h^{-1}}$. This is of the same order as maximum transpiration rates measured with plants when stomata are open (Larcher 1995). Plants regulate their water household by opening or closing stomata. When stomata are closed, transpiration rates are less than one tenth of maximum rates. This cannot be realised with water barriers having properties such as the polymer matrix. Evolution solved this problem by incorporating waxes into cuticles (Sect. 4.6).

4.2 Isoelectric Points of Polymer Matrix Membranes

The polymer matrix (MX) carries immobile (fixed) electrical charges which are contributed by pectins, polypeptides and cutin. Where are the various polar groups located, what are their properties and what are the consequences for water permeability of cuticles? Hydroxyl groups donated by cutin, hydroxyfatty acids, pectins, cellulosic compounds and polypeptides are permanent dipoles, and are not affected by *p*H and cations. Polarity and hydration of carboxyl, amino and phenolic hydroxyl groups depend on *p*H and type of counter ions. In this section, we take a look at location and properties of ionisable groups in the polymer matrix.

Ionised fixed charges electrostatically attract ions of opposite sign, and for this reason carboxyl groups are surrounded by cations such as K^+ or Ca^{2+}. Amino groups are electrically neutralised by anions. Fixed charges in a membrane result in unequal distribution of ions in the membrane and adjacent solutions, and this concentration difference results in an electrical potential between membrane and solution. This potential is called Donnan potential, and the phase containing immobilised charges is the Donnan phase. The sign of the electrical potential in the Donnan phase relative to the surrounding electrolyte solution is the same as the charge of the immobile ion. If fixed charges are homogeneously distributed and identical electrolyte solutions are used on either side of the membrane, the Donnan potentials are in opposite directions and cancel.

The driving force of ion fluxes is the electrochemical potential, which is the sum of the chemical potential (2.13) and the electrical potential ($z\mathscr{F}E$), where z is the charge of the ions (i.e., +1 for K^+ and −2 for SO_4^{2-}), $\mathscr{F}$ is the Faraday constant and E is the electrical potential (V). A concentration difference of ions across a membrane results in a difference in electrochemical potential (diffusion potential), and ions diffuse across the membrane. The mobility of ions in solution differs as they have different sizes. K^+ and Cl^- are an exception as they have identical mobility, and this is the reason they are used as electrolyte solutions in electrodes. In a charged membrane, mobility of cations and anions are also affected by their interactions with fixed charges.

The membrane potential (E_{membrane}) can be measured by placing two electrodes (i.e., calomel or Ag/AgCl) into salt solutions facing the two sides of the membrane. E_{membrane} is the sum of the Donnan potentials (E_{Donnan}) and the diffusion potential ($E_{\text{diffusion}}$). The Donnan potentials enter with opposite signs (Helfferich 1962):

$$\begin{aligned} E_{\text{membrane}} &= E_{\text{Donoan}}^{\text{outside}} + E_{\text{Donoan}}^{\text{inside}} + E_{\text{diffusion}} \\ &= -\frac{RT}{\mathscr{F} z_{\text{counterion}}} \left[\ln \frac{a_{\text{counterion}}^{\text{outside}}}{a_{\text{counterion}}^{\text{inside}}} - (z_{\text{co-ion}} - z_{\text{counterion}}) \right. \\ &\quad \left. \times \int_{\text{inside}}^{\text{outside}} \bar{t}_{\text{co-ion}} \, \mathrm{d} \ln a_{\text{salt}} \right]. \end{aligned} \tag{4.1}$$

Here R and T have the usual meaning, a is the activity of ions or salt, z is the valence of the ions and $\bar{t}$ is the transference number of the co-ion. The transference number of an ionic species is defined as the number of equivalents of ions transferred by 1 faraday of electricity.

The first term on the right hand side of the equation gives the thermodynamic limiting value of the concentration potential, and the second term is the deviation due to co-ion flux. With an ideal permselective membrane ($\bar{t}$ of the co-ion is zero) the second term vanishes, and (4.1) reduces to the so-called Nernst equation. For an ideal cation permselectivity, we have

$$E_{\text{membrane}} = -\frac{RT}{\mathscr{F} z_{+}} \ln \frac{a_{+}^{\text{outside}}}{a_{+}^{\text{inside}}} \tag{4.2}$$

and for an ideal anion permselective membrane

$$E_{\text{membrane}} = -\frac{RT}{\mathscr{F}(-z_{-})} \ln \frac{a_{-}^{\text{outside}}}{a_{-}^{\text{inside}}} \tag{4.3}$$

is obtained. For a monovalent ion ($z = 1$) and 25°C, the term $RT/\mathscr{F}z$ amounts to 25.7 mV. Neglecting activity coefficients, the membrane potential of an ideal permselective membrane is 17.81 mV when the concentration ratio is 2.

Between pH 2 and 9 and at 25°C, membrane potentials measured with identical buffers and a KCl concentration of $4 \times 10^{-3}\,\text{mol}\,\text{l}^{-1}$ on both sides were smaller than 0.5 mV with all cuticles (apricot, pear and *Citrus* CM and *Citrus* MX) tested. When salt concentrations are the same on both sides of the membranes, there is no difference in electrochemical gradient and no driving force. Hence the transference number for anions and the diffusion potential is zero, and any electrical potential measured would be the difference of the Donnan potentials (4.1) on the two surfaces of the membrane. This difference is called asymmetry potential. As it was close to zero at all pH values, it appears that the concentration of fixed charges was the same with all cuticles tested. This is a surprising result in view of the gradient of polarity seen in most TEM pictures (Sect. 1.4).

Membrane potentials across CM and MX membranes were measured with buffered KCl solutions of $4 \times 10^{-3}\,\text{mol}\,\text{l}^{-1}$ and $2 \times 10^{-3}\,\text{mol}\,\text{l}^{-1}$ in contact with the inner and outer surfaces of the membranes respectively. Membrane potentials were strongly pH dependent (Fig. 4.1). At pH 9, potentials between −14 and −16 mV were measured. Potentials decreased with pH, at first slowly but below pH 6 more rapidly, and they assumed positive values at pH 2 ranging from 2 to 13 mV. pH values resulting in zero mV were 2.9 (pear), 3.2 (*Citrus*) and 3.4 (apricot). These pH values mark the isoelectric point of the cuticles. If the pH is higher the membranes carry a net negative charge, at pH values below the isoelectric point membranes have a net positive charge. At the isoelectric point, the membranes have no net charge. This does not mean that they have no fixed charges, rather that the number of positive and negative fixed charges is equal.

Above the isoelectric point, cuticles carry net negative charges which are neutralised by cations, in the present case by K^{+}. Below the isoelectric point they are positively charged, and charges are neutralised by anions (Cl^{-}). At pH 9 membrane

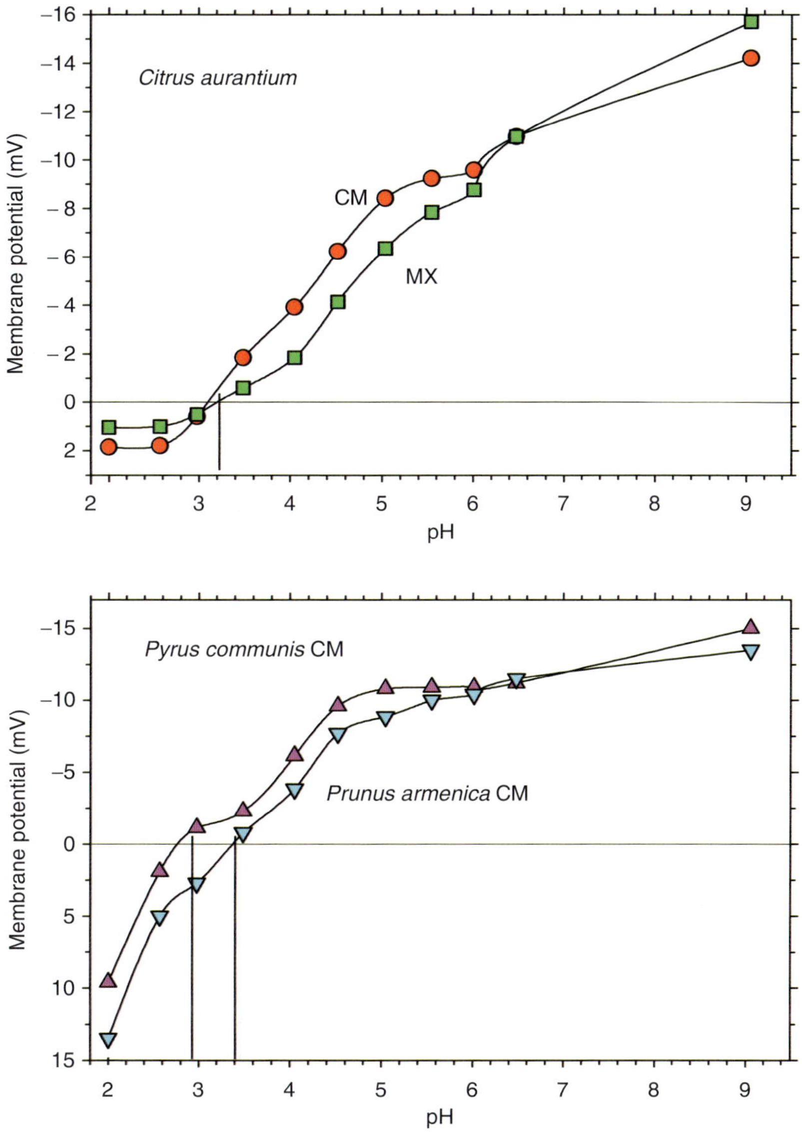

Fig. 4.1 Membrane potentials measured with cuticular membranes isolated from adaxial leaf surfaces of *Citrus aurantium, Pyrus communis* (pear) and *Prunus armeniaca* (apricot) at 25°C and KCl solutions of 4×10^{-3} (inner surface) and 2×10^{-3} mol l^{-1} respectively. (Redrawn from Schönherr and Huber 1977)

potentials were highest, but did not quite reach the Nernst potential of 17.8 mV, hence they were not perfectly permselective for cations. Under natural condition the *p*H at the surfaces of the cuticles will be lower, and exclusion of anions will be far from complete. This means that diffusion of salt can take place, and this has been confirmed experimentally. Self-diffusion of NaBr across *Citrus* MX membranes was studied at *p*H 3 and 8.5 at 25°C, and Na^+ and Br^- permeances were calculated (Schönherr and Huber 1977). The concentration of NaBr and *p*H was the same on both sides of the membrane (4×10^{-3} mol l^{-1}), which implies that there was no net driving force. Radio-labelled ions were used ($^{24}Na^+$ and $^{82}Br^-$), and the fluxes of the ions were not coupled and both ions could diffuse independently. At *p*H 3 the ratio of the permeances of Na^+/Br^- ranged from 0.58 to 0.69, showing that permeance for Br^- was higher because the membranes carried a net positive charge and co-ion (Na^+) diffusion was reduced by relative Donnan exclusion. At *p*H 8.5 the ratio of permeances ranged from 3.83 to 4.39, because the membranes were negatively charged and now Br^- experienced Donnan exclusion. Thus, depending on *p*H, cuticles are either cation or anion exchangers. The nature and concentration of the fixed charges will be dealt with next.

4.3 Ion Exchange Capacity

The concentration of charges covalently bound to the polymer matrix can be determined by potentiometric titration, as can be done with soluble electrolytes. The main difference is the fact that it takes longer to obtain equilibrium, because ions must diffuse into the polymer matrix, where diffusion coefficients are much smaller than in water and mixing is absent. The progressive batch method can be employed (Schönherr and Bukovac 1973). Small amounts (200 mg) of isolated CM or MX are weighed into glass tubes, and a constant volume of degassed water or salt solutions are added. To determine cation exchange capacity, increasing amounts of base is added under a stream on nitrogen, the vessels are closed airtight and agitated at constant temperature (25°C). After 1–4 days, equilibrium is obtained. The *p*H of an aliquot from the supernatant is determined under a stream of nitrogen. If *p*H deviates significantly from 7 the supernatant is titrated back to *p*H 7 with standard acid or base, to determine the amount of H^+ released or base that was not used up in ion exchange. The reaction is stoichiometric and the equilibrium condition can be written as

$$\begin{aligned} &\overline{RH} + NaOH \rightleftarrows \overline{RNa} + H^+ + OH^-, \\ &H^+ + OH^- \rightleftarrows H_2O, \end{aligned} \tag{4.4}$$

where R^- is the polyanion and barred quantities refer to the polymer phase. The exchange capacity at a given *p*H is calculated from the amount of base added, minus the amount of base left at equilibrium. If a solution of a neutral salt is added to the cuticle, some ion exchange takes place and the supernatant becomes acidic:

$$\overline{RH} + NaCl \rightleftarrows \overline{RNa} + H^+ + Cl^-. \tag{4.5}$$

The higher the affinity between R^- and the cation, the lower is the resulting *p*H. For instance, the affinity of the MX for Ca^{2+} ions is very high, and the *p*H obtained with $CaCl_2$ is much lower as with NaCl.

The high affinity for Ca^{2+} can be utilised to determine cation exchange capacity (Schönherr and Huber 1977). A known amount of isolated CM is equilibrated in buffered solutions containing $^{45}CaCl_2$ until equilibrium is obtained. Then, the CM pieces are washed repeatedly with deionised water to remove adhering solution and sorbed electrolyte ($CaCl_2$). With *Citrus* MX four washes sufficed, and radioactivity contained in cuticles no longer decreased when washing was continued. When cuticles are dropped in 1 N HCl, the exchangeable Ca^{2+} is released and can be determined by scintillation counting. This method is simple and very accurate, and it can be used with very small amounts of CM (<1 mg) while with potentiometric titration the batches had to be 200 mg. It is not necessary to work under a nitrogen atmosphere, since solutions are buffered. The method works only with divalent cations. Attempts to determine exchange capacity using monovalent $^{137}Cs^+$ failed, as radioactivity of MX continuously decreased during washing until the MX was free of radioactivity. Cs^+ was exchanged for H^+ contained in equilibrium water having a *p*H of 5.5 due to dissolved CO_2 (Schönherr and Huber 1977).

Before isolated cuticles can be titrated, they must be conditioned by cycling between 1 N HCl and deionised water at room temperature. This removes Ca^{2+} and Mg^{2+} contained naturally in isolated cuticles. However, Cu^{2+}, Zn^{2+} and Fe^{3+} are not completely removed, even though Schönherr and Bukovac (1973) used 6 N HCl. After the third treatment with HCl, cuticles are washed extensively with deionised water until free of chloride ions. This treatment eliminated 0.18 eq kg^{-1} cations contained in isolated tomato fruit cuticles. Before conditioning, large amounts of Ca, Mg, Ba, Fe, Cu and Zn were detected in the ash. Traces of Cu, Fe and Zn were still left in the ash after conditioning (Schönherr and Bukovac 1973).

Cation exchange capacity increased with increasing *p*H (Fig. 4.2). There is a distinct plateau around *p*H 7 and another one around *p*H 9. Above *p*H 9, exchange capacity of tomato fruit (cv. Traveller) MX increased by not much, while with pepper fruit and *Schefflera* MX exchange capacity increased further up to *p*H 11. Higher *p*H values were not used in these experiments, to avoid hydrolysis of ester bonds in cutin. Figure 4.2 suggests the presence of three dissociable groups in the MX differing in acid strength (pK_a). The first group is fully ionised at about *p*H 6, the second at *p*H 9 and the third at *p*H 12 (Schönherr and Bukovac 1973). The exchange capacity of the first group differed among the species and ranged from 0.1 to 0.2 eq kg^{-1}, while the second group had an exchange capacity of around 0.33 eq kg^{-1} with all three species.

Similar results were obtained with *Citrus* and apricot leaf cuticles, except that the plateau around *p*H 6 is not so pronounced (Fig. 4.3). Using the method of paired comparisons, exchange capacity of *Citrus* CM and MX was studied and no difference was detected (Schönherr and Huber 1977). Data for tomato fruit MX seen in Fig. 4.3 were obtained with the red variety "Campbell 17", and at *p*H 9 exchange capacity was about 0.5 eq kg^{-1}. If titration was continued up to *p*H 12 the third group was fully ionised, and exchange capacity reached 1 eq kg^{-1} such that the third group

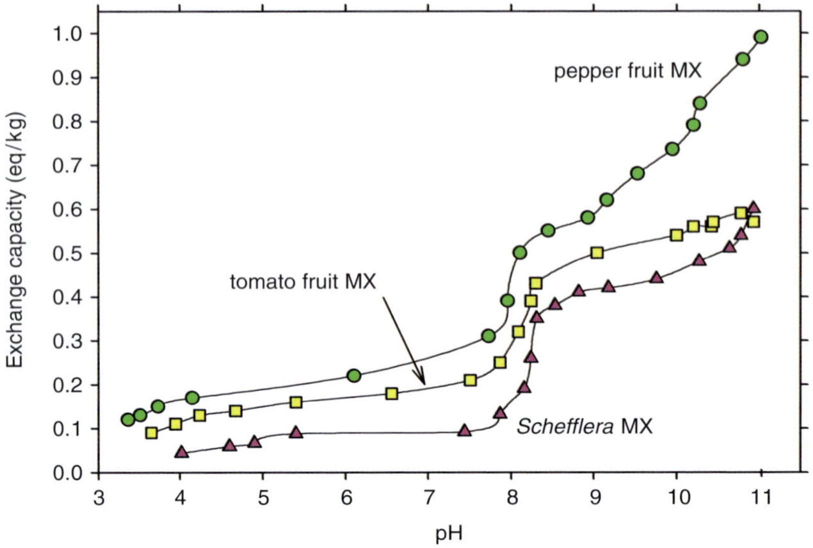

Fig. 4.2 Exchange capacities of the MX obtained from isolated fruit cuticles of ripe pepper (*Capsicum annuum*) and tomato (*Lycopersicon esculentum* cv. "Traveller") and adaxial cuticles from Australian umbrella-tree (*Schefflera actinophylla*) leaves. MX was titrated at 25°C in the presence of 0.1 N $CaCl_2$. (Redrawn from Schönherr and Bukovac 1973)

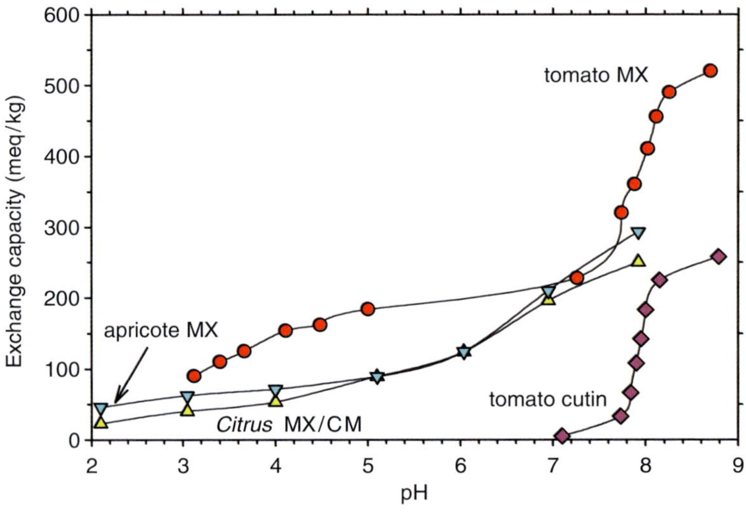

Fig. 4.3 Exchange capacities of MX from leaves of apricot (*Prunus armeniaca*) and *Citrus aurantium*and tomato fruits cv. Campbell 17. Tomato fruit cutin was obtained by treating the MX (cv. Campbell 17) with 6 N HCl for 36 h at 110°C. Exchange capacity was determined at 25°C in presence of 0.1 N $CaCl_2$. Data for tomato MX and cutin were obtained by potentiometric titrations (Schönherr and Bukovac 1973). With apricot and *Citrus*, $^{45}Ca^{2+}$ in buffered solutions was used (Schönherr and Huber 1977). (Redrawn from the original figures)

has an exchange capacity of 0.5 $eq\,kg^{-1}$ (data not shown). Between *p*H 9 and 12, the MX changed colour from yellow/orange to red. The variety "Traveller" (Fig. 4.2) does not turn red when ripe, and isolated cuticles have an opaque white appearance. At *p*H 9 exchange capacity was around 0.5 $eq\,kg^{-1}$, and above *p*H 9 it increased only slightly.

Titration of MX with base results in titration curves indicative of three distinctly different fixed groups whose acid strengths are influenced by *p*H, the nature of counter ions and concentration of neutral salts. This is typical of weak acid groups such as carboxyl and phenolic hydroxyl groups. Based on acid strengths, the first group titrating up to *p*H 6 was assigned to carboxyl groups of pectins and polypeptides, the second group appeared to be donated by non-esterified carboxyl groups of hydroxyfatty acids, and the third group most likely was phenolic in nature (Schönherr and Bukovac 1973). This picture is complicated by the presence of nitrogen in isolated cuticles. Schönherr and Huber (1977) determined amino acids liberated by acid hydrolysis (6 N HCl for 12 h at 110°C). Since amino acid amounts and compositions of apricot leaf MX were the same when isolated with enzyme or Zn/ZCl_2, the polypeptides were not contaminants from the enzymatic isolation. They are located inside the polymer, and were not removed by washing or by the recycling procedure used to condition isolated cuticles.

Basic and acidic amino acids can contribute both acidic and basic fixed charges. Below the isoelectric point, cuticles are positively charged because of the presence of basic amino acids (Table 4.2). Histidine, lysine and arginine are basic amino acids, and K_b values of their free amino groups are 6.0, 10.53 and 12.48 respectively. Below the isoelectric points of cuticles all of them are protonated, and they lose their charges only at *p*H values above 8 (histidine) or higher. Asparaginic and glutamic acid are acidic amino acids, and the pK_a of their free carboxyl groups are 3.65 and 4.25 respectively. They are ionised above the isoelectric points, and contribute to exchange capacities of all three ionisable groups (Table 4.2).

Upon acid hydrolysis the tomato fruit MX lost 13% of its weight and all of its nitrogen. Oxygen content decreased from 22.7% to 19.7%. Water droplets spread on

Table 4.2 Total Exchange capacity ($\mathcal{EC}$) of selected fruit and leaf MX membranes at *p*H 6, 8 and 9, and contribution of amino acids (AA) and pectin to $\mathcal{EC}$

Species	$\mathcal{EC}$ ($meq\,kg^{-1}$) *p*H 6.0	$\mathcal{EC}$ ($meq\,kg^{-1}$) *p*H 8–9	Basic AA ($meq\,kg^{-1}$)	Acidic AA ($meq\,kg^{-1}$)	Pectin ($meq\,kg^{-1}$)	Total AA ($mMol\,kg^{-1}$)
Capsicum fruit	230	580 (*p*H 8)	21	46	184	256
Lycopers. fruit	200	500 (*p*H 9)	13	40	160	190
Citrus aurantium	133	250 (*p*H 8)	12	22	308	107
Prunus armeniaca	120	290 (*p*H 8)	19	33	87	152
Schefflera actinoph.	80	400 (*p*H 8)	14	27	53	139
Pyrus communis	–	–	26	54	–	266

$\mathcal{EC}$ of pectins was estimated by subtracting $\mathcal{EC}$ of acidic amino acids from the total $\mathcal{EC}$ at *p*H 6 (data taken from Schönherr and Bukovac 1973; and Schönherr and Huber 1977)

the morphological inner surface of isolated CM, but after acid hydrolysis the surface turned hydrophobic and water droplets had contact angles around 90°. Obviously, polar polymers were lost from the inner surface, and only cutin was left. This is confirmed by ion exchange properties, since the first and the third ionisable groups were eliminated. Only the second group attributed to cutin remained (Fig. 4.3). With this background, it is clear that exchange capacity of the first group was due to pectins and acidic amino acids, which contributed approximately 20% of the total cation exchange capacity at *p*H 6. The third group was suggested to be phenolic, because dissociation started only above *p*H 9. This is supported by the fact that: (1) ripe red tomato fruits contain large amounts (4.2–5.6%) of bound phenolics such as coumaric acid, naringinin and chalconaringinin (Hunt and Baker 1980), while in mature green tomatoes flavonoids were lacking and only small amounts of coumaric acid (0.8%) were detected, and (2) the green variety "Traveller" lacked the third ionisable group (Fig. 4.2). Coumaric acid has two acidic groups, a carboxyl group with a pK_a around 4.5 and a phenolic hydroxyl group. Naringinin is a trihydroxy flavanon. The pK_a of unsubstituted phenol is 10.0 (Albert and Serjeant 1971). Substitution tends to increase acid strength, such that *p*-hydroxybenzoic acid has two pK_a values of 4.57 and 9.46 respectively. We could not find exact values for the pK_a values of hydroxyl groups of coumaric, naringenin and chalconaringenin, but most likely they are >9, and this perfectly fits the titration curves seen in Figs. 4.2 and 4.3. For the varieties used in titration, we have no exact data concerning which amounts of phenolics were actually present. If coumaric acid was present, it would have made a contribution to the first ionisable group. Naringenin has only phenolic hydroxyl groups, and its equivalent weight is about 91 g eq^{-1}. Since the variety "Campbell 17" had an exchange capacity of 0.5 eq kg^{-1} between *p*H 9 and 12, only 45 g naringenin or 4.5% by weight could account for this. Hunt and Baker (1980) detected 4.2–5.6% phenolics in three other post-climacteric tomato varieties.

4.3.1 Cation Selectivity

As already mentioned, monovalent Cs^+ is not bound very tightly, while Ca^{2+} cannot be washed out with water. At *p*H 9 and in presence of 0.1 N salt, the exchange capacity decreased slightly from 0.40 to 0.35 eq kg^{-1} in the order $Li^+ > Na^+ > Rb^+ > N(CH_3)_4^+$, which is the order of increasing crystal radii of the ions. The same was observed with Ca^{2+} and Ba^{2+}, which at the same conditions had a higher exchange capacity of 0.52 and 0.49 eq kg^{-1}.

The selectivity of the tomato fruit MX for divalent over monovalent cations was further studied using simultaneous exchange of Ca^{2+} and Na^+ as a function of *p*H of the supernatant (Fig. 4.4). The molal concentrations of NaCl and $CaCl_2$ were 0.1 and 5×10^{-3} mol kg^{-1} respectively. Total exchange capacity was determined by back titration, exchange of Ca^{2+} was measured using $^{45}Ca^{2+}$, and Na^+ exchange was calculated by difference.

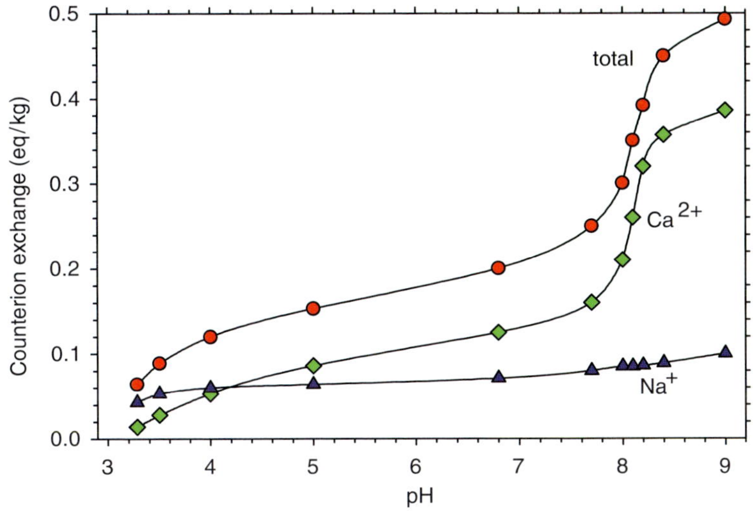

Fig. 4.4 Simultaneous counter ion exchange of MX of *Lycopersicon esculentum* cv. Campbell 17 of Ca^{2+} and Na^{+} as a function of pH at 25°C. The molal concentrations of NaCl and $CaCl_2$ were 0.1 and 5×10^{-3} mol kg^{-1} respectively. (Redrawn from Schönherr and Bukovac 1973)

Above pH 4, more Ca^{2+} than Na^{+} was exchanged, even though Na^{+} concentration was 20 times higher. Most Ca^{2+} was exchanged by the second group (Fig. 4.4). The apparent acid strength of dissociable fixed charges increased with increasing concentration of neutral salt, increasing valence and decreasing crystal radii of counter ions (Schönherr and Bukovac 1973). This behaviour is typical of polyelectrolytes of the weak acid type (Helfferich 1962). The main reason for this is the electrostatic free energy arising from the electrostatic repulsion of neighbouring fixed charges. This repulsion causes the polymer chains to uncoil and stretch, which lowers the configurational entropy and increases the free energy of the polymer. As charge density and electrostatic potential increase during titration, the tendency to form more negative charges in proximity to existing ones diminishes, and the apparent acid strength therefore decreases as the degree of ionisation and exchange capacity increase.

The electrostatic free energy can be reduced by association of fixed charges and counter ions. The smaller the crystal radius of the counter ion, the greater the interaction between fixed charge and counter ions, the lower the electrostatic free energy, and the higher the acid strength. This is exactly what was observed with tomato fruit MX (Schönherr and Bukovac 1973). Increasing the concentration of neutral electrolyte increases the apparent acid strength because the sorbed salt tends to shield neighbouring fixed charges, and it reduces the Donnan potential, i.e., the salt reduces the pH difference between external solution and interstitial fluid in the polymer. The magnitude of the combined effects of electrostatic free energy and Donnan potential can be seen from the fact that the intrinsic pK_a of hydroxyfatty acids is about 5 (Albert and Serjeant 1971). In tomato fruit MX, apparent pK_a was about 8 at 0.1 N

$CaCl_2$ (Fig. 4.3). At zero and 0.01 N $CaCl_2$, the apparent pK_a of the second group was 8.75 and 8.5 respectively (Schönherr and Bukovac 1973).

Similar arguments can explain counter ion selectivity. If two ions are available, the polymer prefers that cation which results in the minimum electrostatic free energy. Hence, the polymer prefers the counter ion that associates more closely with the fixed charges (minimising the electrostatic free energy) and which results in the smallest polymer volume (minimising the free energy of stretching and maximising the configurational entropy). In comparison with sodium, the divalent calcium associated more closely with the fixed charges and reduced swelling, because half the number of osmotically active particles were present. Exchangeable Ca^{2+} ions also tend to have a lower osmotic coefficient than sodium (Howe and Kitchner 1955). In epidermal cell walls there is no excess of K^+ over Ca^{2+}, and most carboxyl groups in cuticles will be neutralised with calcium ions.

High selectivity for Ca^{2+} over Na^+ is evidence that in the MX negative charges occur in close proximity. It follows that negative charges are not randomly distributed, but clustered. Likewise, a random distribution of COOH groups in a CH_2 environment would be energetically unfavourable und thus improbable. This also applies to other polar functions such as hydroxyl and amino groups. This leaves us with the question as to the shape and location of polar clusters and their possible role in permeability to water and ions. This will be considered next.

4.4 Water Vapour Sorption and Permeability as Affected by *p*H, Cations and Vapour Pressure

Synthetic polymeric membranes can be classified as lipophilic or polar polymers (Sect. 4.1). In lipophilic polymers such as polyethylene, water sorption is proportional to vapour pressure, and permeability does not depend on partial vapour pressure. Polar polymers sorb more water, their sorption isotherms are non-linear, and permeability increases with water vapour pressure. A few examples are shown in Fig. 4.5.

Silicon rubber is an example of a highly lipophilic matrix. Permeance does not depend on partial pressure (p/p_0), and it is high because diffusion coefficients are exceptionally high due to very high chain mobility (Barrie 1968). Ethyl cellulose (EC) is a cellulose ether, with about half of the hydroxyl groups of glucose monomers linked to ethyl groups. Water is sorbed mainly by free hydroxyl groups. Polymethyl methacrylate (PMA) contains carboxyl groups as dipoles, and under experimental conditions they were neutralised by Ca^{2+} ions. Permeance of both polar polymers (EC and PMA) increased with partial pressure, which is most pronounced when $p/p_0 > 0.6$. At the lowest p/p_0 (0.22), permeances were smaller by factors of 0.28 and 0.72 with PMA and EC respectively.

The polymer matrix (MX) carries immobile (fixed) electrical charges, which are contributed by pectins, polypeptides, coumaric acid and cutin (Sect. 4.2). Hydroxyl groups donated by cutin, hydroxyfatty acids, pectins, cellulosic compounds and

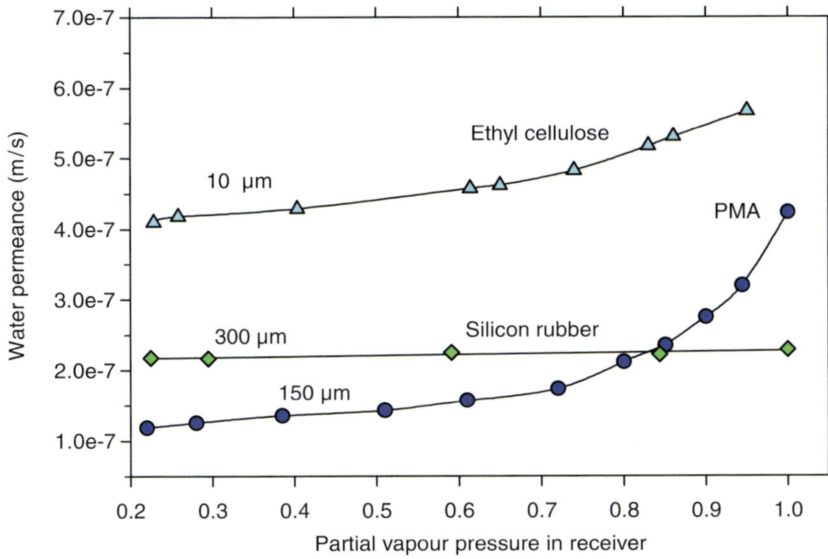

Fig. 4.5 Water permeance (P_w) at 25°C of polymer membranes as affected by partial vapour pressure. With ethyl cellulose, water vapour in the donor was varied, and in the receiver vapour pressure was close to zero. Membrane thickness was 1 mm, and P_{Hg} (Wellons and Stannett 1966) was converted to P_w for a 10 μm-thick membrane as described in Chap. 3. With silicon rubber and polymethyl methacrylate (PMA) an aqueous buffer of *p*H 6 containing 0.1 mol l^{-1} $CaCl_2$ was used, and partial pressure of the receiver was varied. Data for silicon rubber and PMA were taken from Schönherr and Schmidt (1979) and Schönherr and Ziegler (1980) respectively

polypeptides are permanent dipoles, and their hydration is not affected by *p*H and cations. Polarity and hydration of carboxyl, phenolic hydroxyl and amino groups depend on *p*H and type of counter ions. From these facts, one would expect permeance to depend on *p*H, type of counter ion and p/p_0 — and in fact it does (Fig. 4.6).

Permeance of polymer matrix membranes from *Citrus aurantium* increased with increasing partial vapour pressure of the receiver. With NaCl in the donor, permeance was lowest at *p*H 3 and increased with *p*H, but dependence on p/p_0 was similar at all *p*H values used (Fig. 4.6). With $CaCl_2$ in the donor, dependence on p/p_0 was similar as when MX-membranes were in the Na^+-form, but *p*H had no effect on permeance. This indicates that hydration of carboxyl groups is similar when non-dissociated or neutralised with Ca^{2+}. When the partial pressure was 0.22, permeance was lower by a factor of 0.5 compared to *p*H 9. All four plots had a plateau at a partial pressure of about 0.5.

Citrus leaf and tomato fruit cuticles have been classified as "all regions reticulate". In anticlinal pegs of *Citrus limon*, a histochemical test (Thiéry reaction) indicated the presence of polysaccharides in the reticulum (Holloway 1982a). Most water vapour sorbed in MX is sorbed by polar polymers (Fig. 4.7) and not by cutin. The dependence of P_w on p/p_0, *p*H and inorganic salts (Fig. 4.6) suggests that

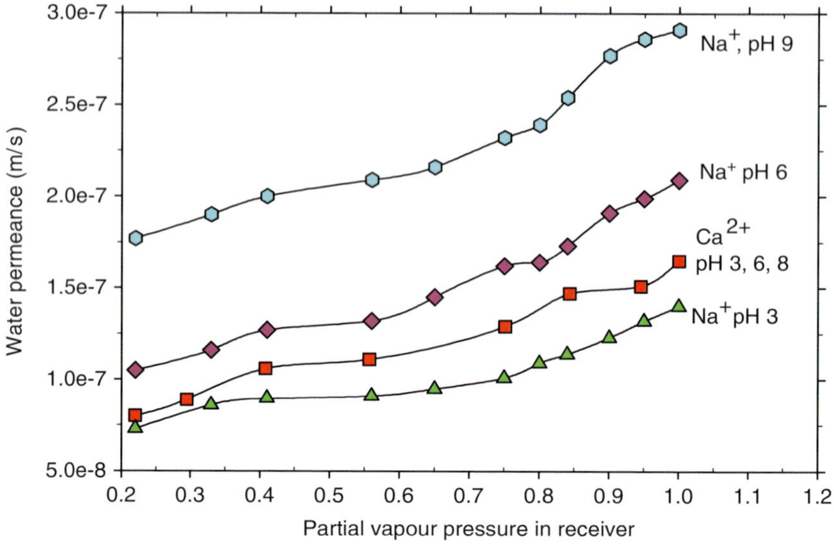

Fig. 4.6 Effect of partial pressure of water vapour in the receiver on water permeance (P_w) of *Citrus aurantium* polymer matrix membranes at 25°C. Data were taken from Schönherr and Schmidt (1979). The aqueous donor contained $CaCl_2$ or NaCl at 0.01 $mol\,l^{-1}$, and was buffered at *p*H 3, 6, 8 or 9. The same set of membranes was used with all *p*H values and salt solutions. Permeance was measured using tritiated water in the donor

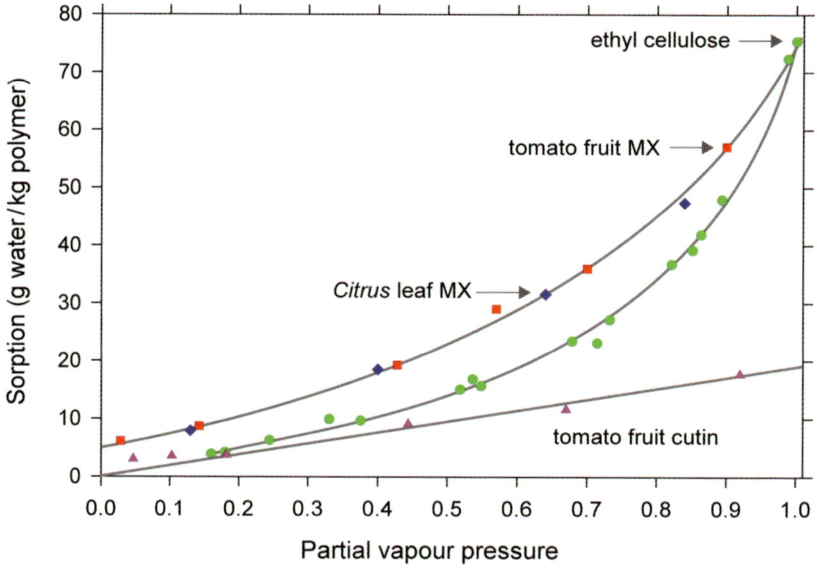

Fig. 4.7 Sorption of water at 25°C by ethyl cellulose, MX from *Citrus aurantium* leaves, tomato (*Lycopersicon esculentum*) fruit and tomato cutin, as affected by partial pressure of water vapour. Data were taken from Wellons and Stannett (1966) and Chamel et al. (1991) for EC and MX respectively

in *Citrus* MX the reticulum contains carboxyl groups and is continuous across the entire polymer matrix, including the cuticle proper. This is convincing evidence, even though such a reticulum is rarely seen in TEM (Sect. 1.4). The aqueous pores across the polymer matrix formed by the reticulum are further characterised in Sect. 4.5. Comparable data for MX from other plant species, including those having a lamellated cuticle proper, are not available.

Polar polymers sorb much more water than hydrophobic ones, and permeance increases with increasing partial pressure (Fig. 4.5) because of increased sorption of water (Barrie 1968). Sorption in the polymer matrix was measured with carboxyl groups in the hydrogen form, that is, in absence of inorganic cations (Chamel et al. 1991). Sorption isotherms are not linear (Fig. 4.7) and resemble B.E.T. type II isotherms. Water vapour sorption in tomato and *Citrus* MX was similar to that in EC. For the MX, sorption close to 100% humidity is not available, but extrapolation leads to figures somewhere between 70 and 80 $g\,kg^{-1}$, which is 7–8% by weight. The plateau seen in Fig. 4.6 for permeance of MX is strikingly absent in sorption data (Fig. 4.7). This may be due to the fact that sorption in MX was measured with carboxyl groups in the hydrogen form and in the absence of inorganic counter ions. COOH and OH groups probably sorb similar amounts of water, because characteristic dipole moments for COOH and OH groups are similar and amount to 1.7 and 1.65 Debye respectively (Israelachvili 1991).

Sorption in tomato fruit cutin was considerably lower, and the isotherm was linear. Sorption at 100% humidity was 19 $g\,kg^{-1}$, which is only about 25% of the amount sorbed in the tomato MX. Cutin was generated by acid hydrolysis of tomato fruit MX (6 N HCl, 110°C, 12 h). This hydrolysis eliminates polysaccharides and polypeptides, and probably also liberates phenolic compounds bound covalently to the MX (Schönherr and Bukovac 1973). The bulk (75%) of the water in the MX is sorbed by dipoles contributed by these compounds.

Polyethylene also has a linear sorption isotherm, and maximum sorption at 100% humidity is 0.65 $kg\,m^{-3}$ (Table 4.1). With a specific gravity of 950 $kg\,m^{-3}$ maximum sorption is 0.68 $g\,kg^{-1}$. Hence, cutin sorbed 28 times more water than PE. Most of this water is probably bound to free hydroxyl groups of cutin acids. When 19 g water is sorbed in 1 kg cutin, this amounts to 1.06 $mol\,kg^{-1}$. Tomato fruit cutin is made up mainly of C_{16}-dihydroxyfatty acids (Baker et al. 1982), which have a molecular weight around 300 $g\,mol^{-1}$. This yields a concentration of 3.33 mol hydroxy fatty acids per kg of cutin. Hence, only a third of the mid-chain hydroxyl groups sorbed a water molecule at 100% humidity. Those hydroxyl groups that did not sorb water were probably engaged in intermolecular hydrogen bonds, and this indicates that they were not distributed at random but are arranged close to each other.

At a partial pressure of 0.22, permeance was smaller by a factor of 0.5 than at $p/p_0 = 1$. Sorption in MX at $p/p_0 = 0.22$ is about 10 $g\,kg^{-1}$, while at $p/p_0 = 1$ sorption amounted to about 70–80 $g\,kg^{-1}$. With ethyl cellulose the difference is even larger (Fig. 4.7). It follows that permeance and water content of the membranes were not proportional. Apparently, not all water sorbed in the MX participated in transport.

4.5 Diffusion and Viscous Transport of Water: Evidence for Aqueous Pores in Polymer Matrix Membranes

In the previous sections, we have considered diffusion of water in membranes and characterised it by permeability, diffusion and partition coefficients. However, water transport across a membrane can be diffusional, viscous or both. If a membrane separates two identical solutions, and in the absence of a pressure difference, the mechanism of transport is always diffusion and that is individual water molecules move by random motion. In a porous membrane subject to a difference of hydrostatic pressure, the flow can be viscous or mass flow. In this case water flows as bulk, not as individual molecules. Ticknor (1958) found that the type of flow depended on the size of the permeating molecule, the radii of the capillaries in the membrane and interactions of the permeating species with the pore walls. He estimated rates of diffusional and viscous flows to be equal when the pore radius was about twice the radius of the penetrating water molecule. In much larger pores, the flow would be mostly viscous. This formed the basis of subsequent attempts to characterize water flows and estimate pore sizes in natural and artificial membranes (Nevis 1958; Robbins and Mauro 1960; Solomon 1968; Lakshminarayanaiah 1969; Schönherr 1976a).

Pores in a membrane matrix can be permanent, as in filter membranes (sintered glass or in Millipore® filters), or they arise by swelling of polar polymers (agar, gelatine, dialysis membranes). Pores in cuticles belong to the last type. Prerequisites for their formation have been recently discussed (Schönherr 2006). Our pore model, on which estimation of pore size is based, assumes a polymer matrix which resembles polyethylene and is made up of ethylene and methyl groups. As a first approximation, this matrix is considered lipophilic and impermeable to water and polar solutes. All water is contained in capillaries which are continuous and cross the polymer matrix. The walls of the capillaries are lined with polar functional groups such as hydroxyl, carboxyl, phenolic hydroxyl and amino groups. In presence of water these groups are hydrated, that is, they are surrounded by water molecules. In the literature (cf. Schönherr 2006) aqueous pores were invoked to explain certain experimental data on permeability of polar non-electrolytes and ions, but they were never characterised. Schönherr (1976a) was the first to estimate radii of aqueous pores in MX membranes.

Having recognised the existence of two types of water flow across membranes, theories have been developed (Renkin 1954; Nevis 1958; Robbins and Mauro 1960) that permit estimating the average pore size from measurements of diffusional flux ($J_{\mathrm{diffusion}}$) and viscous flux (J_{viscous}). We shall refer to this approach as model I.

Diffusion of water in MX membranes can be measured using tritiated water (THO) and identical aqueous buffers or salt solutions on both sides of the membrane. Under these conditions, there is no difference in hydrostatic or osmotic pressure, and water activity is the same on both sides of the membrane. There is no driving force and hence no net flux of water. The diffusional flow of water ($F_{\mathrm{diffusion}}$ in Bq s^{-1}) can only be measured as isotope flux. This process is called self-diffusion. According to

Fick's law, we can write for the steady state

$$A_{\text{membrane}} \times J_{\text{THO}} = -D_{\text{THO}} \times A_{\text{pore}} \frac{\Delta C_{\text{THO}}}{\ell}, \tag{4.6}$$

where C_{THO} is the concentration of tritiated water (Bq m^{-3}), D_{THO} is the diffusion coefficient (m^2 s^{-1}) in the pore liquid, and ℓ is the thickness of the membrane. J_{THO} is the flux per unit membrane area and driving force (Bq m^{-2} s^{-1}). By hypothesis, water flows exclusively in pores assumed to be circular, and the total pore area is

$$A_{\text{pore}} = n\pi r_{\text{pore}}^2. \tag{4.7}$$

The permeability coefficient for diffusion of THO is

$$A_{\text{membrane}} \times J_{\text{THO}} = -P_{\text{diffusion}} \times A_{\text{membrane}} \times \Delta C_{\text{THO}}. \tag{4.8}$$

Combining (4.6)–(4.8) we obtain after rearranging

$$P_{\text{diffusion}} = \frac{A_{\text{pore}}}{A_{\text{membrane}}} \times \frac{D_{\text{THO}}}{\ell} = \frac{D_{\text{THO}} n\pi r^2}{\ell A_{\text{membrane}}}. \tag{4.9}$$

According to Poiseuille's law, the viscous water flux (J_{viscous} in m^3 m^{-2} s^{-1}) in capillaries is proportional to the fourth power of the radius and inversely proportional to viscosity (η in Pa s):

$$A_{\text{membrane}} \times J_{\text{viscous}} = \frac{n\pi r_{\text{pore}}^4}{8\eta \overline{V_{\text{w}}}} \times \frac{RT\Delta C_{\text{solute}}}{\ell} = P_{\text{viscous}} \times A_{\text{membrane}} \times \Delta C_{\text{solute}}, \tag{4.10}$$

where $\overline{V_{\text{w}}}$ is the partial molar volume of water (18 cm^3 mol^{-1}). The driving force for viscous flux is the hydrostatic pressure difference (in Pa) across the membrane. Cuticular membranes are fragile, and an osmotic pressure difference must be used in determining J_{viscous}. For this reason, the difference of hydrostatic pressure was substituted by the difference in osmotic pressure, which according to Van't Hoff equals $RT\Delta C_{\text{solute}}$. Hydrostatic pressure and osmotic pressure cause identical viscous fluxes if the membrane is impermeable to the solute. Dividing (4.10) by (4.9) and solving for r^2 yields

$$r_{\text{pore}}^2 = \frac{8 D_{\text{w}} \eta \overline{V_{\text{w}}}}{RT} \times \frac{P_{\text{viscous}}}{P_{\text{diffusion}}} \tag{4.11}$$

and after substituting the literature values for the constants and taking the square root, we have at 25°C

$$r_{\text{pore}} = \sqrt{\frac{8\,(2.44 \times 10^{-9}\,\text{m}^2\,\text{s}^{-1})(9 \times 10^{-4}\,\text{Pa s})(18 \times 10^{-6}\,\text{m}^3\,\text{mol}^{-1})}{2.48 \times 10^3\,\text{Pa mol}^{-1}\,\text{K}^{-1}}} \times \sqrt{\frac{P_{\text{viscous}}}{P_{\text{diffusion}}}}$$
$$= 0.36\sqrt{(P_{\text{viscous}}/P_{\text{diffusion}})}. \tag{4.12}$$

Accounting for the diffusional component present in $P_{viscous}$, Nevis (1958) wrote the above equation as

$$r_{pore} = 0.36\sqrt{\frac{P_{viscous} - P_{diffusion}}{P_{diffusion}}}, \tag{4.13}$$

which gives the average or equivalent pore radius in nm. In deriving (4.13), the bulk quantities for the diffusion coefficient (D_w) and viscosity (η) of water are used.

These equations have been employed to estimate equivalent pore radii in the polymer matrix from *Citrus* leaves (Schönherr 1976a) and onion bulb scales (Schönherr 1976b). Beyer et al. (2005) published some data from which pore radii of pepper and tomato fruit CM can be calculated. Data for other species are not available.

Diffusion of water across polymer matrix membranes from *Citrus* leaves was measured using tritiated water. When both buffer solutions contained 0.1 N, NaCl permeance ($P_{diffusion}$) was higher above *p*H 4 than in presence of 0.1 N $CaCl_2$. The effects of *p*H and type of counter ions demonstrate the involvement of three different weakly acidic groups in the MX, as was already discussed (Sect. 4.3). Above *p*H 9, phenolic hydroxyl groups are responsible for the effect of *p*H on permeance of MX membranes in Na^+ form. Data for MX membranes in Ca^{2+} form above *p*H 8 are not available. Under natural conditions *p*H in the MX will certainly be <8, and due to high selectivity of carboxyl groups to Ca^{2+} ions they will be in the Ca^{2+} form. This minimises swelling, and permeance is similar at all *p*H values, no matter if carboxyl groups are not ionised or when neutralised with Ca^{2+} (Fig. 4.8).

Using (4.9) we can calculate the fractional pore area ($A_{pore}/A_{membrane}$), if we assume that D_{THO} in the pore liquid is the same as in bulk water ($2.44 \times 10^{-9}\,m^2\,s^{-1}$) and the length of the pores is the same as the thickness of the membrane ($2.66 \times$

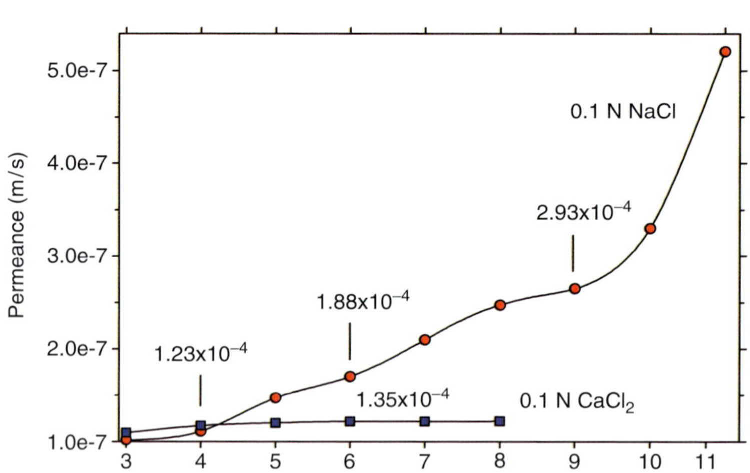

Fig. 4.8 Permeance (P_w) of *Citrus aurantium* MX membranes measured at 25°C using tritiated water. The same MX membranes were employed for studying the effect of salts and *p*H on permeance. Numbers are fractional pore areas. (Plotted using data from Schönherr 1976a, b)

10^{-6} m). This is not a good assumption, as pointed out in Sect. 4.1. D_{THO} in the pore fluid is definitely lower, and the path length is greater than ℓ due to tortuosity. As the fractional pore area is $P_{diffusion}\ell/D_{THO}$, the ratio ℓ/D_{THO} in the pore liquid will be much larger than in a water film having the same thickness as the MX membranes. However, ℓ/D_{THO} is probably not affected by pH, and for the sake of argument we have added selected fractional pore areas to Fig. 4.8. Since permeance of MX membranes in Na^+ form increased with pH, $A_{pore}/A_{membrane}$ also increased. $A_{pore}/A_{membrane}$ is proportional to the fractional volume of water in the membrane (volume of water/total volume of membrane), and this is a quantitative measure of swelling (Kedem and Katchalsky 1961). The absolute values of $A_{pore}/A_{membrane}$ are in error, but the increase of $A_{pore}/A_{membrane}$ with pH reflects the change in water content of MX. The fractional volume of water in the MX is independent of pH when the MX is in Ca^{2+} form, or when carboxyl groups are not ionised (Fig. 4.8).

Having established that total pore area increases with increasing pH, as long as the MX is in Na^+ form, we can now test if this is due to larger pore radii or to an increase in number of pores. Size of pores can be estimated using (4.13) when $P_{diffusion}$ and $P_{viscous}$ are known. Volume flux of water was measured using an apparatus made from glass (Schönherr 1976a) and a number of solutes differing in size. All measurements were made with identical buffers on both sides and with the osmotic solutes in the outer compartment facing the morphological outer surface of the MX. The volume flux was measured in a calibrated capillary (0.24 µl mm^{-1}) connected to the outer compartment. The entire apparatus was submerged in a water bath maintained at 25 ±0.02°C, and only the tips of the capillaries protruded over the surface of the water bath. Temperature control is critical, since water volume of water varies greatly with temperature. A 0.01 moll^{-1} citric acid and Na_2HPO_4 buffer was used in the pH range of 3–7 and 0.01 moll^{-1} disodiumtetraborate (borax) adjusted with HCl was used at pH 9. With these buffers in donor and receiver, the MX membranes are in the Na^+ form. Solute concentrations were 0.5 molkg^{-1} with urea, glucose and sucrose, and with raffinose 0.25 molkg^{-1} were used, which is close to the solubility limit.

Viscous or volume fluxes of water were determined at pH 3, 6 and 9 with urea glucose, sucrose and raffinose, and $P_{viscous}$ was calculated from (4.10). The same set of membranes was used for all pH values and solutes. $P_{viscous}$ increased with increasing pH and solute size, and asymptotically approached the maximum value of $P_{viscous}$ (Fig. 4.9). As the differences in $P_{viscous}$ between sucrose and raffinose were small, Schönherr (1976a) assumed that at all pH values MX membranes were impermeable to raffinose, and permeance measured with raffinose represented maximum permeance. Solutes larger than raffinose were not included in the work. Here we use an approach for estimating maximum $P_{viscous}$ that is superior to that which would be obtained with hydrostatic pressure or with solutes to which the membranes are impermeable. By fitting a parabola to the data points, $P_{viscous}^{maximum}$ can be obtained and the above assumption can be tested. The curves fitted to the data points (Fig. 4.9a) represent the hyperbola where θ is a constant, and $P_{viscous}^{maximum}$ is the maximum permeance that would be obtained when solute radius (r_{solute}) approaches

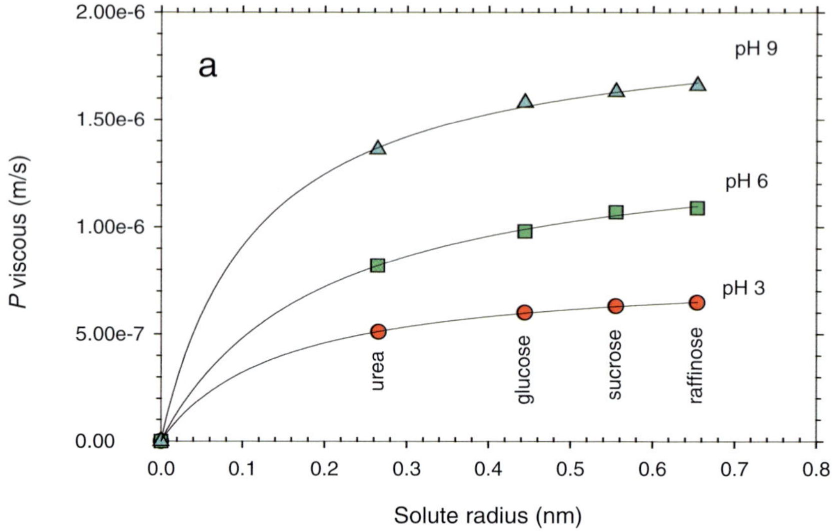

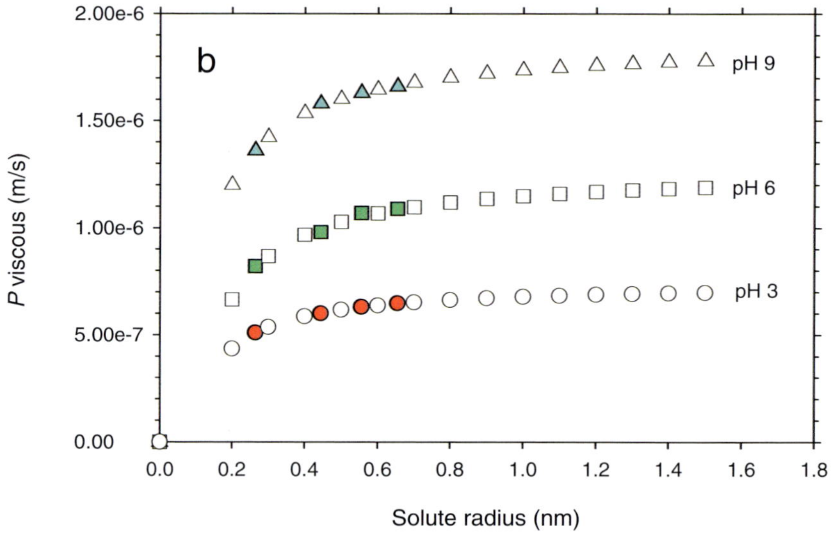

Fig. 4.9 Dependence on solute radius of permeance for viscous flow of water at 25°C across polymer matrix membranes obtained from *Citrus aurantium* leaves. Experimental data are shown in colour. *Curves* shown in **(a)** represent the hyperbola $(ax/b + x)$ and *empty symbols* in **(b)** were calculated from (4.15). Data were taken from Schönherr (1976a)

infinity

$$P_{\text{viscous}} = \frac{P_{\text{viscous}}^{\text{maximum}} \times r_{\text{solute}}}{\theta P_{\text{viscous}}^{\text{maximum}}}. \tag{4.14}$$

Data can also be analysed assuming that P_{viscous} varies linearly with the reciprocal of the solute radius. Dependence was in fact linear, as the coefficients of determination were 0.99 or better at all three pH values. The following regression equations were obtained:

$$\begin{aligned}
\text{at } p\text{H 3:} \quad P_{\text{viscous}} &= -1.34 \times 10^{-7} \frac{1}{r_{\text{solute}}} + 1.87 \times 10^{-6} (r^2 = 0.997), \\
\text{at } p\text{H 6:} \quad P_{\text{viscous}} &= -1.21 \times 10^{-7} \frac{1}{r_{\text{solute}}} + 1.27 \times 10^{-6} (r^2 = 0.985), \\
\text{at } p\text{H 9:} \quad P_{\text{viscous}} &= -6.05 \times 10^{-8} \frac{1}{r_{\text{solute}}} + 7.38 \times 10^{-7} (r^2 = 0.999).
\end{aligned} \tag{4.15}$$

The constants on the right hand side of the equations are the maximum values of P_{viscous} when r_{solute} approaches infinity, that is, when $1/r_{\text{solute}}$ is zero. These equations were used to calculate P_{viscous} for selected solute radii and results (empty symbols) were plotted in Fig. 4.9b. Experimental data are included as filled symbols.

P_{viscous} increased in the order urea $<$ glucose $<$ sucrose, showing that these solutes penetrated the MX membranes to some degree (Fig. 4.9). Inspection of the figure might suggest that membranes were impermeable to raffinose, but the $P_{\text{viscous}}^{\text{maximum}}$ values obtained from the hyperbola and inverse functions were slightly larger than the experimental values (Table 4.3). The difference is smallest at pH 3 (0.06×10^{-6}), but at pH 6 and 9 the difference amounts to 0.27×10^{-6} and 0.12×10^{-6} respectively. This indicates that MX membranes were not totally impermeable to raffinose.

According to our hypothesis, water and polar solutes can penetrate the membrane only in aqueous pores. This is confirmed by the fact that viscous flow increased with solute size. Solutes which dissolve in the polymer matrix, that is, solutes having a high partition coefficient (2.17) would not have induced viscous water flux. Clearly, large polar solutes are discriminated, and membranes should be impermeable to solutes larger than the size of the pores (Fig. 4.9).

This was tested by measuring permeability at pH 9 of MX membranes to radio-labelled water, urea and glucose (Schönherr 1976a). Sucrose, raffinose or larger

Table 4.3 Values of $P_{\text{viscous}}^{\text{experimental}}$ (m s^{-1}) obtained experimentally with raffinose, and values calculated by regression analysis ($P_{\text{viscous}}^{\text{maximum}}$) based on a hyperbola or an inverse function. In the last column, the means of the fitted values are given

pH	$P_{\text{viscous}}^{\text{experimental}}$	$P_{\text{viscous}}^{\text{maximum}}$ hyperbola	$P_{\text{viscous}}^{\text{maximum}}$ inv. function	$P_{\text{viscous}}^{\text{maximum}}$ mean
3.0	0.68×10^{-6}	0.80×10^{-6}	0.68×10^{-6}	0.74×10^{-6}
6.0	1.08×10^{-6}	1.43×10^{-6}	1.27×10^{-6}	1.35×10^{-6}
9.0	1.80×10^{-6}	1.97×10^{-6}	1.87×10^{-6}	1.92×10^{-6}

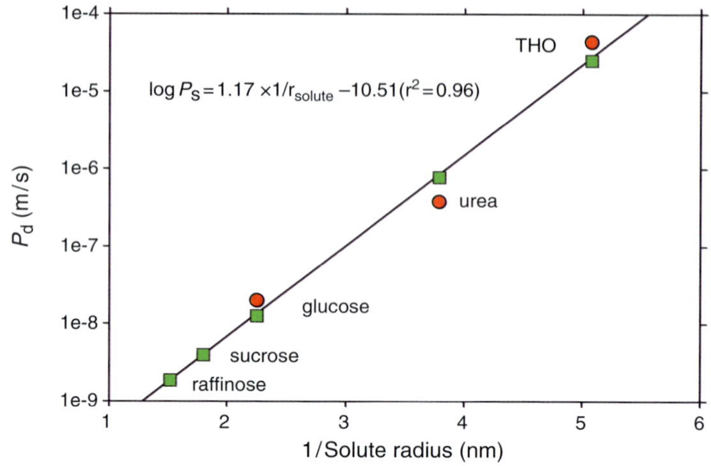

Fig. 4.10 Permeances for diffusion of radio-labelled tritiated water (*THO*) and ^{14}C-labelled polar solutes across *Citrus* MX membranes at *p*H 9 and 25°C. *Red dots* represent experimental data, *Green squares* were calculated based on the linear regression equation shown in the graph. Solute size was taken from Longsworth (1953)

solutes were not included, but the data available indicate that membranes are nearly impermeable to solutes larger than raffinose (Fig. 4.10). Linear regression shows that log P is a function of $1/r_{\text{solute}}$, and it decreases by a factor of 1.17 when $1/r_{\text{solute}}$ increases by 1.0. This corresponds to a decrease in P by a factor of 14.8. P_{urea} and P_{glucose} are smaller than P_{THO} by factors of 32 and 2,008 respectively. Predicted permeances for sucrose and raffinose are 3.94×10^{-9} and $1.85 \times 10^{-9}\,\text{m s}^{-1}$ respectively. As the y-intercept is -10.51, the limiting permeance is $3.0 \times 10^{-11}\,\text{m s}^{-1}$. A solute having twice the molecular weight of raffinose, that is $1{,}008\,\text{g mol}^{-1}$, has a radius of 0.9 nm, and its permeance would be $6.0 \times 10^{-10}\,\text{m s}^{-1}$, which is very low and not too far from the limit. It would be very difficult to measure it precisely.

The above arguments can be used to illustrate the difficulties encountered in determining permeability to polar solutes using a steady state experiment. According to (2.2), permeance is the ratio of flux (J) and driving force (ΔC_{solute}). With raffinose the predicted permeance is $1.85 \times 10^{-9}\,\text{m s}^{-1}$. When working with radio-labelled solutes, which is the most sensitive method, one needs a flux of at least $100\,\text{Bq h}^{-1}$ for reasonable counting statistics. The membrane area of cuticles is rarely larger than $1\,\text{cm}^2$. Under these conditions, we would need a donor concentration of $1.5 \times 10^5\,\text{Bq cm}^{-3}$. Radiochemical purity is rarely better than 98%. To make sure that the radioactivity in the receiver is not mainly due to impurities, the total flux should at least be 10% of the activity of the donor, which in our example would be 1.5×10^4 Bq. Since the minimum flux should be $100\,\text{Bq cm}^{-2}\,\text{h}^{-1}$, the experiment would have to last 150 h or 6.25 days for penetration of 10% of the solute from the donor to the receiver. This can be done, but it is close to the limit, because all experimental variables must be closely controlled, all instruments must work properly and there must be no power failure and the like. This is one of the

Table 4.4 Estimating pore size and pore number in MX-membranes of *Citrus aurantium* in Na^+ form. Data for $P_{diffusion}$ were taken from Schönherr (1976a), and $P^{maximum}_{viscous}$ values were obtained from Table 4.3

pH	$P_{diffusion}$ ($m\,s^{-1}$)	$P^{maximum}_{viscous}$ ($m\,s^{-1}$)	P_{visc}/P_{diff}	r_{pore} (nm)	A_{pore} (m^2)	Number of pores/m^2
3.0	2.56×10^{-7}	0.74×10^{-6}	2.89	0.50	2.79×10^{-4}	3.55×10^{14}
6.0	4.16×10^{-7}	1.35×10^{-6}	3.25	0.54	4.16×10^{-4}	4.54×10^{14}
9.0	7.30×10^{-7}	1.92×10^{-6}	2.63	0.46	7.96×10^{-4}	11.97×10^{14}

r_{pore} was calculated using (4.13); A_{pore} is the total pore area per m^2 membrane calculated according to (4.9) using $D = 2.44 \times 10^{-9}\,m^2\,s^{-1}$ and $\ell = 2.66 \times 10^{-6}\,m$; number of pores $= A_{pore}/\pi r^2_{pore}$.

reasons why we have no data for large solutes, even for the MX. Since P_{solute} for cuticular membranes is at least 100 times smaller, it is clear that accurate measurements in the steady state are not possible. The problem can be overcome using the SOFP technique (Sect. 6.4).

From $P_{diffusion}$ and $P^{maximum}_{viscous}$, equivalent pore radii can be calculated using (4.13). The ratio $P_{viscous}/P_{diffusion}$ is larger than 2 (Table 4.4), indicating the presence of aqueous pores (Nevis 1954). Their radii range from 0.46 to 0.54 nm. Since radii are calculated from the ratio of two empirical permeances, the differences are not significant (Schönherr 1976a). The pore radii did not depend on pH, and the mean is 0.50 nm. The increase in permeance with pH can be attributed to an increase in number of pores rather than to larger pore radii. The average pore size estimated from diffusional and viscous permeability (0.5 nm) is larger than that calculated by Schönherr (1976a), who obtained radii of 0.44–0.46 nm because he used $P_{viscous}$ obtained with raffinose, which is a little smaller than that estimated by curve fitting (Table 4.3). This estimated pore radius of 0.5 nm is too small, since with sucrose ($r_{solute} = 0.555$ nm) and raffinose ($r_{solute} = 0.654$ nm) viscous permeance was smaller than $P^{maximum}_{viscous}$. Clearly, the MX membranes were not totally impermeable to these solutes. Some reasons for this discrepancy are considered next.

There are a number of assumptions inherent in the above calculations. The determination of $P_{diffusion}$ and $P_{viscous}$ is straightforward and no assumptions are needed. However, in calculating the pore radius from the permeances the diffusion coefficient and viscosity enter (4.11), and we used bulk properties (4.12). Activation energies of diffusion of water and of viscous flow in MX membranes are 54 and 46 $kJ\,mol^{-1}$ respectively, while for bulk liquid the level is only about 19 $kJ\,mol^{-1}$ (Schönherr 1976a). The activation energy of a hydrogen bond is about 20 $kJ\,mol^{-1}$ (Nobel 1983). Hence, when water molecules diffuse in bulk only one of the hydrogen bonds is broken at a time, while in the pore liquid more than two H-bonds need to be broken for a water molecule to move. This suggests that D in the pore liquids is considerably smaller than in bulk water. This is not too surprising, since pores are very narrow and the radius of a water molecule is about 0.19–0.197 nm, depending on source (Durbin 1960; Renkin 1954).

Only about 2–3 water molecules fit into the diameter of these pores and since pore walls are made of permanent dipoles and ionised groups, many of these water molecules represent hydration water. Water molecules bound to permanent dipoles or to ions are not completely immobilised. They exchange with bulk water, and at room temperature the mean residence time of water in the primary hydration shell of monovalent cations is about $10^{-9}\,s^{-1}$, and with hydrogen-bonded water residence time is shorter ($10^{-11}\,s^{-1}$) (Israelachvili 1991). This means that many water molecules jump from one dipole to the next. Thus, viscosity of water in the pores is much higher than in bulk. Fortunately, in calculating pore radii the product $D\eta$ enters rather than D alone (4.11). For diffusion in liquids, the Stokes–Einstein relationship states that D_{solute} is proportional to the Boltzmann constant (κ) and temperature (T) and inversely proportional to viscosity (η) and the radius of the solute:

$$D_{solute} = \frac{\kappa T}{6\pi\eta \times r_{solute}}. \tag{4.16}$$

This implies that at constant temperature (T in Kelvin) the product of the diffusion coefficient and viscosity is constant. It is reasonable to assume this to hold also for aqueous pores, and deviation of D and η from bulk properties should not greatly affect pore size estimates. However, it could account in part for the fact that estimated pore size is somewhat smaller than pore size expected from viscous permeance (Fig. 4.9). Another factor might also have contributed. For a solute to enter the pore by a diffusional jump it must find the opening and not hit the pore wall, from which it would be reflected. This steric hindrance increases as the solute approaches the size of the pore opening and may be estimated. Empirical corrections are discussed by Lakshminarayanaiah (1969).

Since we have data on sorption of water in MX membranes in H^+ form (Fig. 4.7), which were measured gravimetrically, we can compare them with fractional volume of water at *p*H 3 calculated from diffusion of THO (Table 4.4). At *p*H 3, most carboxyl groups are not ionised and fractional water content was about 2.8×10^{-4}. At 100% humidity ($p/p_0 = 1.0$) the weight fraction of water in *Citrus* MX was about 0.08 (Chamel et al. 1991), which can be taken to be the volume fraction, since specific gravity of water and the MX do not differ much (Schreiber and Schönherr 1990). This figure is 285 times larger than fractional volumes of water derived from diffusion of water. In calculating fractional pore area and number of pores, the diffusion coefficient of water in bulk was used and the tortuosity was disregarded. Absolute values of the fractional pore areas given in Fig. 4.8 could be multiplied by 285, which is the factor by which ℓ/D is larger in the pore liquid than in bulk. Such a calculation results in a fractional pore area and a fractional pore volume of 0.0385 for the MX in Ca^{2+} form. Thus, the average volume of water in the MX would be 3.85% of its total volume. As total water content is about 8% (Chamel et al. 1991), 48% of the total water would be located in aqueous pores, while 52% should be sorbed in other compartments, possibly in cutin. In tomato fruit, cutin water sorption amounted to $19\,g\,kg^{-1}$, which is 1.9% by weight. Data for *Citrus* cutin are not available.

In Sect. 1.4 we introduced the terms cuticle proper and limiting skin, which refer to a thin outer layer of the CM. It has been suggested that the thickness of the limiting skin is only 1/10 of the total thickness of the MX (Schönherr and Riederer 1988). From transmission electron microscopy we know that the electron-dense reticulum which marks the location of polar functional groups decreases in density from the cell wall side towards the outer surface of the CM (Jeffree 2006). Often it is very hard to detect this reticulum in the cuticle proper of non-laminated cuticles, and it appears to be absent in cuticles having lamellated cuticle proper. This suggests that water content of the cuticles decreases from the inner to the outer surface.

Above, we have demonstrated the existence of aqueous pore in the polymer matrix membranes of *Citrus* leaves, and we estimated the equivalent pore radius. A similar study was conducted using onion bulb scale and a different set of *Citrus* MX membranes (Schönherr 1976b). Pore radii were calculated from P_{viscous} and $P_{\text{diffusion}}$ measurements at *p*H 7 using a slightly different approach. P_{viscous} was determined using 0.25 mol kg^{-1} raffinose, and it was assumed that values so determined represented $P^{\text{max}}_{\text{viscous}}$. The pore radii in both types of MX membranes were similar (0.41 nm), even though both permeances for onion MX were 5.8 times larger than for *Citrus* MX.

In the study of Schönherr (1976b) water permeances measured with MX were around 500 times higher than with CM, both with *Citrus* and onion bulb scale. The volumetric apparatus used in these studies of Schönherr (1976a, b) was not sensitive enough to accurately measure viscous water flux across CM. Hence, we have no information concerning how waxes affect the ratio $P_{\text{viscous}}/P_{\text{diffusion}}$ or pore radii. Beyer et al. (2005) measured P_{viscous} and $P_{\text{diffusion}}$ with pepper and tomato fruit cuticular membranes at 25°C. P_{viscous} was determined gravimetrically, and $P_{\text{diffusion}}$ was determined with THO. Polyethylene glycol with a molecular weight of 6,000 was used as solute in the receiver facing the morphological inner surface of the CM, and *p*H was not controlled. Hence, the water flux was from the outer surface to the inner surface of the CM. Permeances and estimated pore radii are given in Table 4.5.

These permeances are very high compared to permeances measured with astomatous leaf CM, but they are similar to P_{w} of *Citrus* and pear leaf MX (Table 4.1). P_{viscous} was about ten times larger than $P_{\text{diffusion}}$, and pore radii were much larger than those measured in *Citrus* MX. The reason(s) for this difference are not known, and it would be futile to speculate about it. Appropriate data for CM and MX of other species are not available, but they are badly needed to get a better picture of pore structure in CM and MX.

Table 4.5 Water permeance for viscous flux and diffusional flux (P_{w}) and pore radii calculated from (4.13). (Data taken from Beyer et al. 2005)

Species	P_{viscous} (m s^{-1})	$P_{\text{diffusion}}$ (m s^{-1})	Pore radius (nm)
Pepper fruit CM	1.31×10^{-8}	0.19×10^{-8}	3.85
Tomato fruit CM	1.52×10^{-8}	0.13×10^{-8}	2.12

We have demonstrated the presence of aqueous pores in some cuticles for the first time. Theories and methods are available. For a better understanding as to how cuticles function, we need systematic studies using cuticles from many more species. Occurrence of aqueous pores in MX and CM of plant cuticles is not an academic problem. Aqueous pores have been speculated about during the last 5 decades, but the issue was not approached experimentally. Aqueous pores are involved in humidity effects on transpiration, and they are a prerequisite for penetration of hydrated ionic species, which are insoluble in cutin and waxes and require aqueous pores to cross cuticles (Schönherr 2006). We shall return to this topic in Chap. 5.

4.5.1 Lipophilic and Hydrophilic Pathways in the Polymer Matrix

The pore model discussed in the previous section postulated that the polymer matrix is composed of aqueous pores surrounded by a lipophilic matrix. Diffusion of water, solutes and viscous flow is limited to aqueous pores, while the lipophilic matrix was assumed to be impermeable to water and solutes. In this section, we shall evaluate this assumption and discuss consequences.

According to our model I, most polar functional groups lined the pore walls. Riederer and Schönherr (1984) estimated cutin composition of CM for various plant species (Table 1.1). Correcting for the weight fraction of waxes, the fraction of cutin in the MX of *Citrus aurantium* was estimated as 0.77, the remainder (0.23) being the hydrolysable fraction (polysaccharides, peptides, phenolic constituents). Most of these polar polymers should be located in the pores, while cutin would be filling the volume between pores.

Citrus cutin is composed of esterified C_{16} dihydroxyfatty acids, and in a linear polyester one of the two hydroxyl groups is free. With a molecular weight of around $300\,\mathrm{g\,mol^{-1}}$ this hydroxyl group amounts to about 5.7% by weight, while the remainder (94.3%) is made up of ester bonds, CH_2 and CH_3 groups. The permeances (P_{wv}) for synthetic polymers having $3\,\mu\mathrm{m}$ thickness (Table 4.1) range from $6.3 \times 10^{-6}\,\mathrm{m\,s^{-1}}$ (polypropylene) to $6.3 \times 10^{-3}\,\mathrm{m\,s^{-1}}$ (ethyl cellulose). We have no data for permeance of ester cutin, but since it has a similar composition to the synthetic polymers in Table 4.1 it is not very likely that it is completely impermeable to water, as was assumed in model I (Sect. 4.5). Water transport in cutin polyester would be by diffusion, and polar solutes are excluded in all probability. If this is accepted, we can conclude that $P_{diffusion}$ across the MX membrane is the sum of two independent permeances, while $P_{viscous}$ is not affected. What would be the magnitude of the contribution of the polymeric pathway to total diffusional flux of water, and what would be the consequences for estimating pore size?

4.5.2 Permeability of the Pore and Cutin Pathways

Our new improved model (model II) depicted in Fig. 4.11 shows the limiting skin of *Citrus aurantium* polymer matrix. Schönherr (1976a) suggested that the pores resemble thin tubes. The walls of these tubes were formed by hydroxyl and carboxyl groups which are surrounded by hydration water. This poses the problem of the location of the remainder of the polar polymers. In model II, we assume that polar polymers (peptides, pectins, phenols) constitute a separate continuous phase, because during synthesis it would be thermodynamically unfavourable to mix lipophilic cutin with hydrophilic polar polymers. In fact, thin sections of cuticles show two separate phases. A lipophilic matrix with little contrast is traversed by a reticulum of heavy contrast (Sect. 1.4). Hydroxyl, carboxyl and amino groups react with OSO_4, MnO_4 and heavy metal dyes. These reaction products are visible in TEM as dark filaments. In this polar polymer phase, aqueous pores arise by hydration of polar groups and counter ions. These pores are not like tubes and they are not circular, but the interstitial water phase can still be characterised by equivalent pore radii. Water can cross the limiting skin by moving either in the pore liquid or in cutin. Thus, we have two resistances arranged in parallel, the cutin polymer and aqueous pores within strands of polar cuticular polymers.

Resistance ($\mathscr{R}$) is the reciprocal of permeance (Chap. 2), and according to Ohm's law the reciprocal of the total resistance of a group of resistors in parallel is the sum of the reciprocals of the individual resistances, that is, conductances (permeances) in parallel are additive

$$\frac{1}{\mathscr{R}_{\mathrm{diffusion}}} = \frac{1}{\mathscr{R}_{\mathrm{pore}}} + \frac{1}{\mathscr{R}_{\mathrm{cutin}}} = P_{\mathrm{diffusion}} = P_{\mathrm{pore}} + P_{\mathrm{cutin}}. \quad (4.17)$$

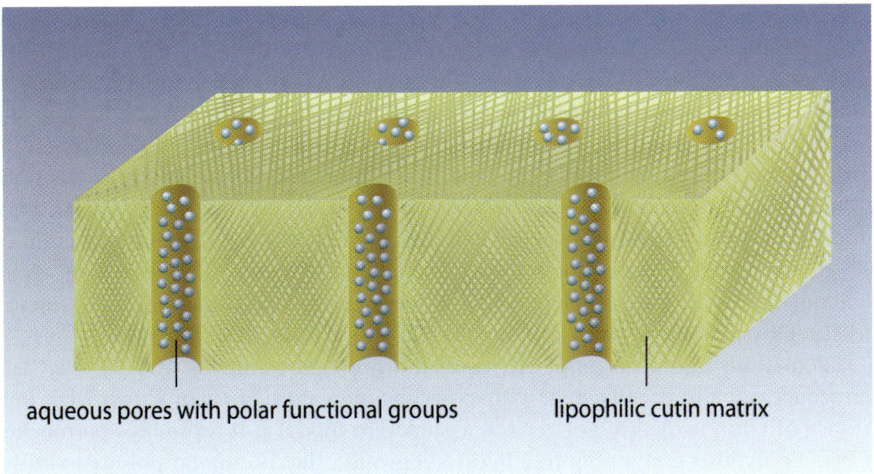

Fig. 4.11 Schematic drawing of the limiting skin of *Citrus* MX, composed of the lipophilic cutin matrix and polar pores arranged in parallel (not to scale)

$P_{\text{diffusion}}$ is calculated as the flux per unit membrane area and driving force (4.8), and it is the only known permeance. Permeance of cutin membranes have never been determined, and the magnitudes of P_{pore} and P_{cutin} must be estimated based on model II. In model I, it was assumed that all water diffuses in aqueous pores. We now loosen that restriction by assuming the water can diffuse in pores and in cutin, while laminar flow is restricted to aqueous pores.

$P_{\text{diffusion}}$ was determined experimentally, and it characterises the total diffusional water flux across the MX membranes. The fraction of water which diffuses in aqueous pores is determined by P_{pore}, which is calculated by multiplying $P_{\text{diffusion}}$ by the weight fraction ($W_{\text{polar polymers}}/\text{W}_{\text{membrane}}$) of polar polymers. P_{cutin} characterises the diffusional flux across the cutin polymer, and it is obtained by difference, that is $P_{\text{cutin}} = P_{\text{diffusion}} - P_{\text{pore}}$, because the two resistances are arranged in parallel (4.17). Model calculations were restricted to pH 3, since pore size is independent of pH (Table 4.4) and most carboxyl groups are not ionised, such that counter ions do not affect the magnitude of permeances at pH 3. It should be remembered that $P_{\text{viscous}}/P_{\text{diffusion}}$ did not depend on pH (Table 4.4).

In Fig. 4.12 the effect of $P_{\text{pore}}/P_{\text{diffusion}}$ on $P_{\text{cutin}}/P_{\text{diffusion}}$ and on pore radius is shown. If $P_{\text{pore}}/P_{\text{diffusion}} = 1$, the total diffusional flux takes place in pores and $P_{\text{cutin}}/P_{\text{diffusion}}$ is zero. As the fraction of water which diffuses across the pore decreases, more water diffuses across cutin, and when only 10% flows across the pores the ratio $P_{\text{cutin}}/P_{\text{diffusion}}$ is equal to 9. When more water penetrates the cutin phase, less water diffuses in pores, and the ratio of viscous flux over diffusion flux (which is proportional to P_{pore}) increases, and hence pore radii increase (Fig. 4.12b).

Pore radii estimated using model I (0.50 nm) are too small, because raffinose having a solute radius of 0.654 nm did penetrate MX membranes. This indicates that the pores are larger than 0.5 nm. Some of the reasons for this deviation have been pointed out in Sect. 4.5. Taking into account steric hindrance at the pore entrance, the "real" average equivalent radius of the pores should be in the range 0.75–0.80 nm.

Such radii would be obtained if $P_{\text{pore}}/P_{\text{diffusion}}$ is 0.46–0.51 (Fig. 4.12b), that is, about half of the diffusional water flux measured would take place across the cutin, filling the space between the pores. At $P_{\text{pore}}/P_{\text{diffusion}} = 0.5$, both P_{pore} and P_{cutin} are $1.28 \times 10^{-7}\,\text{m s}^{-1}$, since $P_{\text{diffusion}}$ is $2.56 \times 10^{-7}\,\text{m s}^{-1}$. This estimate of permeances is based on membrane area. Assuming that weight and area fractions are numerically identical, corrected permeances based on fractional area of polar cuticular polymers and cutin can be calculated. With a weight fraction of cutin of 0.77, the permeance of a cutin membrane having a thickness of $2.7 \times 10^{-6}\,\text{m}$ would be about $1.66 \times 10^{-7}\,\text{m s}^{-1}$. This value is identical to the P_{w} of $1.60 \times 10^{-7}\,\text{m s}^{-1}$ measured by Schönherr and Lendzian (1981) with *Citrus aurantium* MX.

Calculations based on model II result in improved estimates of pore size, and permeance of cutin is consistent with experimental values of P_{w} of *Citrus* MX. Permeance of cutin is clearly not zero as assumed in model I. It resembles permeance of EC, a polyether containing free hydroxyl groups, that is, a polar polymer similar to cutin. A 3 μm-thick EC membrane has a P_{w} of $1.45 \times 10^{-7}\,\text{m s}^{-1}$ (Table 4.1). In *Citrus* MX, water diffuses across two pathways arranged in parallel, and permeances

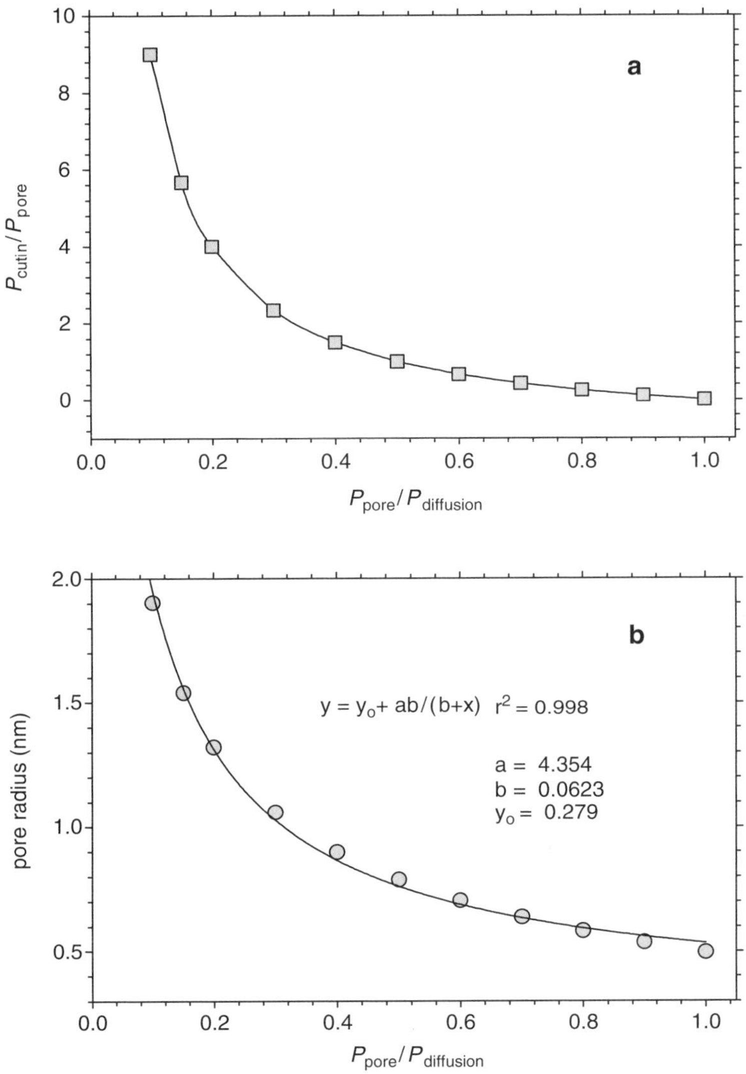

Fig. 4.12 Diffusion of water across cutin and pores and pore radius as effected by $P_{cutin}/P_{diffusion}$. Model calculations for *Citrus* MX at *p*H 3. $P_{diffusion}$ and $P_{viscous}$ data were taken from Table 4.4, and values for $P_{cutin}/P_{diffusion}$ were assumed. Pore radius was calculated using (4.13), but substituting $P_{diffusion}$ by P_{pore}

corrected for the different area fractions are 1.66×10^{-7} and $5.56 \times 10^{-7}\,\mathrm{m\,s^{-1}}$ for P_{cutin} and P_{pore} respectively. Hence, permeance of pores is larger by a factor of 3.35 than permeance of cutin, but water fluxes across the two pathways are the same, since the weight fraction of cutin is larger by a factor of 3.35 than the weight fraction of polar polymers. In this calculation, the area fraction of polar polymers was used rather than area fraction of pores. This neglects the volume occupied by polar polymers, which is not known.

Model II is superior to model I, as it provides more realistic pore size estimates. No assumptions concerning the diffusion coefficient and viscosity of water in the pores are needed, but it is still assumed that $D\eta$ is the same in pores as in bulk water. Calculations are based on $P_{\mathrm{diffusion}}$ which reflects permeance of the cuticle proper, but pore size estimates do not depend on weight fraction of polar polymers or distribution of polar polymers between cuticular layer and cuticle proper or limiting skin. The ratios between P_{pore} and $P_{\mathrm{diffusion}}$ are modelled, and the real contribution of P_{cutin} to $P_{\mathrm{diffusion}}$ should be measured experimentally. This can be done, since after acid hydrolysis intact cutin membranes are obtained. Dependence on partial vapour pressure of P_{cutin} can also be determined.

4.5.3 Effect of Partial Pressure of Water Vapour on Permeances of the Pore and Cutin Pathways

In models I and II, $P_{\mathrm{diffusion}}$ and P_{viscous} were estimated for fully swollen MX membranes, as the membranes were in contact with water on both sides. In vivo, cuticles are wet by water only on their inner surfaces, which are in equilibrium with water of the epidermal cell wall. The outer surfaces face more or less dry air, except during fog and rain. When humidity on the outer surface of cuticles is below 100% ($p/p_0 < 1$) permeances of MX decrease (Fig. 4.6), and this was attributed to reduced vapour sorption and swelling (Sect. 4.3). It is not known if partial vapour pressure affects permeances of cutin and polar polymers similarly, but it is likely that the effect is larger with polar cuticular polymers than with cutin. When partial pressure was reduced from 1 to 0.22, permeance of ethyl cellulose decreased by a factor of 0.28 (Fig. 4.5), while permeance of MX membranes decreased by a factor of about 0.5 (Fig. 4.6).

There are two further complications. P_{viscous} was larger than $P_{\mathrm{diffusion}}$, and during transpiration there can be a large difference in water activity or water potential between the inner and the outer surfaces of the cuticles. In cutin, water transport is probably solely by diffusion, but in polar cuticular polymers viscous flow has been demonstrated, and with decreasing partial vapour pressure the contribution of viscous flow to total water transport is likely to decrease. When humidity is only 20%, sorption of water is very small (Fig. 4.7) and pores may be absent, such that water flux is limited to diffusion. The ratio $P_{\mathrm{pore}}/P_{\mathrm{cutin}}$ would decrease from 3.35 to 1.0, and the change in nature of water flux is likely to account for most of the effect on permeance of partial vapour pressure of MX membranes seen in Fig. 4.6.

Permeance of cuticular membranes (CM) was also affected by partial vapour pressure (Sect. 4.6). CM contain cuticular waxes which occur as intracuticular and surface waxes (Sects. 1.2 and 1.4). It is not very probable that waxes are deposited within the polar polymer phase. This leaves the cutin matrix as site for deposition of intracuticular waxes. This might affect water sorption in cutin, but this is not known.

Since the effect of partial vapour pressure on permeance of cutin is probably small, the wax-incrusted cutin may not respond at all to partial pressure of water vapour. In this case, water vapour pressure would mainly affect permeance of the polar polymer phase. This aspect is treated more comprehensively in the following section.

4.6 Water Permeability of Isolated Astomatous Cuticular Membranes

4.6.1 Comparing Water Permeability of CM with that of MX

Cuticles are composed of a polymer matrix (Sect. 1.1) and various amounts of intra- and epicuticular waxes (Sect. 1.2). Permeances (P_w) of *Citrus* and pear CM are 8.3×10^{-10} and $4.9 \times 10^{-10}\,m\,s^{-1}$ respectively (Table 4.6). Jeffree (2006) has classified these as Type 3 cuticles, which have an amorphous cuticle proper of low electron density but lacking lamellae. The other species in Table 4.6 have a lamellated CP, and P_w was lower and ranged from 1 to $3.2 \times 10^{-10}\,m\,s^{-1}$.

Extracting cuticular waxes increased permeance by factors of 142–2,031 which demonstrates that waxes play a major role in water permeability of CM. Inspection of Table 4.6 shows that wax amounts, permeance and effect of extraction are not correlated. A similar observation was made by Buchholz (2006), who reported mobility of the lipophilic solute bifenox in CM from 22 different species. Mobility was not related to thickness of CM or amounts of wax. The onion bulb scale cuticle is about 1 μm thick or less, and only $1\,\mu g\,cm^{-2}$ of wax could be extracted from these thin cuticles. The waxes consisted mainly of C_{31} and C_{33} n-alkanes (together 79% of total) and C_{44} and C_{46} n-alkyl esters (18% of total wax). The effect of extraction of waxes on P_w of onion bulb scales is mainly due to the relatively high P_w of the

Table 4.6 Effect of extraction of waxes on water permeance at 25°C of astomatous isolated cuticular membranes from leaves (L) or bulb scales (BS)

Species	P_w (CM) ($m\,s^{-1}$)	P_w (MX) ($m\,s^{-1}$)	$\frac{P(MX)}{P(CM)}$	Wax mass ($\mu g\,cm^{-2}$)	CM mass ($\mu g\,cm^{-2}$)
Citrus aurantium L[a]	8.3×10^{-10}	1.6×10^{-7}	193	12[d]	316[f]
Pyrus communis L[a]	4.9×10^{-10}	1.1×10^{-7}	224	133[e]	343[f]
Allium cepa BS[b]	3.2×10^{-10}	6.5×10^{-7}	2,031	1[b]	110[b]
Clivia miniata L[c]	1.2×10^{-10}	1.7×10^{-8}	142	113[d]	715[f]
Hedera helix L[a]	1.0×10^{-10}	2.7×10^{-8}	270	64[d]	476[f]

[a]Schönherr and Lendzian (1981)
[b]Schönherr and Mérida (1981)
[c]Schmidt et al. (1981)
[d]Schreiber and Riederer (1996a)
[e]Riederer and Schönherr (1984)
[f]Becker et al. (1986)

MX, rather than to a particularly low P_w of the CM. *Pyrus* CM had $130\,\mu g\,cm^{-2}$ wax, and permeance was higher than that of onion bulb scale CM.

Extraction of waxes increased P_w by orders of magnitude, even though they contribute little to total mass of CM (Table 4.6). With onion bulb scale, waxes amounted to less than 1% of the total mass of the CM. With *Citrus* 3.8% wax decreased P_w by a factor of 193. Pear leaf CM contained 38% wax ,while with ivy and *Clivia* waxes amounted to 13% and 16% of the total mass respectively. Clearly, a mechanistic analysis of water permeability of cuticles must focus on the contribution of waxes to the water barrier.

4.6.2 Water Permeability of CM

According to model II introduced in Sect. 4.5.2, two parallel paths for water exist in the polymer matrix. Water can move either in aqueous pores of the polar polymers or diffuse in cutin, which fills the space between the polar polymers. CM contain as additional component epi- and intracuticular waxes (Sect. 1.2), and CM are highly asymmetrical membranes (Sects. 1.4 and 6.3). The transport-limiting barrier is located at the morphological outer surface. Waxes contribute to this limiting barrier, but it is not known how exactly waxes embedded in the cutin of the limiting skin and waxes deposited as thin continuous wax film on the cuticle cooperate in forming this limiting barrier (Sect. 1.4). Waxes embedded in cutin of cuticular layers (Fig. 1.6) apparently do not contribute much to barrier properties of CM. This conclusion is based on the observation that diffusion coefficients of lipophilic solutes in the limiting skin are smaller by orders of magnitude than D in cuticular layers (Sect. 6.3). Based on these considerations, we can develop a more refined model of water transport in CM (model III). There are three different options how waxes can be incorporated in model II.

Model III A: A thin layer of wax is deposited on the outer surface of the MX and forms the limiting barrier. In this case, permeability of the CM would exclusively depend on the transport properties of the wax, since this wax layer would be a resistance in series with cutin, which reduces water permeability of the MX by 2–3 orders of magnitude.

Model III B: A transport-limiting barrier could be made of waxes embedded in the cutin and at the outer surface of the MX. Wax is soluble in cutin but not in polar polymers, which amount to about 20% of the mass of the MX. This would reduce water permeability of the cutin, but polar polymers containing the aqueous pores would still exist in parallel.

Model III C: This option is a mixture of the above alternatives, except that the superficial wax layer covers a fraction of the aqueous pores, which reduces the amount of water that moves in aqueous pores.

These three models are depicted schematically in Fig. 4.13. In the following, we shall test which options of model III are consistent with published data on water permeability of CM and MX membranes. This is complicated by the fact that for water

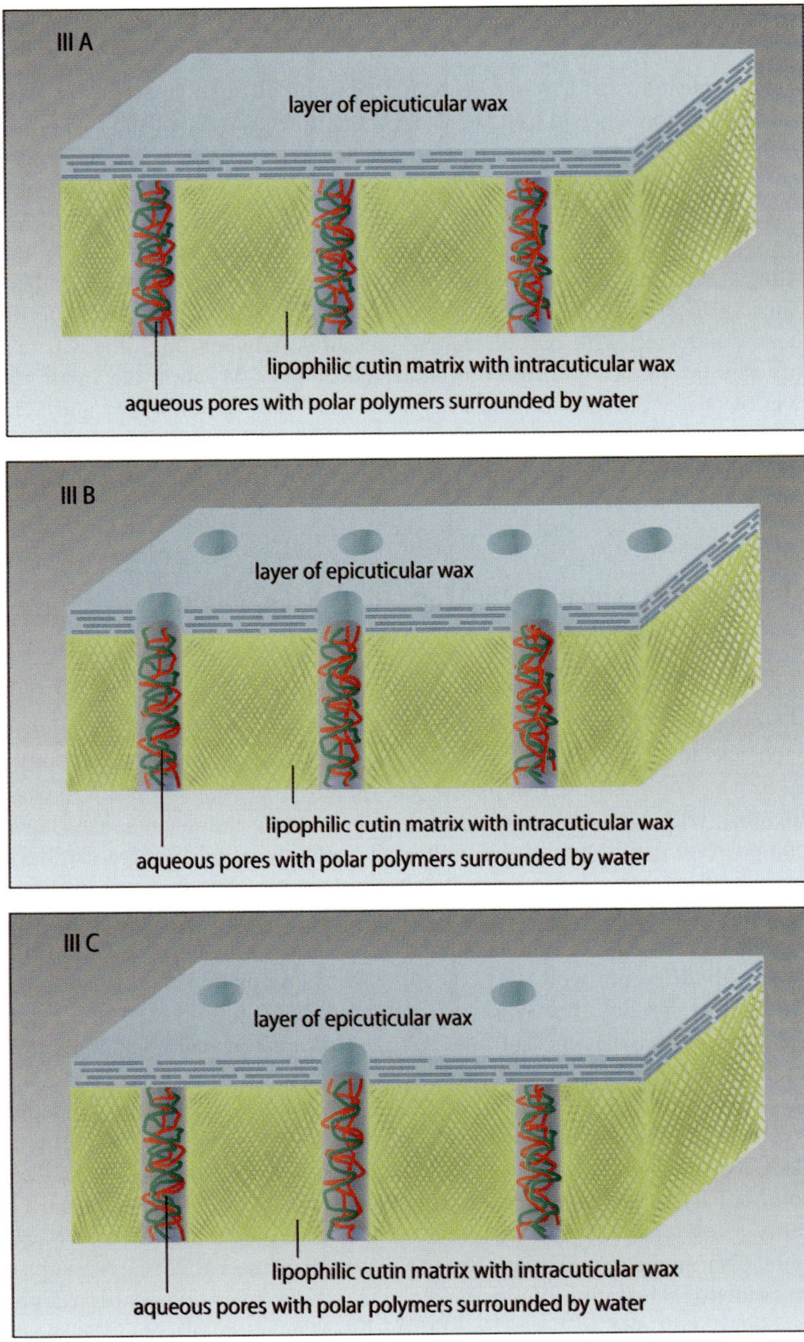

Fig. 4.13 Schematic drawings for the three models IIIA, IIIB and IIIC showing the water pathways across cuticular membranes composed of polar polymers, cutin and waxes embedded in the outer cutin layer of the MX and on the surface of the MX (not to scale)

only permeances have been measured, but not diffusion and partition coefficients in the transport-limiting barrier.

Water permeability of CM has been studied using different approaches. With the cup method (Schönherr and Lendzian 1981) the inner surface of the CM is in contact with water, and the outer surface is exposed to dry air of nearly 0% humidity (Sect. 9.7). The amperometric method (Becker et al. 1986) used nitrogen gas saturated with water vapour as donor at the morphological inner surface of the CM, and dry nitrogen served as receiver facing the morphological outer surface of the CM. Permeances obtained with these two methods are minimum permeances, because the outer surface of the CM is not or only weakly hydrated. Other experiments had been conducted with partial vapour pressures between 0.02 and 1.0 or with liquid water in contact with the outer surface of the CM while the inner surface was wet by water (Niederl et al. 1998; Schönherr 1976b; Schönherr and Schmidt 1979; Schreiber 2001). These experiments show if permeance increases with partial vapour pressure. With 100% humidity or liquid water on the receiver side, maximum permeances are measured.

4.6.2.1 Chemical Composition of Wax and Its Relationship to Water Permeability

Since neither thickness of the CM nor amounts of wax provide a satisfactory explanation for the large variability in water permeability of CM from different plant species, some workers in the field have postulated that permeability might be related to wax composition. Qualitative and quantitative wax composition can vary considerably with leaf development and with growing conditions. Wax amounts and composition can be measured with high accuracy using modern capillary gas chromatography/mass spectrometry. Each plant species has its own specific wax composition, with homologues varying in chain lengths distribution and substance class composition. This might account for differences in cuticular transpiration observed with different species. By analysing wax composition and measuring water permeability in parallel, this hypothesis can be tested. Suitable data sets are available for astomatous isolated CM from *Citrus aurantium* (Geyer and Schönherr 1990; Riederer and Schneider 1990) and *Hedera helix* (Hauke and Schreiber 1998).

Wax composition of *Citrus aurantium* leaves varied with leaf age and depended on growing conditions (Riederer and Schneider 1990). Increased day temperatures (15–35°C) resulted in smaller wax amounts (alkanes, alcohols, acids and esters) by factors of 2–3, whereas increased night temperatures had the opposite effect. Humidity (50% or 90%) had no effect on wax composition. Water permeability determined using the cup method (Sect. 9.7) was measured at 25°C with subsets of CM from identical isolations. Even though wax coverage and composition depended on growing conditions, water permeability of CM was not affected. Permeability of all sets of CM decreased during storage of CM at 8°C. During about 50 days of storage, permeance decreased to about 40% of the initial P_w. This reduction in permeance was attributed to a change in wax structure during storage. Possibly the ratio between crystalline and amorphous fractions of the wax changed and reduced water permeability. We return to this point when we discuss water permeability of paraffin waxes (Sect. 4.6.4, subsection: "Water Permeance of Polyethylene and Paraffin Wax").

Wax composition and water permeability as related to exposure to sun light and leaf age have been studied using *Hedera helix* leaf CM (Hauke and Schreiber 1998). Sun and shade leaves of at least eight different developmental stages were harvested from bud break to leaf senescence, and leaf area, cuticle weight, wax amount, mean chain lengths and cuticular transpiration were measured (Fig. 4.14).

Leaf area increased rapidly during the first 30 days, and more slowly for another 30 days when maximum leaf size had been reached (Fig. 4.14a). Cuticle weight continuously increased for 60 days (Fig. 4.14b). Wax amounts increased 1.5-fold within the first 40 days, and decreased again until senescence at the end of the season (Fig. 4.14c). Mean chain length of the wax homologues continuously increased from C_{27} to C_{33} within the first 60 days (Fig. 4.14d). Water permeability decreased about

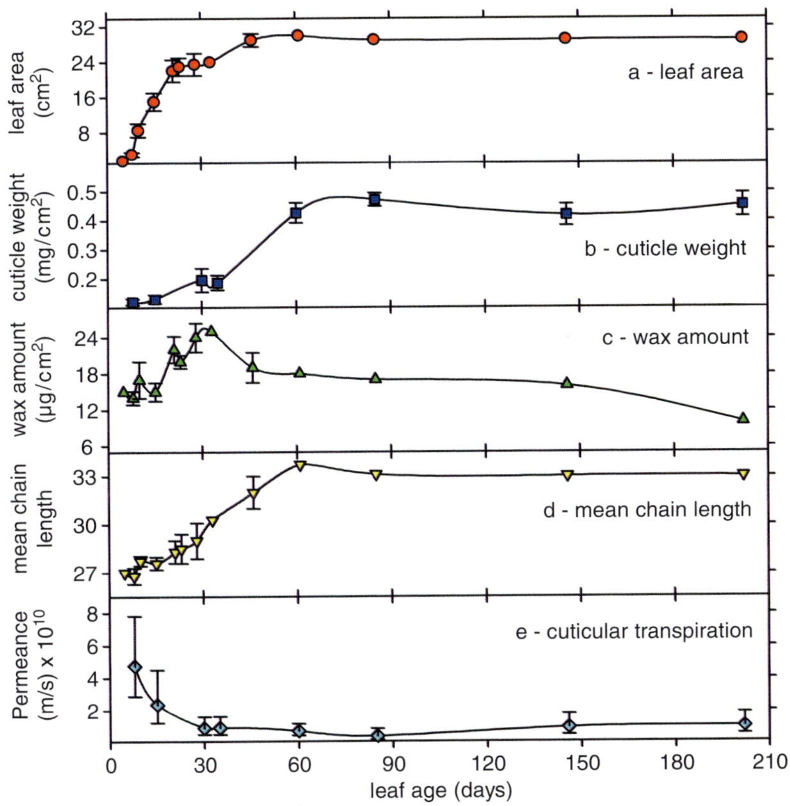

Fig. 4.14 Leaf and cuticle development of *Hedera helix* shade leaves from bud break to leaf senescence. **(a)** The variation of leaf area. **(b)** Cuticle weight. **(c)** Wax amount. **(d)** Mean chain length of the wax molecules. **(e)** The decrease of cuticular permeability during about 7 months from end of April to beginning of November is shown. (Data from Hauke and Schreiber 1998)

seven-fold during the first month and remained constant during the remaining part of the season (Fig. 4.14e). There is no obvious relationship between water permeability and the other properties studied. These studies give no hints as to how waxes reduce water permeability.

4.6.2.2 Water Permeability of CM and Diffusion of Stearic Acid in Wax

Schreiber and Riederer (1996a, b) studied water permeance P_w (m s^{-1}) of isolated CM from leaves and fruits of 24 plants using the cup method. From the same lots of CM, cuticular waxes were extracted and diffusion coefficients of ^{14}C-labelled lipophilic tracer stearic acid (SA) in reconstituted cuticular wax (D_{SA}) were determined (for the method, see Sect. 6.5 and 9.6). It was tested whether P_w of the intact CM is correlated with D_{SA} in cuticular wax from the same species.

P_w varied among species by more than two orders of magnitude (Table 4.7), from 1.7×10^{-11} m s^{-1} (*Vanilla planifolia*) to 2.07×10^{-9} m s^{-1} (*Malus domestica*). D_{SA} covered a similar range of two orders of magnitude ranging from 2.7×10^{-19} m^2 s^{-1} (*Hedera helix*) to 290×10^{-19} m^2 s^{-1} (*Malus domestica*). Mass of CM varied from 68 µg cm^{-2} (*Maianthemum bifolium*) to 3,217 µg cm^{-2} (*Malus domestica*), and mass of cuticular wax varied from 11.8 µg cm^{-2} (*Citrus aurantium*) to 1,317 µg cm^{-2} (*Malus domestica*).

Plotting P_w as a function of D_{SA} resulted in a fairly good correlation (Fig. 4.15), whereas P_w was not correlated with either mass of the CM or wax. Equation (4.18) states that P_w in CM is directly proportional to stearic acid mobility D_{SA} in reconstituted wax

$$P_w = 6.6 \times 10^7 D_{SA} - 4.5 \times 10^{-11} (r^2 = 0.90). \quad (4.18)$$

Species having high water permeance have a wax in which stearic acid mobility is high, and vice versa. We take this as evidence that water in CM and stearic acid in reconstituted waxes both diffuse along the waxy pathway characteristic for each species. Data points above the regression line could indicate that some water transport in aqueous pores might have been involved, but direct evidence is lacking. Besides, P_w was measured using the cup method, and swelling of CM was minimum. Wax structure, that is, physical arrangement of amorphous and crystalline fractions, appears to be similar in CM and reconstituted waxes (Sect. 6.5). Once again, chemical composition appears not to be directly related to barrier properties.

What is the physical meaning of the *y*-intercept? It is statistically not different from zero, but we should still consider its possible role in model III. The *y*-intercept must be subtracted from the product $6.6 \times 10^7 \times D_{SA}$, which overestimates P_w. This correction is more important for CM with low P_w. With *Vanilla* CM, P_w calculated from the regression equation is 3.6×10^{-11} m s^{-1}, which is reasonably close to the value measured of 1.7×10^{-11} m s^{-1}. Had the *y*-intercept not been subtracted, P_w calculated would have been 8.1×10^{-11} m s^{-1}. The corresponding values for *Citrus aurantium* are 2.1×10^{-10} or 2.5×10^{-10} m s^{-1}. Both figures are not far from

Table 4.7 Water permeance P_w of CM, diffusion coefficient D_{SA} of stearic acid in reconstituted wax, mass m_{cm} of the CM, and mass m_{wax} of the wax obtained from the investigation of 24 species. (Data from Schreiber and Riederer 1996a, b)

Species	$P_w \times 10^{11}$ ($m\,s^{-1}$)	$D_{SA} \times 10^{19}$ ($m^2\,s^{-1}$)	m_{CM} ($\mu g\,cm^{-2}$)	m_{WAX} ($\mu g\,cm^{-2}$)
Leaf CM				
Vanilla planifolia (vanilla)	1.7	12.3	359	122
Monstera deliciosa (breadfruit-vine)	4.3	25.8	808	242
Philodendron selloum	6.6	8.9	335	54
Ficus elastica (Assam rubber plant)	9.4	25.1	510	87
Ficus benjamina (Java fig)	13.0	29.3	306	64
Hedera helix (ivy)	5.70	2.7	337	114
Clivia miniata	15.7	36.0	1,020	285
Camellia sinensis (tea-plant)	10.8	8.1	252	11.8
Prunus laurocerasus (cherry laurel)	13.3	63.0	333	83
Nerium oleander (oleander)	52.2	55.7	664	113
Olea europaea (olive-tree)	12.6	59.8	661	88
Citrus aurantium (bitter orange)	12.8	38.3	369	32
Citrus limon (lemon)	47.0	77.9	1,373	381
Euonymus japonicus (evergreen e.)	35.8	70.0	403	64
Liriodendron tulipifera (tulip-poplar)	42.0	74.6	233	72
Juglans regia (English walnut)	45.8	99.8	125	27
Ginkgo biloba (ginkgo)	52.2	100.5	342	40
Cydonia oblongata (quince)	62.9	34.1	191	51
Ligustrum cf. vulgare (prim)	43.4	42.7	227	39
Forsythia suspensa (golden bells)	38.7	80.2	955	137
Maianthemum bifolium (false lilly-of-the valley)	111.0	152.3	68	36
Fruit CM				
Lycopersicon esculentum (tomato)	62.2	121.0	1,554	54
Capsicum annuum (bell-pepper)	134.5	222.3	2,162	96
Malus domestica (apple)	207.0	290.5	3,217	1,317

the measured value of $1.3 \times 10^{-10}\,m\,s^{-1}$. Since the y-intercept lowers calculated permeance, it is not related to water permeation in a hypothetical parallel aqueous pathway. These data are consistent with model III A.

The good correlation between P_w of CM and D_{SA} in reconstituted wax is amazing, because permeance is a composite quantity and depends on mobility and solubility of water in wax and thickness of the CM. According to (2.18), $P_w = D_w \times K_{ww}/\ell$, while D_{SA} is a pure mobility parameter. In Chap. 2 we defined ℓ as thickness of a membrane, while in reality the lengths of the diffusion path can be much larger due to the tortuosity factor τ. Thus, linearity implies that $K_{ww}/\ell\tau$ is equal to the slope of $6.6 \times 10^7\,m^{-1}$, which is the same with all species. However, K_{ww} and $\ell\tau$ are likely to vary independently among species, and it is not very likely that they are the same for all species. High permeance could be the consequence of large K_{ww} or a short diffusion path, that is, thin CM and low τ. Cuticular wax is composed of crystalline and amorphous fractions (Reynhardt and Riederer 1991; 1994), and both sorption and diffusion of water should exclusively take place in

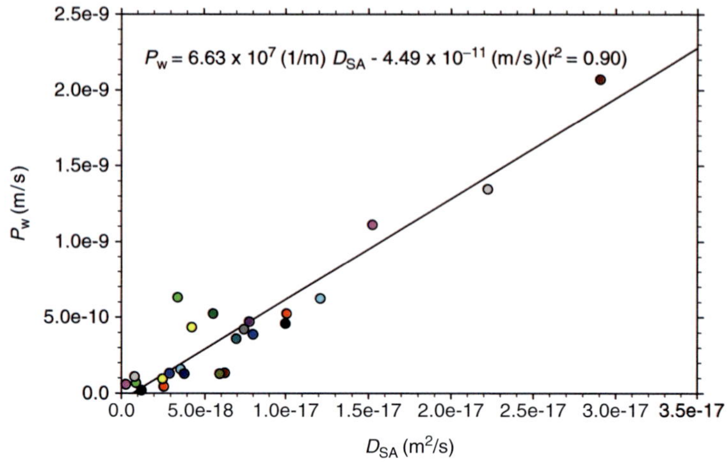

Fig. 4.15 Water permeances P_w of isolated CM plotted as a function of D_{SA} (diffusion coefficient of stearic acid) in reconstituted cuticular waxes of 24 plant species. (Data from Table 4.7)

the amorphous phase (Riederer and Schreiber 1995), whereas the crystalline phase is not accessible. With an increasing amorphous fraction in the wax, sorption is expected to increase. Since diffusion of water and stearic acid takes place in the same amorphous wax phase, path length of diffusion might increase as well, and thus the ratio of $K_{ww}/\ell\tau$ could be very similar for all investigated species. This hypothesis could help to explain why P_w is correlated with D_{SA}.

Diffusion coefficients for lipophilic solutes in CM and wax (Sect. 6.3.2.3: "Variability of Solute Mobility with Size of Solutes" and Sect. 6.5.2) differ among species because tortuosity of the diffusion path differs. This implies that differences in D_{SA} (Table 4.7) are caused by differences in length of the diffusion paths ($\ell\tau$). Since water in CM of all species diffuses in the waxy pathway, the path lengths of stearic acid and water should be the same for both. Hence, partition coefficients (K_{ww}) may vary little among species, and since $K_{ww}/\ell\tau$ is constant ($6.6 \times 10^7\,\mathrm{m}^{-1}$) differences in P_w should mainly be due to differences in lengths of the diffusion path.

Data on sorption and diffusion of water in cuticular waxes could help us to better understand the waxy barrier. There are data on water concentration in polyethylene and liquid hydrocarbons. Water concentration in polyethylene at 25°C and 100% humidity is about $7 \times 10^{-4}\,\mathrm{g\,g}^{-1}$ (Table 4.1). According to Schatzberg (1965) the solubility of water in a liquid alkane (hexamethyl tetracosane) at 25°C is $5.5 \times 10^{-5}\,\mathrm{g\,g}^{-1}$ or $55\,\mathrm{mg\,kg}^{-1}$. Sorption of water in isolated and reconstituted solid cuticular waxes has not been studied, for good reasons as the following example will show. Starting with the assumption that the above paraffin/water partition coefficient ($5.5 \times 10^{-5}\,\mathrm{g\,g}^{-1}$) is valid for cuticular waxes, we can estimate the concentration of tritiated water (THO) needed to detect THO in a wax sample of 1 mg. For a reasonable counting statistic, we aim at $100\,\mathrm{Bq\,mg}^{-1}$ wax, and we can calculate the concentration of THO in water from the definition of K (2.12). The ratio of the radioactivity in the wax ($100\,\mathrm{Bq\,mg^{-1}}$) and the partition coefficient K_{ww} (5.5×10^{-5}) is $1.82 \times 10^{6}\,\mathrm{Bq\,mg^{-1}}$ or $1.82 \times 10^{9}\,\mathrm{Bq\,g^{-1}}$ water. Transfer from the vapour phase into the scintillation cocktail requires special equipment to prevent desorption of water from the thin wax film. Since wax platelets cannot be handled without support (for instance aluminium foil), water sorption must be measured with different wax amounts, to be able to correct for sorption on the support.

4.6.2.3 Co-Permeation of Water and Lipophilic Solutes

Co-permeability experiments of lipophilic solutes and water were conducted to identify the diffusion paths of water and lipophilic solutes in the CM (Niederl et al. 1998). The cuticle is a lipophilic polymer membrane (Chap. 1), and permeability of lipophilic non-electrolytes increases with increasing partition coefficients (Sect. 6.2). This is due to the fact that the concentration of solutes in CM and wax increase with increasing K. Using ^{3}H-labelled water (THO) and ^{14}C-labelled organic solutes (benzoic acid, salicylic acid), simultaneous penetration of each of the two pairs of THO and ^{14}C-labelled organic solutes across *Prunus laurocerasus* CM was measured (Fig. 4.16). Donor solutions were buffered at pH 3, and the concentration of non-ionised solutes was used as driving force. The receiver contained a phospholipid suspension (1%) in which lipophilic solutes are sorbed such that their concentration in water is essentially zero.

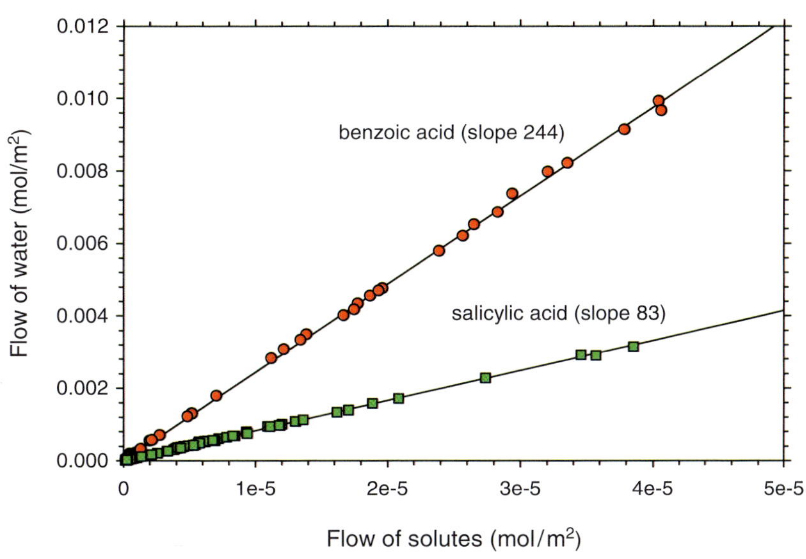

Fig. 4.16 Simultaneous penetration of ^{3}H-labelled water and ^{14}C-labelled benzoic acid and salicylic acid across isolated cuticular membranes of *Prunus laurocerasus*. Data points represent the amounts of labelled compounds which had diffused across each of the 15 investigated cuticles at each sampling time (1, 2, 3, 4, 5 and 6 h). (Data from Niederl et al. 1998)

Plotting flows of THO vs flows of lipophilic solutes resulted in straight lines which go through the origin (Fig. 4.16). When flows of benzoic acid or salicylic acid were high, water flow was also high. This was the case with all CM, and indicates that water and lipophilic solutes diffused along the same path. Since this path is accessible to lipophilic solutes, simultaneous co-permeation took place in the waxy path, not in aqueous pores. With benzoic acid, the slope is steeper by a factor of 2.9 than with salicylic acid. This is explained by larger driving force for benzoic acid, since diffusion coefficients in *P. laurocerasus* wax are similar (Sect. 6.5.2; Kirsch et al. 1997).

Similar experiments were conducted using the pair THO/benzoic acid and CM from 12 species. Water permeances and benzoic acid permeances were calculated. Plots log P_w vs log P_{BA} were linear, with a coefficient of determination of 0.95 (Fig. 4.17). Fluxes of water and benzoic acid were coupled in all species. Species having high P_{BA} also have a high P_w, and vice versa. This indicates that benzoic acid and water used the same pathway in all species. Benzoic acid is lipophilic, as its K_{ww} in wax from a number of species is around 20 (Kirsch et al. 1997). Hence, benzoic acid diffused along the waxy path and so did water (model III A). The regression equation for the data in Fig. 4.17 is

$$\log P_w = 0.86 \times \log P_{BA} - 1.32. \quad (4.19)$$

The y-intercept of the regression equation (4.19) is -1.32, that is, P_w is $4.79 \times 10^{-2}\,\mathrm{m\,s^{-1}}$ (i.e., $10^{-1.32}$) when $P_{BA} = 1.0$. As the slope is 0.86, P_w increases only by a factor of 7.24 (i.e., $10^{0.86}$) when P_{BA} increases tenfold. When permeance is

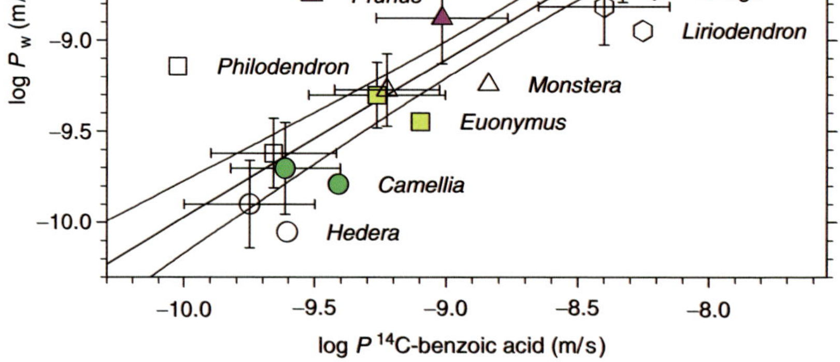

Fig. 4.17 Correlation of water permeance (log P_w) with permeance of benzoic acid (log P_{BA}) across isolated cuticular membranes of 12 species. (Data from Niederl et al. 1998)

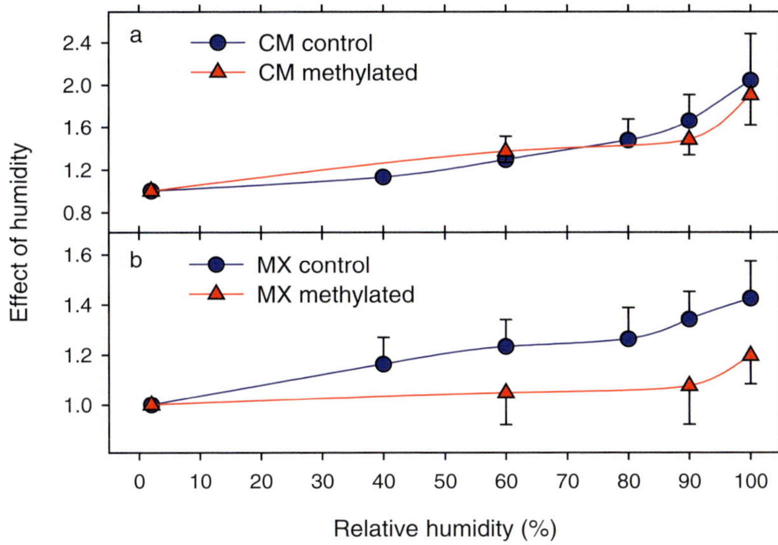

Fig. 4.18 Effect of relative humidity in the receiver on water permeance (P_w) of *Prunus laurocerasus* CM and MX at 25°C. Permeance was measured using tritiated water in the donor and isolated astomatous cuticular membranes (CM), or polymer matrix membranes (MX). Methylation of membranes was obtained by diazomethane. *Error bars* are 95% confidence intervals. Data were taken from Schreiber et al. (2001)

equal to $9.72 \times 10^{-10}\,\mathrm{m\,s^{-1}}$, P_w and P_{BA} are numerically equal. If P_{BA} is lower, then $P_w > P_{BA}$, and if P_{BA} is larger than $9.72 \times 10^{-10}\,\mathrm{m\,s^{-1}}$, $P_w < P_{BA}$. For example, when $P_{BA} = 10^{-8}\,\mathrm{m\,s^{-1}}$, P_w is $6.3 \times 10^{-9}\,\mathrm{m\,s^{-1}}$. With species having P_{BA} of $10^{-10}\,\mathrm{m\,s^{-1}}$, the calculated P_w is a little higher ($1.2 \times 10^{-10}\,\mathrm{m\,s^{-1}}$). This could be taken as evidence that in CM from species having a low permeance for benzoic acid, some water diffused in aqueous pores, while in CM having a high BA permeance, all or most water diffuses in the waxy path, and the contribution of a parallel aqueous paths goes unnoticed. In view of the large error bars seen in Fig. 4.17, this conclusion may very well be wrong.

Both permeances are related to D, K and ℓ, and this can be written as

$$\frac{\log P_w}{\log P_{BA}} = \frac{\log(D_w K_{ww})/\ell}{\log(D_{BA} K_{BA})/\ell} = 0.86. \tag{4.20}$$

If water and benzoic acid use the same waxy pathway, ℓ is the same and cancels. K_{ww} for benzoic acid is around 20 for waxes of *P. laurocerasus*, *G. biloba* and *J. regia*. The respective D_{BA} values in waxes of these species are 3.5×10^{-17}, 4.4×10^{-17} and $5.6 \times 10^{-17}\,\mathrm{m^2\,s^{-1}}$. This is the sequence seen on the regression line in Fig. 4.17. $D_w \times K_{ww}$ is not known for these three species, but their products should be smaller by a factor of 7.2 ($10^{0.86}$). The wax–water partition coefficient for water is much smaller than for benzoic acid, which implies that D for water must be considerably larger than for benzoic acid.

4.6.2.4 Effect of Partial Vapour Pressure (Humidity) on Permeability of CM

All polymers sorb water when exposed to water vapour or liquid water. Sorption is greater in polar polymers than in non-polar polymers (Table 4.1). When dealing with effects of water vapour pressure on permeability of membranes, effects on fluxes and permeance must be distinguished. With lipophilic polymers (polyethylene, silicon rubber) sorption isotherms are linear, and permeance is the same at all partial pressures, while permeance of polar polymers increases with partial pressure (Figs. 4.5 and 4.6). When measuring the effect of humidity on permeance of CM, the donor faced the inner side of the CM while partial pressure was varied in the receiver facing the outer surface of the CM. With increasing partial pressure of the receiver, the driving force (Δa_{w} or ΔC_{wv}) decreases, and with 100% humidity in the receiver, the driving force is zero and no net flux of water can be observed. For this reason, the effect of humidity on P_{w} was studied using tritiated water (THO) as donor, and the flux of the tritiated water was measured. The driving force was the concentration of tritiated water in the donor, because THO concentration in the receiver was practically zero. In this way, water fluxes at very high humidity in the donor can be measured with high accuracy.

Extracting cuticular waxes increases permeance by orders of magnitude (Table 4.6), and the question arises whether waxes modify the effect of p/p_0 on permeance of MX as seen in Fig. 4.6. In several investigations, it was shown that an effect of partial water vapour pressure (p/p_0) on permeance can also be seen in CM (Schönherr and Schmidt 1979; Schönherr and Mérida 1981; Schreiber et al. 2001). Water permeability of CM increased by a factor of 2.3 for *Vinca major*, 3.2 for *Citrus aurantium* with humidity increasing from 2% to 100%, and factors of 2.4, 2.8 and 2.9 were found for *Hedera helix*, *Prunus laurocerasus* and *Forsythia intermedia* respectively (Schreiber et al. 2001). There was a weak increase in permeability of up to 70% humidity, and a much more pronounced increase between 70% and 100% humidity (Fig. 4.18). Similar effects of humidity on permeability were obtained with MX (Fig. 4.18b).

The effect of humidity on permeance demonstrates the involvement of polar functional groups in cuticular wax and/or the MX (Sects. 4.1 and 4.5). In waxes, polar groups are mainly contributed by fatty acids and alcohols, but compared to MX cuticular waxes have very few polar groups (Riederer and Schneider 1990). Methylation of carboxylic groups significantly decreased the effect of humidity on permeability of MX by 50%. This demonstrates the involvement of carboxylic groups of the MX. Since the effect of humidity did not totally disappear after methylation, other functional polar groups (i.e., –OH) linked to proteins, phenolics and carbohydrates which cannot be methylated apparently contributed to the effect of p/p_0 on P_{w}. There was no effect on P_{w} after methylation of *Prunus laurocerasus* CM. This can be interpreted as evidence that penetration of water was limited by cuticular waxes lacking carboxyl groups, or in which carboxyl groups did not contribute to water permeability. Perhaps carboxyl groups in waxes were not accessible to diazomethane. As there was a significant effect of partial pressure on P_{w} of CM, hydroxyl groups which formed water clusters and polar pores may have caused it.

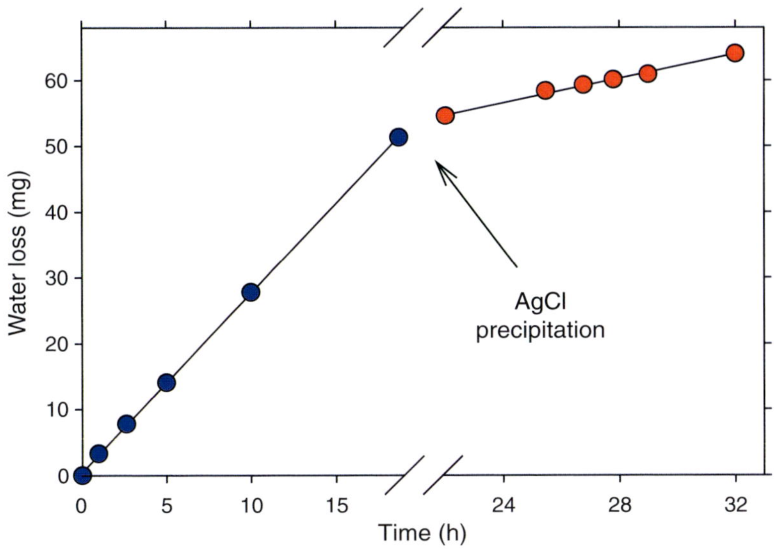

Fig. 4.19 Water permeability of an isolated *Populus canescens* CM before and after counter diffusion of NaCl and $AgNO_3$. The slope of the linear transpiration kinetic is significantly decreased by 64% after the formation of AgCl precipitations in the CM. Data were taken from Schreiber et al. (2006)

4.6.2.5 Effect of AgCl Precipitates on Water Permeance

Polar functional groups contribute to the sorption of water to the MX (Sect. 4.3) and to the formation of aqueous pores (Sect. 4.4). Water permeability of some CM increases with humidity (Sect. 4.6.2, subsection: "Effect of Partial Vapour Pressure (Humidity) on Permeability of CM") and if they are traversed by aqueous pores they are permeable to hydrated ions (Chap. 5). As hydrated ions diffuse exclusively in aqueous pores, it was attempted to selectively block these pores in CM membranes with insoluble AgCl precipitates. CM were inserted in the transpiration cups (Fig. 9.2), and water permeance was determined gravimetrically with $0.01\,\mathrm{mol\,l^{-1}}$ NaCl as donor solution. Partial pressure of the receiver was practically zero. With NaCl still in the donor, the outer surface of the CM was treated with $0.01\,\mathrm{mol\,l^{-1}}$ $AgNO_3$ in deionised water for 24 h. This was done by adding $AgNO_3$ solution to the outer side of the CM, sealing the chambers with adhesive tape and incubating them on a rolling bench, which guaranteed homogeneous wetting of both sides of the CM. After blotting off the silver nitrate solution, water permeability was again determined gravimetrically (Schreiber et al. 2006). When Ag^+ and Cl^- diffusing towards each other in aqueous pores of the CM meet, insoluble precipitates are formed. These precipitates could at least partially block aqueous pores, and water permeability should decrease. This is exactly what happened (Fig. 4.19). If water loss was plotted vs time, slopes decreased after the formation of AgCl precipitates.

Similar counter-diffusion experiments were conducted with CM isolated from 15 species (Table 4.8). Water permeability of 13 species significantly decreased; the exceptions were *Nerium oleander* and *Hedera helix*. These data suggest that in most species water moved in two parallel pathways, aqueous pores and a lipophilic pathway composed of cutin and waxes (model III B). After precipitation of AgCl, aqueous pores are blocked and most water transport observed after treatment occurred in the lipophilic pathway. If AgCl precipitates completely blocked aqueous pores, 15% (*Stephanotis*) to 64% (*Populus canescens*) of the water moved along aqueous pores and 85–36% diffused in the waxy path.

Water permeance of *Citrus* MX membranes as determined using the cup method is $1.6 \times 10^{-7}\,\mathrm{m\,s^{-1}}$ (Schönherr and Lendzian 1981). It was estimated that about half of this penetrated in cutin and the other half in aqueous pores (Sect. 4.5.2). The waxy pathway in *Citrus* CM had a permeance of $1.23 \times 10^{-10}\,\mathrm{m\,s^{-1}}$ (Table 4.8), which suggests that waxes associated with cutin reduced permeance by a factor of 1,300. Permeance of the aqueous path in fully swollen *Citrus* MX is $1.3 \times 10^{-7}\,\mathrm{m\,s^{-1}}$ (Sect. 4.5.2), while in *Citrus* CM $P_w^{aqueous}$ is $0.64 \times 10^{-10}\,\mathrm{m\,s^{-1}}$, which is smaller than in the MX by a factor of 2,031. It appears that most of the aqueous pores observed in the MX were eliminated or covered in the CM by waxes. From the original 4.54×10^{14} aqueous pores (Table 4.5), only 2.23×10^{11} survived incrustation of the limiting skin with waxes. These surviving pores respond to partial vapour pressure just like those in the MX, because P_w of *Citrus* CM increased with humidity. These data are compatible with model III C.

Table 4.8 Permeances ($\mathrm{m\,s^{-1}}$) of cuticular membranes isolated from 15 different species measured before P_w^{total} and after P_w^{AgCl} the formation of AgCl precipitates. Results are means of at least 12 CM. $P_w^{aqueous}$ is the difference between P_w^{total} and P_w^{AgCl}. (Data from Schreiber et al. 2006)

Species	$P_w^{total} \times 10^{10}$ ($\mathrm{m\,s^{-1}}$)	$P_w^{AgCl} \times 10^{10}$ ($\mathrm{m\,s^{-1}}$)	$P_w^{aqueous} \times 10^{10}$ ($\mathrm{m\,s^{-1}}$)	$\frac{P_w^{aqueous}}{P_w^{total}}$
Leaf CM				
Nerium oleander	0.71	0.73	0	0
Hedera helix	1.49	1.46	0	0
Stephanotis floribunda	4.71	4.01	0.70	0.15
Forsythia intermedia	1.97	1.58	0.39	0.20
Ligustrum vulgare	4.03	3.19	0.84	0.21
Vinca major	1.20	0.95	0.25	0.21
Prunus laurocerasus	1.10	0.76	0.34	0.31
Citrus aurantium	1.87	1.23	0.64	0.34
Juglans regia	4.51	2.96	1.55	0.34
Syringia vulgaris	3.51	2.06	1.09	0.41
Pyrus communis	8.30	3.92	4.38	0.53
Populus canescens	26.8	9.64	17.16	0.64
Fruit CM				
Malus domestica	3.20	2.17	1.03	0.32
Lycopersicon esculent..	24.8	15.7	9.10	0.37
Prunus domestica	33.2	12.2	21.0	0.63

CM of most species had permeances which increased with humidity by factors of 2–3, and *Hedera helix* belongs to these species This humidity effect has been attributed to the presence of aqueous pores, which shrink when humidity is lowered and approaches 0% (Schönherr and Schmidt 1979; Schönherr and Mérida 1981; Schreiber et al. 2001). Results shown in Table 4.8 were taken as indicative of pore closure by AgCl crystals, but *Hedera helix* did not respond. This inconsistency is difficult to explain. Possibly this lot of CM was not porous, and it would have been desirable to test humidity effects prior to treatment with $AgNO_3$ and not simply to determine P_w with zero humidity in the receiver. The same applies to the CM after treatment. Unfortunately, P_w^{AgCl} was not measured at different humidities. It was assumed that pores were completely blocked. Thus, the treatment with $AgNO_3$ should have completely eliminated the humidity effect seen in Fig. 4.6 and discussed in Sect. 4.6.2, subsection: "Effect of Partial Vapour Pressure (Humidity) on Permeability of CM". Since this test has not been made, the above considerations are speculative, although they are plausible. There is another problem with the interpretation of AgCl effects on permeance. Data in Figs. 4.15–4.17 had been interpreted as evidence that lipophilic solutes and water diffuse in the waxy pathway, and no evidence for significant involvement of aqueous pores was detected. Some of the species were also used in the study with AgCl blockage, and a significant contribution of aqueous pores to total permeances was calculated.

4.6.3 Diffusion Coefficients of Water in CM and Cuticular Wax

Diffusion and permeation of water across isolated MX membranes, including the transport of water across polar pores, is discussed in Sects. 4.1–4.5. In the previous sections we have discussed the nature of the waxy barrier, based on experimental data obtained with MX and CM. We have also pointed out the lack of data on wax/water partition coefficients and diffusion coefficients in cuticular waxes. There is only one study which tried to estimate D_w in CM (Becker et al. 1986) and this will be considered next.

4.6.3.1 Measurement of D_w for Water in CM from Hold-up Times

Using an amperometric method, Becker et al. (1986) measured P_{wv} of and D in CM isolated from a number of plant species. Nitrogen gas saturated with water vapour served as donor on the morphological inner surface of the CM. The receiver was dry N_2. The calculation of P_{wv} from the flux and the vapour concentration is straightforward and requires no assumptions. However, in calculating D using the hold-up time (t_e) it must be decided which thickness ℓ should be employed (2.5). Becker et al. (1986) used the total thickness calculated from the mass per area of the CM and a specific weight of 1.1 g cm^3. Data are shown in Table 4.9.

Table 4.9 Water permeance (P_{wv}) and diffusion coefficients (D_w) of cuticular membranes at 25°C

Species	t_e (s)	$\ell(\mu m)$	$P_{wv}(m\,s^{-1})$	$D_w(m^2\,s^{-1})$	$C_w(g\,kg^{-1})$	$C_w(g\,kg^{-1})$[a]
Schefflera L.	930	2.96	8.19×10^{-7}	1.57×10^{-15}	33	–
Clivia L.	770	6.50	1.14×10^{-6}	9.14×10^{-15}	17	–
Hedera L.	933	4.33	2.66×10^{-6}	3.35×10^{-15}	73	–
Nerium L.	960	12.81	3.25×10^{-6}	2.84×10^{-14}	34	24
Ficus L.	512	5.68	4.25×10^{-6}	1.05×10^{-14}	48	30
Citrus L.	264	2.87	1.20×10^{-5}	5.20×10^{-15}	139	63
Pyrus L.	550	3.12	1.22×10^{-5}	2.95×10^{-15}	270	43
Solanum F.	640	6.47	2.23×10^{-5}	1.08×10^{-14}	244	–
Capsicum F.	361	8.03	9.28×10^{-5}	2.98×10^{-14}	523	38
Lycopersicon F.	215	8.00	1.42×10^{-4}	4.96×10^{-14}	477	43

D was calculated from the hold-up time and the total thickness (ℓ) of CM; C_w is sorption of water at 25°C and 100% humidity calculated from $\mathscr{P}/D$ [(3.17) and (3.20)] using the data by Becker et al. (1986). L is leaf CM and F is fruit CM. (Data from Becker et al. 1986)
[a]Data taken from Chamel et al. (1991) derived from water vapour sorption

Permeance varied by a factor of 173, and ranged from 8.19×10^{-7} (*Schefflera* leaf) to $1.42 \times 10^{-4}\,m\,s^{-1}$ (tomato fruit). Diffusion coefficients varied by a factor of only 32, and ranged from 1.57×10^{-15} (*Schefflera*) to $4.96 \times 10^{-14}\,m^2\,s^{-1}$ (tomato fruit). According to theory [Chap. 2; (2.18)] one might conclude that variations in membrane thickness and partition coefficient caused these differences. The D_w values are much lower than those for synthetic polymers, PVA being the only exception (Table 4.1). With some CM (*Nerium* and *Ficus*), water sorption (C_w) derived from the $\mathscr{P}_{wv}/D_w$ ratio is low and similar to sorption in CM when determined gravimetrically (Table 4.9, last column). With the other CM, water concentration calculated from $\mathscr{P}_{wv}/D_w$ is unreasonably high, and much larger than that determined by a sorption experiment. This indicates structural heterogeneity of these membranes.

When calculating D_w from hold-up time, the thickness of the CM enters into consideration. Becker et al. (1986) used the weight average thickness, while the waxy limiting skin or the cuticle proper is the limiting barrier both in water and solute diffusion (Sects. 4.6 and 6.5). The thickness of the limiting skin in CM of various plant species is not known, but plausible estimates are 100–500 nm (Sect. 1.4). In calculating D_w from (2.5), thickness of the limiting barrier enters as ℓ^2, and using a thickness less than total thickness of the CM would result in lower diffusion coefficients than those in Table 4.9. In Sect. 6.5.2 we estimate the diffusion coefficient of water in cuticular wax as $1.2 \times 10^{-16}\,m^2\,s^{-1}$. This is about a factor of 10 lower than D_w values in leaf CM estimated from the hold-up time (Table 4.9). With *Citrus* CM, a D_w of $1.2 \times 10^{-16}\,m^2\,s^{-1}$ would have been obtained with ℓ equal to 0.44 μm. This amounts to 15% of the total thickness of the CM and appears plausible. This neglects the fact that the waxy diffusion path of water is tortuous, but in calculating D_w for synthetic polymers membrane thickness (ℓ) is used (Table 4.1) and tortuosity is also disregarded.

We can further test these cuticles for homogeneity by plotting P_{wv} vs $1/\ell$ and D_w vs $1/6\times$ hold-up time. In homogenous membranes, these plots should be linear. The

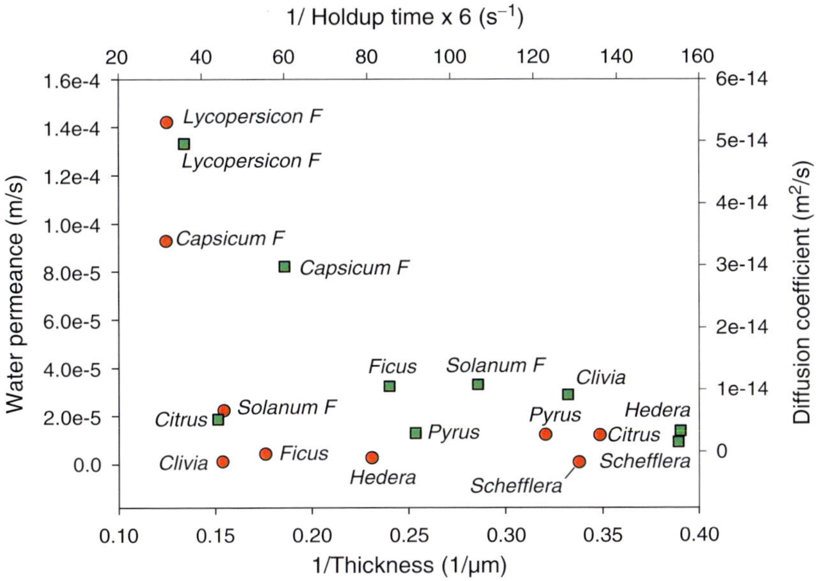

Fig. 4.20 Test for homogeneity of cuticular membranes. P_{wv} was plotted vs $1/\ell$ (*circles*), and D_w was plotted vs $1/6 \times t_e$ (*squares*). (Data from Becker et al. 1986)

plot P_{wv} vs $1/\ell$ should have a slope of $D_w K_{wv}$ (2.18), while a plot D_w vs $1/6 \times t_e$ should have a slope equal to ℓ^2 (2.5). These plots are shown in Fig. 4.20.

The two plots are not linear, and there is considerable scatter. P_{wv} is highest with tomato and pepper fruit CM, and no dependence on $1/\ell$ is detectable. Similarly, no dependence of D_w on $1/6 \times t_e$ can be seen. With leaf CM and *Solanum* fruit CM, data are clustered and no dependence of P_{wv} on thickness or D_w on hold-up time is detectable. It is obvious that no simple relationship between P_{wv} and D_w on total thickness of CM exists. Such a dependence can not really be expected, because extracting small amounts of waxes increases permeance by orders of magnitude, and the weight fraction of waxes in cuticles is by no means constant (Table 4.6). Furthermore, even MX membranes are not homogeneous (Sect. 1.4), and CM contain waxes in addition to polar polymers and cutin. Polar polymers form a separate phase (Sects. 4.5 and 4.6.2, subsection: "Effect of Partial Vapour Pressure (Humidity) on Permeability of CM"), while waxes occur in cutin and on the surface of the cuticles (Sect. 1.3).

4.6.3.2 Estimation of D_w from Diffusion of Lipophilic Neutral Molecules

Alternatively, the diffusion coefficients of water in wax can be estimated from diffusion coefficients of neutral solutes in cuticular waxes. Diffusion coefficients of a series of ^{14}C-labelled non-electrolytes in cuticular wax of different species were measured (for details see Sect. 6.5). Plotting the logarithm as a function of the molar volume, good linearity was obtained for the wax of the three species *Prunus laurocerasus*, *Ginko biloba* and *Juglans regia* (Kirsch et al. 1997). Thus, with the molar volume known, D_w in cuticular waxes of any of other compound can be estimated. Using the molar volume of water ($18\,cm^3\,mol^{-1}$) and the equations in Table 6.10, diffusion coefficients of $1.19 \times 10^{-16}\,m^2\,s^{-1}$, $1.60 \times 10^{-16}\,m^2\,s^{-1}$ and $1.06 \times 10^{-16}\,m^2\,s^{-1}$ can be calculated for the three species *Prunus laurocerasus*, *Ginko biloba* and *Juglans regia* respectively. These are fairly low values, and compared to D_w obtained from extrapolated hold-up times (Table 4.9) they are by 1–2 orders of magnitude lower. However, D_w values in Table 4.9 were calculated using the thickness of the CM and not the thickness of the wax layer. Assuming that cuticular wax forms the transport-limiting barrier of the CM and not the thickness of the CM itself, D_w of most of the species shown in Table 4.9 can be recalculated (Table 4.10) if wax coverage of the CM is known. The thickness of the wax layer is calculated dividing the wax amount per area by wax density ($0.9\,g\,cm^{-3}$).

Depending on the wax amounts used for calculating the thickness of the wax layer, this estimation of diffusion coefficients results in values for leaf CM ranging from 0.11×10^{-16} (*Citrus*) to 56.6×10^{-16} (*Nerium*). D_w calculated for *Nerium* leaf CM and fruit CM from tomato and pepper clearly differ from the other CM of the data set. The reasons are not known. Most of the estimated D_w are around $10^{-16}\,m^2\,s^{-1}$, and this agrees fairly well with the D_w estimated from diffusion of organic non-electrolytes in reconstituted cuticular waxes. If D_w calculated for CM is much higher than D_w values in reconstituted wax, some water flow in aqueous pores

Table 4.10 Estimated diffusion coefficients (D_w) of water in cuticular waxes calculated from extrapolated hold-up times of water permeation across the CM. Thickness of the wax layers were calculated from amounts of wax per unit area (coverage)

Species	t_e(s)[a]	Wax coverage ($\mu g\,cm^{-2}$)	ℓ(μm)[d]	D_w in wax[e] ($m^2\,s^{-1}$)[e]
Leaf CM				
Clivia miniata	770	106[b]–113[c]	1.17–1.25	2.96×10^{-16}–3.38×10^{-16}
Hedera helix	933	64[c]–85[b]	0.71–0.94	0.90×10^{-16}–1.58×10^{-16}
Nerium oleander	960	381[c]–514[b]	4.23–5.71	31.1×10^{-16}–56.6×10^{-16}
Ficus elastica	512	87[c]–114[b]	0.97–1.26	3.06×10^{-16}–5.16×10^{-16}
Citrus aurantium	264	12[b]–32[c]	0.13–0.35	0.11×10^{-16}–0.77×10^{-16}
Pyrus communis	550	133[b]	1.44	6.28×10^{-16}
Fruit CM				
Solanum melongena	640	48[b]	0.53	0.73×10^{-16}
Capsicum annuum	361	96[c]–197[b]	1.06–2.18	5.18×10^{-16}–21.9×10^{-16}
Lycopersicon esculentum	215	54[c]–152[b]	0.60–1.69	2.79×10^{-16}–22.1×10^{-16}

[a] Data from Becker et al. (1986)
[b] Data from Riederer and Schönherr (1984)
[c] Data from Schreiber and Riederer (1996a)
[d] Wax coverage divided by wax density of $0.9\,g\,cm^{-3}$
[e] D calculated from (2.5)

could be responsible for the difference. Unfortunately, in the two sets of experiments different species were used. Nevertheless, it is interesting that these two independent approaches yield estimates for D_w of the same order of magnitude. It is an advantage of these approaches that solubility of water in cuticular wax (wax/water partition coefficients) are not needed for estimating D_w. Unfortunately, assumptions concerning the thickness of the limiting skin must be made, and these assumptions enter as the square.

4.6.4 Water Permeability of Paraffin Waxes

Model III A postulates that wax in and on top of the CM completely determine water permeability, while aqueous pores are not involved. This model is difficult to test, because there are no data on water permeability of reconstituted waxes. The evidence presented is indirect, and relies on co-penetration in CM of lipophilic solutes and water (Figs. 4.16 and 4.17). The observation that cuticular transpiration is correlated with diffusion of stearic acid in cuticular wax (Fig. 4.15) agrees with model III A, and the correlation between water and benzoic acid permeances is also consistent with model III A, even though (4.17) indicates that, in CM of species with relatively high P_w, water transport in aqueous pores may have been involved. On the other hand, data of Table 4.8 are consistent with model IIIC, which assumes that a significant amount of water crosses the CM via aqueous pores, which can be blocked by silver chloride precipitates. There is no reason why CM of all species investigated so far should meet model III A, B or C criteria. The question arises which thickness of a wax barrier could account for observed water permeances. Since we have no data for reconstituted wax, we shall take a look at data obtained with paraffin waxes.

4.6.4.1 Water Permeance of Polyethylene and Paraffin Wax

In Table 4.7, data on wax coverage for CM of several species are given. They range from 11.8 $\mu g\,cm^{-2}$ (*Citrus aurantium*) to 1,317 $\mu g\,cm^{-2}$ (*Malus*). For *Allium* (Table 4.6), only 1 $\mu g\,cm^{-2}$ has been determined using gas chromatography. Multiplying the wax amounts by specific weight (0.9 $g\,cm^{-3}$), we obtain wax layers having a thickness from 11 nm (*Allium*), 131 nm (*Citrus*) and 14.6 μm (*Malus*). For most species in Table 4.7, wax amounts varied from 30 to 100 $\mu g\,cm^{-2}$, resulting in wax layers of 333 nm to 1.1 μm. Can we estimate P_{wv} for such wax layers?

Waxes are partially crystalline, hydrophobic and they consist mainly of CH_2 and CH_3 groups. This is also true for polyethylene, which is partially crystalline but exclusively composed of CH_2 and CH_3 groups. $\mathscr{P}_{wv}$ of polyethylene is $3.3 \times 10^{-11}\,m^2\,s^{-1}$ (Table 4.1). A membrane of a comparable physico-chemical structure is parafilm (polyethylene covered with paraffin) and water permeability P_{wv} of 127 μm-thick parafilm is $3.33 \times 10^{-7}\,m\,s^{-1}$ (Santrucek et al. 2004), which results in a $\mathscr{P}_{wv}$ of $4.23 \times 10^{-11}\,m^2\,s^{-1}$. This is nearly identical to polyethylene. Polyethylene and parafilm of 333 nm thickness would have a permeance ($P_{wv} = \mathscr{P}_{wv}/\ell$) of $1–1.27 \times 10^{-4}\,m\,s^{-1}$, and for a 1.1 μm thick membrane permeance would be $3.0–3.84 \times 10^{-5}\,m\,s^{-1}$. Comparing these values with those shown in Table 4.9 leaves no doubt that a hydrophobic membrane having P_{wv} similar to PE or parafilm cannot account for most of the permeances observed especially not for those of leaf CM.

$\mathscr{P}_{wv}$ measured for a liquid paraffin (hexamethyl tetracosane) is even 100 times higher ($3.3 \times 10^{-9}\,m^2\,s^{-1}$) (Schatzberg 1965), and it can be ruled out that liquid cuticular waxes might contribute to barrier properties. *Citrus* wax is in the fluid state when $T > 70°C$ (Reynhardt and Riederer 1991). At 25°C P_w was $5.9 \times 10^{-10}\,m\,s^{-1}$, and at 55°C it had dropped to $2 \times 10^{-8}\,m\,s^{-1}$ (Haas and Schönherr 1979). This is not too far from P_w measured for the polymer matrix, which is $2.5 \times 10^{-7}\,m\,s^{-1}$. Clearly, we need a solid wax to account for water permeability of cuticular membranes.

Fox (1958) studied water vapour transmission of thin cellophane films coated with paraffin waxes. Waxes were applied at temperatures above the melting point, and solidified at various temperatures. The composite membranes were stored at 23°C or 35°C before vapour transmission rates were determined at 23°C and 50% humidity using a cup method (Sect. 9.7). Data obtained with a paraffin wax having a melting point of 62°C are shown in Table 4.11.

In the original paper, wax load was given as lb/ream. These figures were converted to $g\,m^{-2}$ using 453.6 g per pound and 278.71 m^2 per ream. Depending on type of paper the ream can have variable size. The ream of glassine paper consists of 500 sheets of 24 × 36 inches (Scott et al. 1995). Water vapour transmission rates reflect permeability of the wax films because cellophane has very high water permeability. WVTR depended on temperature of water used for cooling the melt and on duration of storage of solidified wax. Rapid cooling at 1.7°C resulted in high initial WVTR, which decreased to about one half during 4–5 days storage. At higher cooling temperatures initial WVTR were lower, and decreased to $0.02\,g\,m^{-2}\,day^{-1}$. Longer storage (38 days) had little effect, but with some waxes WVTR dropped to $0.01\,g\,m^{-2}\,day^{-1}$. After 4–5 days storage at 35°C, permeance (P_{wv}) ranged from 2.2×10^{-8} to $8.9 \times 10^{-8}\,m\,s^{-1}$. This is lower by at least one order of magnitude than P_{wv} for CM (Table 4.9).

Table 4.11 Water vapour transmission rate (WVTR) of cellophane membranes coated with molten paraffin wax and cooled at various temperatures either in water or air. WVTR rates measured at 23°C and 50% humidity

Cooling T(°C)	Wax load ($g\,m^{-2}$)	ℓ of wax (μm)	WVTR (first day) ($g\,m^{-2}\,day^{-1}$)	WVTR (after 4–5 days) ($g\,m^{-2}\,day^{-1}$)	P_{wv} ($m\,s^{-1}$)
1.7 (water)	11.7	12.9	0.15	0.08	8.9×10^{-8}
7.2 (water)	10.9	12.0	0.12	0.03	3.3×10^{-8}
12.8 (water)	9.6	10.6	0.05	0.03	3.3×10^{-8}
18.3 (water)	9.8	10.8	0.07	0.02	2.2×10^{-8}
23.0 (air)	10.7	11.8	0.09	0.02	2.2×10^{-8}

By microscopic inspection Fox (1958) detected significant changes in the structure of the wax films. After waxing, paraffin films exhibited numerous cracks. During storage at 35°C these cracks and fissures healed, and grain size of the crystals increased. Apparently, the paraffin recrystallised. When storage at 35°C was long enough, the crystals became so large and compacted that the wax film began to resemble one large crystal. Furthermore, very thin layers of recrystallised wax formed on the top surface of the film. X-ray diffraction studies showed that the air-cooled films contained crystals predominantly oriented with their *c*-axes (long axes) perpendicular to the cellophane sheet. Such wax crystals oriented parallel to the sheet gave a low WVTR. Formation of such plates was aided by slow air cooling, or by recrystallisation and reorientation of the water-cooled films during storage. Wax crystals in cuticles are also oriented parallel to the surface of the CM, and *c*-axes are perpendicular (Sect. 1.4).

Fox (1958) concluded that vapour transmission occurred through defects between crystals and that a high WVTR was the result of more numerous defects, while the crystals themselves were impermeable to water vapour. However, paraffinic waxes are complex mixtures of alkanes varying in chain lengths. Le Roux and Loubser (1980) estimated for paraffinic wax having a melting point of 59°C that 83% of the molecules were incorporated into crystals at room temperature, 16% formed a rigid amorphous phase and 1% was liquid oil (1%). The technical paraffin used by Fox most likely contained amorphous wax which is permeable to water, and water vapour penetrated across defects and in amorphous wax. It is unfortunate that wax coverage was not varied systematically. Thus, we do not know if number and healing of defects depend on film thickness. If it is assumed that this is not the case, $\mathscr{P}_{\mathrm{wv}}$ should be of the order of $2.6 \times 10^{-13}\,\mathrm{m^2\,s^{-1}}$, which is two orders of magnitude smaller than $\mathscr{P}_{\mathrm{wv}}$ for polyethylene (Table 4.1) or parafilm. Since crystals are impermeable, a very thin wax layer should suffice to build an excellent water barrier.

We can calculate the thickness of a paraffinic film on the CM (2.4) necessary to obtain the permeances measured (Table 4.9). For instance, with *Schefflera* a wax layer of 317 nm would suffice, and with *Hedera* 98 nm, while with *Pyrus* and *Citrus* a wax layer of 22 nm thick would be enough if we disregard the contribution of the MX to P_{wv}. With *Lycopersicon* a wax layer of only 2 nm would be necessary, but with such a thick CM the contribution of cutin to P_{wv} cannot be neglected. Since a wax layer of 11 nm thickness has a weight of $1\,\mu\mathrm{g\,cm^{-2}}$, the coverages of wax shown in some of the previous tables (Tables 1.1, 4.6, 4.7 and 4.11) are far in excess of the amount needed for a wax barrier on top of the cuticle.

In these calculations, we have assumed that cuticular waxes have water permeability and structure similar to that of the paraffin wax used by Fox (1958). However, this is not the case since cuticular wax is characterised by functional groups (Sect. 1.3.2), and this must be considered in models of the cuticular transpiration barrier. The methods of Fox (1958) should also work with isolated cuticular waxes, and it is amazing that experiments using this approach have not yet been carried out. Permeability across cuticular wax films could be measured. Since cellophane is isotropic, molecular orientation and degree of crystallinity of these wax films can be investigated using polarised light and X-ray diffraction.

4.6.4.2 Water Permeability of Lipid Monolayers

Using atomic force microscopy, Koch et al. (2004) demonstrated the formation of highly ordered mono- and bilayers on cuticles from which surface waxes had been stripped. They suggested that they are composed of long chain lipids. Water permeability of such monolayers can be estimated from data obtained with monolayers of fatty acids or n-alcohols (Archer and LaMer 1955; LaMer et al. 1964). At high surface pressure, resistance (1/permeance) doubled when two methylene groups were added (Fig. 4.21). Resistances of monolayers of alcohols having 16, 18, 20 or 22 carbon atoms were 200, 400, 800 and 1,600 $\mathrm{s\,m^{-1}}$ respectively. The fatty alcohols are somewhat inclined, and the thickness ℓ of a monolayer (nm) can be calculated (Barrow 1961) from (4.21)

$$\ell = 0.154\,n\mathrm{C} \times \cos 35^{\circ}, \tag{4.21}$$

where nC is the number of carbon atoms. A monolayer with 40 carbon atoms has a thickness of 5 nm, and P_{wv} would be $1.22 \times 10^{-6}\,\mathrm{m\,s^{-1}}$. With nC equal to 34, which corresponds to the mean chain length of ivy wax (Fig. 4.14), thickness would be 4.3 nm and P_{wv} would be $9.77 \times 10^{-6}\,\mathrm{m\,s^{-1}}$ (Fig. 4.21). The monolayers observed by Koch et al. (2004) had a thickness of 3–5 nm. Permeances P_{wv} measured for ivy CM in different studies range from $2.66 \times 10^{-6}\,\mathrm{m\,s^{-1}}$ (Table 4.10) to $4.3 \times 10^{-6}\,\mathrm{m\,s^{-1}}$ (Table 4.6) or $1.55 \times 10^{-5}\,\mathrm{m\,s^{-1}}$ (Table 4.7). Comparing these estimated permeances of monolayers with measured permeances of ivy, we see that a single monolayer could in theory account for water permeability of the CM.

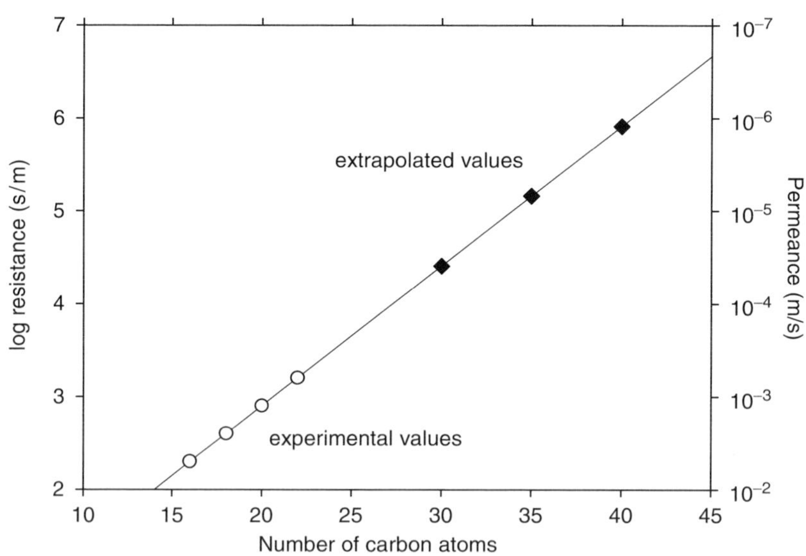

Fig. 4.21 Resistances and permeances (P_{wv}) of monolayers of fatty alcohols as a function of number of carbon number. (Data taken from LaMer et al. 1964)

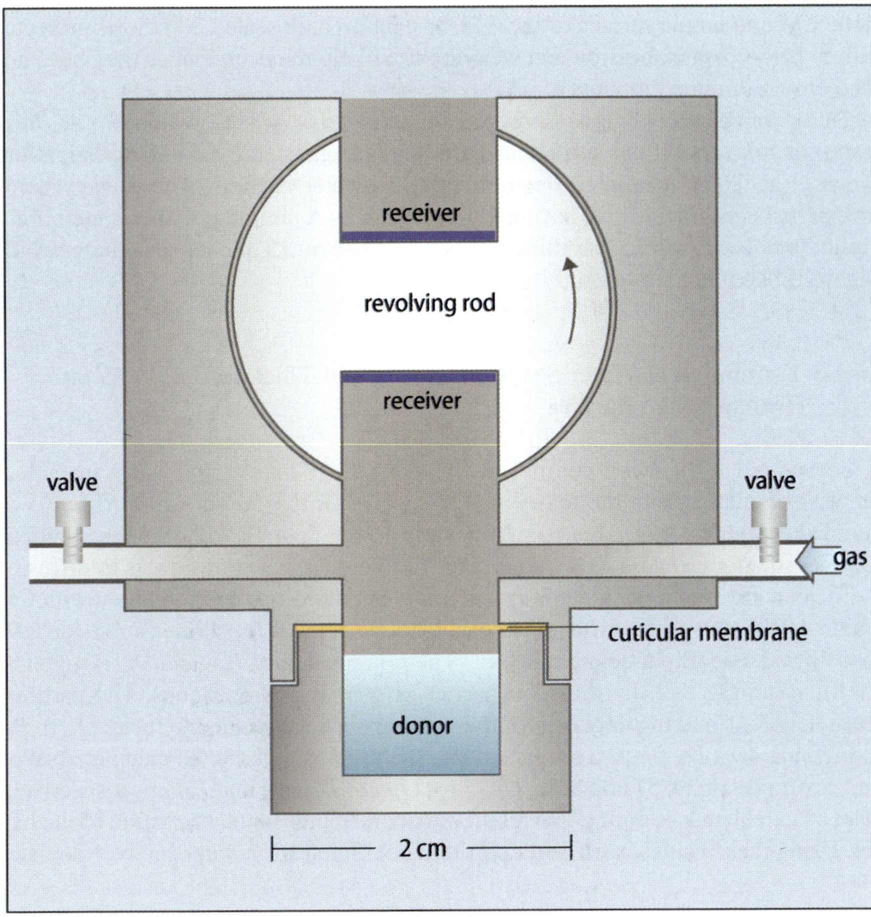

Fig. 4.22 Transport chamber used for measuring the flux of ^{3}H-labelled water from the donor across stomatous cuticles into the receiver. In the receiver was a water-saturated filter paper (depicted in *dark blue*) absorbing ^{3}H-labelled water which diffused across the CM. The revolving rod allows sampling of the receiver without changing the atmosphere in the receiver. (Modified from Santrucek et al. 2004)

The fact that resistance of fatty alcohols is a function of log *n*C could explain why evolution selected very long chain wax monomers for building a water barrier. At ambient temperatures these monomers are in the solid state, and very small amounts of 1–2 $\mu g\,cm^{-2}$ suffice. We are again confronted with the question why wax coverage is much higher with most plants (Tables 1.1, 4.6, 4.7 and 4.11). Leaves can be subjected to considerable abiotic (e.g., wind, rain, ice and storm) and biotic forces (e.g., insects, fungi and bacteria). This should lead to a continuous disturbance of a highly ordered wax structures on the leaf surface, and losses of wax might even occur during the season. A fast and efficient compensation of this disturbance, keeping the cuticular transport barrier intact, might require fairly high amounts of wax in the CM and on the surface of the CM. The onion bulb scale CM is well-protected within the bulb, and abrasion and wear are not likely to occur. Hence they have and need only minimum amounts of wax.

These model calculations show convincingly that waxes which would form mono- or bilayers on the surface of cutin, as is suggested by AFM investigations (Koch et al. 2004), could be very effective as water barriers. These waxy barriers can be very thin and still have a low permeance. A simple polymeric membrane similar to polyethylene or parafilm would have to be much thicker, and energetically this would be much more costly.

4.6.4.3 Estimation of Water Sorption in Wax and Thickness of the Waxy Transpiration Barrier

In Sect. 4.6.3 a diffusion coefficient of about $10^{-16}\,m^2\,s^{-1}$ for water in cuticular waxes isolated from the leaves of the species *Ginko*, *Juglans* and *Prunus* was calculated (Table 4.12). This is significantly lower than D_w of water in synthetic polymer membranes and MX membranes (Table 4.1), but it is similar to diffusion coefficients for water calculated from extrapolated hold-up times obtained with CM (Table 4.10). Can these diffusion coefficients be related to P_w measured for CM, paraffin wax and aliphatic monolayers? The relationship of P_{wv} and D_w is given by (2.18), assuming that the transport barrier of wax is homogeneous. The partition coefficient K_{ww} and the thickness of the wax barrier ℓ are needed to relate D_w to P_w.

In Table 4.12 D_w and P_w are listed, and the ratio K_{ww}/ℓ can be calculated. With wax coverages of 39, 51 and 83 $\mu g\,cm^{-2}$ for *Ginko*, *Juglans* and *Prunus* respectively, ℓ can be calculated assuming that all the wax contributes to the formation of the barrier. Using these values, wax/water partition coefficients for water can be calculated

Table 4.12 Diffusion coefficients in wax (D_w), permeances (P_w) of water across the CM, and thickness ℓ of the wax layers in *Ginkgo biloba*, *Juglans regia* and *Prunus laurocerasus*. Wax/water partition coefficients K_{ww} are calculated by multiplying K_{ww}/ℓ by ℓ. Thickness ℓ of the wax layer was calculated dividing K_{ww} by K_{ww}/ℓ

	Ginkgo	*Juglans*	*Prunus*
D_w ($m^2\,s^{-1}$)[a]	1.60×10^{-16}	1.06×10^{-16}	1.19×10^{-16}
P_w ($m\,s^{-1}$)[b]	4.3×10^{-10}	6.3×10^{-10}	1.33×10^{-10}
ℓ (nm) calculated from wax amounts[c]	433	566	922
$P_w/D_w = K_{ww}/\ell$	2.69×10^6	5.94×10^6	1.11×10^6
K_{ww} calculated[d]	1.16	3.36	1.03
ℓ (nm) calculated for $K_{ww} = 0.01$	3.7	1.7	9.0
ℓ (nm) calculated for $K_{ww} = 0.04$	14.9	6.7	36.1

[a]Calculated from equations listed in Table 6.10
[b]Data from Table 4.7
[c]Calculated from wax overages given in Table 4.7
[d]K_{ww} calculated here have the units gram per volume (g water/cm^3 polymer)/(g water/cm^3 water) and not gram per gram as given in Sect. 2.4

from K_{ww}/ℓ. These calculated partition coefficients vary between 1 and 3. This is definitively too high, since this would imply that solubility of water in lipophilic wax is the same as in the polar water phase.

Alternatively, K_{ww} can be assumed and ℓ calculated. Partition coefficients of water in different lipophilic phases and water vary depending on the chemistry of the lipid phase. Solubility of water at 25°C in C_{16} hexadecane and in a branched liquid C_{30} hydrocarbon (hexamethyl tetracosane = squalene) is 4.2×10^{-5} and $4.4 \times 10^{-5}\,g\,cm^{-3}$ (Schatzberg 1965) respectively. For squalene this amounts to 5.5×10^{-5} g water per g hydrocarbon, and thus the partition coefficient K_{ww} is 5.5×10^{-5}. Much higher partition coefficients of 0.0014, 0.013 and 0.04 have been published for olive oil, ether and octanol respectively (Wolosin and Ginsburg 1975). Even higher values of 0.06 (Potts and Francoeur 1991) and 0.162 (Schwindt et al. 1998) were estimated for stratum corneum, the transport-limiting lipid barrier of the mammalian skin which also contains squalene. Cuticular waxes contain small but significant amounts of polar functional groups which are expected to raise the partition coefficients somewhat. On the other hand, a large fraction of the waxes is in the crystalline state and probably sorbs no water in crystals and little on crystal surfaces. If these factors cancel, K_{ww} for cuticular waxes is likely to be similar to that for the liquid paraffin. In Table 4.1 K_{wv} values for solid polymers are given, and from these K_w can be obtained by multiplying them with 43,394 (3.11), (3.16), (3.17). This calculation gives K_w for polyethylene and polyethylene terephthalate of 4.5×10^{-4} and 0.011 respectively. These K_w are higher by factors about 10–200 than that for the liquid paraffin. For model calculations they may be accepted to mimic water contents of cuticular waxes.

Water sorption in cuticular waxes may also be estimated based on concentrations of polar functional groups. If we assume that polar functional groups amount to 4% of the wax and each oxygen atom would bind one water molecule, the wax/water partition coefficient for water would be 0.041. If only 10% would sorb a water molecule due to intermolecular hydrogen bonding, the partition coefficient would be 0.0041. Working with partition coefficients of 0.01 and 0.04 for model calculations seems reasonable. Using (2.18), the thickness of the wax barrier can be estimated (Table 4.12). Depending on magnitude of K_{ww}, the required thickness of the waxy barrier ranges from 1.7 to 9 nm ($K_{ww} = 0.01$) or 6.7 to 36.1 nm ($K_{ww} = 0.04$). Since 1 μg wax per cm^2 gives a wax film having 11 nm thickness, the required thickness of the waxy barriers ranges from about 0.2 to $4\,\mu g\,cm^{-2}$.

Wax coverage of most plant species is much higher (Tables 4.7 and 4.11) and the question is why? In many cuticles substantial amounts of waxes occur as embedded waxes (Fig. 1.6), which contribute little to barrier properties (Sects. 6.3 and 6.4). The function of these "useless" and hidden waxes is obscure. Not all CM have such high wax coverage as shown in Table 4.7. *Allium* epidermis (Table 4.6) has a wax coverage of only $1\,\mu g\,cm^{-2}$, which results in a wax layer of only 11 nm. A similar low wax coverage of about $1.6\,\mu g\,cm^{-2}$ was reported for *Arabidopsis* leaves (Jenks et al. 1996). According to our model calculations (Fig. 4.13), a functional transpiration barrier can be built in *Allium* and *Arabidopsis* leaves. The average mean chain length of the linear long-chain aliphatic molecules in *Arabidopsis* leaf wax is about 31 carbon atoms (Jenks et al. 1996), and using (4.21) the average thickness of a monolayer would be 4.0 nm. With a wax amount of $1.6\,\mu g\,cm^{-2}$ at most, 4–5 monomolecular layers of wax molecules could be established.

4.7 Permeances of Adaxial and Abaxial Cuticles

The analysis of water permeance of cuticles presented in this chapter relies on data obtained with isolated astomatous cuticular membranes and with reconstituted waxes. However, most leaves have stomata at least on one side, and due to this lateral heterogeneity we have at least three parallel pathways of water: (1) diffusion across stomatal pores, (2) diffusion across cuticles over guard cells and accessory cells, and (3) diffusion across cuticles over ordinary pavement cells.

A more sophisticated approach permits the quantification of the fluxes of water vapour across cuticles and across stomatal pores (Santrucek et al. 2004). It is based on the fact that water diffusion in the vapour phase depends on density of the vapour, whereas the water diffusion through the solid phase of the cuticle remains unaffected. The total flux F_{total} across an isolated cuticle with stomatal pores is the sum of the two fluxes occurring across the solid cuticle $F_{cuticle}$ and the stomatal pore F_{stoma}

$$F_{total} = F_{cuticle} + F_{stoma}. \tag{4.22}$$

Transpiration experiments were carried out using the experimental setup depicted in Fig. 4.22 and ^3H-labelled water vapour in either a pure helium (He) or pure nitrogen (N) atmosphere at ambient pressure. Radio-labelled water diffuses from the donor across the membrane into the receiver, where it is trapped and sampled in regular time intervals. Different gas phases can be obtained by flushing the receiver chamber with either helium of nitrogen gas.

The two fluxes of water $F_{totalHe}$ in helium and F_{totalN} in nitrogen can be measured using isolated stomatous cuticles

$$F_{totalHe} = F_{cuticle} + F_{stomaHe}, \tag{4.23}$$

$$F_{totalN} = F_{cuticle} + F_{stomaN}. \tag{4.24}$$

This experimental approach is based on the fact that the diffusion coefficient of water in helium D_{wHe} is 3.6 times higher than the diffusion coefficient of water in nitrogen D_{wN} (Cussler 1984). Thus, using membranes with pores (e.g., stomatous cuticles) higher fluxes will be measured in a helium atmosphere compared to a nitrogen atmosphere (4.25). However, no differences occur with a non-porous membrane (e.g., astomatous cuticle) and fluxes in both gases (helium and nitrogen) will be the same

$$\frac{F_{stomaHe}}{F_{stomaN}} = \frac{D_{wHe}}{D_{wN}} = 3.6. \tag{4.25}$$

The ratio D_{wHe}/D_{wN} will be called d, and y is $d/(d-1)$. Substituting $F_{stomaHe}$ in (4.23) by (4.25) and combining (4.23) and (4.24), the following solution is obtained describing the flux of water across the solid phase of a stomatous cuticle

$$F_{cuticle} = y \times F_{totalN} - (y-1) \times F_{totalHe}. \tag{4.26}$$

Using this approach and (4.26) with stomatous and astomatous CM isolated from ivy leaves, it was found that P_{wv} of the astomatous cuticle was $3.12 \times 10^{-6}\,\mathrm{m\,s^{-1}}$, whereas P_{wv} of the stomatous cuticle was $3.57 \times 10^{-5}\,\mathrm{m\,s^{-1}}$ (Santrucek et al. 2004). The ratio of both permeances is 11.4. This indicates that water permeability of the abaxial CM was 11-fold higher than water permeability of theadaxial CM. So far, this new approach was used only with ivy CM, and additional studies with different plant species are necessary. We expect this ratio to be similar in other plant species. Much higher permeability to ions of cuticles from abaxial leaf surfaces compared to adaxial surfaces has also been observed (Chap. 5).

4.8 Water Permeability of Isolated Cuticular Membranes as Compared to Intact Leaves

Traditionally cuticular transpiration was studied by gravimetrically measuring water loss from detached leaves (Kamp 1930). Lower stomatous sides were either covered with grease or it was assumed that stomata were completely closed when leaves had reduced turgor and started to wilt. With amphistomatous leaves, this approach is the only one possible. Rates of cuticular transpiration measured with detached leaves are in general significantly higher (by factors between 3 and 10) than those for isolated astomatous CM (Kerstiens 1996b).

Kerstiens (1996b) suggested using the term minimum conductance for water loss measured with intact leaves, whereas the term cuticular transpiration should be reserved for water transport across astomatous cuticles. Imperfect closure of stomata was taken to be responsible for the higher water permeability of stomatous leaf surfaces; however, even when diffusion of water vapour across stomatal pores was eliminated, P_{wv} for the cuticular pathway was still 11.4 times higher with adaxial leaf CM (Sect. 4.7). This implies that rates of transpiration of leaves under water stress would be limited by permeability of cuticles rather than by incomplete closure of stomata. Results obtained with ivy, comparing intact leaves with isolated cuticles, supports this conclusion (Burghardt and Riederer 2003).

In spite of the fact that water permeability of CM is generally lower than minimum conductance of intact leaves, there is often concern that barrier properties of cuticles might be adversely affected during isolation. Solute permeability (Kirsch et al. 1997) and water permeability (Schreiber et al. 2001) of isolated CM and cuticles still attached to leaves has been compared, and identical permeances were observed in the two studies (Fig. 4.23). In this study, fully expanded but only several weeks old leaves had permeances which were 2 to 3 times lower than those

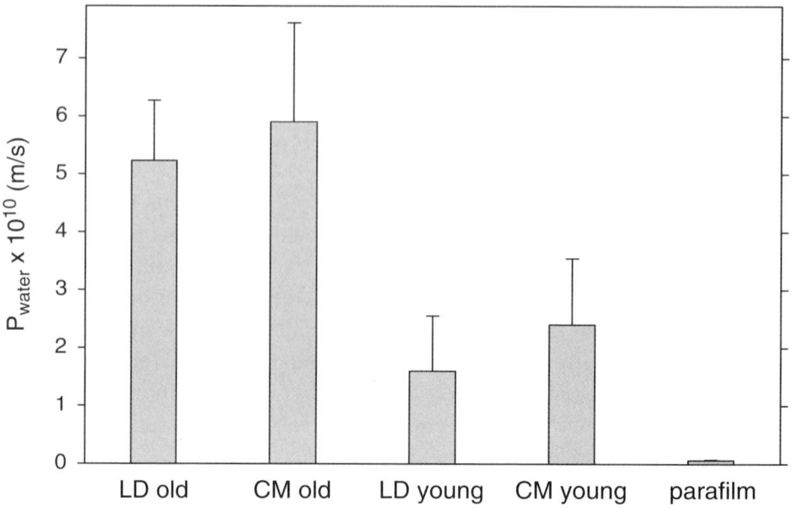

Fig. 4.23 Permeance P_w measured with old and young *Prunus laurocerasus* leaf disks (LD) and isolated cuticles (CM), using ^{3}H-labelled water. (Data from Schreiber et al. 2001)

of several months old leaves. This indicates that cuticular permeability increases slightly when leaves age. A similar observation was also reported for ivy (Fig. 4.14e; Hauke and Schreiber 1998).

4.9 The Shape of the Water Barrier in Plant Cuticles

The data presented in the preceding Sects. 4.6.1–4.6.4 have to be considered when trying to establish a model for the structure of the cuticular transpiration barrier. While there is no doubt that cuticular waxes significantly reduce cuticular transpiration (Sect. 4.6.1), the question still remains how this is accomplished. In Sect. 4.6.2 three possible models have been suggested, and in Sects. 4.6.2–4.6.4 it was shown that not all experimental data can be fitted to only one model.

In model III A, all water must diffuse across a cuticular wax layer which is the transport-limiting barrier. No alternative pathway exists. Data analysing co-permeation and correlation of water permeability with diffusion of stearic acid in cuticular waxes are consistent with this model. In addition, water transport across paraffin wax and fatty acid monolayers fits this model.

Experimental results analysing the effect of humidity on cuticular transpiration and the effect of salt precipitates in cuticles favour model III B, which includes aqueous pores and provides evidence that some water also diffuses across polar aqueous pores in parallel to the lipophilic waxy pathway. Model III C is similar to model III B, but it takes into account the fact that in CM only a fraction of polar pores in the polymer matrix contributes to water transport because a layer of cuticular wax covers and closes aqueous pores to some extent.

Our present knowledge does not permit us to decide which model best describes water permeability of all cuticles. In fact, there are indications that different models describe permeability of cuticles of different species adapted to different habitats. We have presented a number of hypotheses regarding the shape of cuticular barriers. The current data base is insufficient to rigorously test these hypotheses. Additional data are need to test these hypotheses. For instance, sorption and diffusion of water in cuticular waxes should be measured directly, and permeation of water across wax films using the approach of Fox (1958) should be carried out. The observation that monolayers and bilayers can spontaneously and rapidly form on the smooth stripped cuticle at the plant/atmosphere interface (Koch et al. 2004) is exciting, and must be analysed in more detail. Species should be used for which data on water permeability are available or can be measured accurately.

Characterisation of the aqueous pathway during leaf development of many more plant species is badly needed. Pore size must be estimated, and the contribution of water flux across pores to total water permeability should be estimated using the approach of blocking polar aqueous pores. More information on the structure and spatial arrangement of the polar polymers in the MX should be collected. This can be obtained using immunological techniques and high-resolution microscopy. With these data available, our models could be refined and it would be possible to test which experimental data can be fitted to these models.

It is likely that different species follow very different strategies in adapting to their specific environmental conditions. There is no reason why the cuticular transpiration barrier should be the same with all species. Species with thin cuticles and very low wax amounts, like *Allium* or *Arabidopsis*, might in fact have to rely on a fragile monolayer deposited on top of the MX. Species with thick cuticles and much higher wax loads, like *Hedera helix* or *Prunus laurocerasus*, might follow the different strategy of impregnating and covering the MX with large wax amounts, with some polar pores still contributing to the water permeability.

Problems

1. What is the permeance at 25°C of a polyethylene bag having a thickness of 20 μm and $\mathscr{P}_{\mathrm{wv}}$ of $4 \times 10^{-11}\,\mathrm{m^2\,s^{-1}}$? How much water (g) penetrates across m^2 per day if the bag is filled with water and stored at 50% humidity?
2. Using the data given in Table 4.1, calculate the partition coefficients K_{wv} and K_{w} for ethyl cellulose. What is the concentration of water (C_{w}) in the polymer (g water/cm^3 polymer) at 20% humidity?
3. (a) How many meq kg^{-1} of carboxyl groups are ionised in apricot leaf MX at *p*H 3.4, which is the isoelectric point? (b) What percentage is this compared to total exchange capacity at *p*H 8?
4. Which condition must be fulfilled for all carboxyl groups of cutin acids to ionise?

5. What is the predominant counter ion in the cuticle at physiological *p*H values of 5–6, and why?
6. Why is water permeance of cuticles higher in Na^+ than in Ca^{2+} form?
7. Which kind of measurements would you conduct to test for presence of aqueous pores in cuticles?
8. Why are aqueous pores not seen in transmission electron micrographs?
9. You have experimentally established that $P_{viscous}$ of a membrane is the same when you use osmotic pressure of hydrostatic pressure of the same magnitude as driving force; what does this mean?
10. What is the diffusion coefficient (D) in water at 25°C of a solute having a molecular having radius of 2 nm? For this calculation you need the Boltzmann constant ($1.38 \times 10^{-23}\,\mathrm{J\,K^{-1}}$ per molecule), and you should remember that $\mathrm{Pa} = \mathrm{J\,m^{-3}}$.
11. Which hypothetical models describing the structure of the cuticular transpiration barrier have been proposed?
12. How do thickness of the CM, thickness of the wax layer, and chemical composition of cuticular waxes affect cuticular transpiration?
13. A diffusion coefficient D of $2.92 \times 10^{-17}\,\mathrm{m^2\,s^{-1}}$ was measured in a sample of reconstituted cuticular wax. What is the permeance P_w of an isolated cuticle of this species?
14. A permeance P_{BA} of $1.34 \times 10^{-10}\,\mathrm{m\,s^{-1}}$ was measured for an isolated cuticle and benzoic acid. What is the permeance P_w for water?
15. In a steady state experiment measuring 2,4-D permeability across an isolated cuticle, the hold-up time t_e was 1,300 s. What is the diffusion coefficient D of 2,4-D in the cuticular barrier, assuming a thickness of the cuticle of 3 µm. What would be D assuming that the transport-limiting barrier of the cuticle is formed by the layer of cuticular wax, which corresponds to about one tenth of the thickness of the cuticle.
16. The average mean chain length of the constituents of a wax sample is 28 carbon atoms. What is the thickness of a monolayer formed by this number of carbon atoms? How many monolayers could be formed on the surface of leaf having a wax amount of $1.0\,\mathrm{\mu g\,cm^{-2}}$?

Solutions

1. P is $2 \times 10^{-6}\,\mathrm{m\,s^{-1}}$, and J is $1.99\,\mathrm{g\,m^{-2}}$ per day.
2. K_{wv} is 1,000, while K_w is 0.023. C_w at 100% humidity is $4.6\,\mathrm{mg\,cm^{-3}}$.
3. (a) At the isoelectric point, positive and negative net charges are equal. Hence, $19\,\mathrm{meq\,kg^{-1}}$ carboxyl groups must be ionised to balance the ionised basic groups. (b) 6.55%.
4. The *p*H in the epidermal wall must be 8.0, which is two *p*H units higher than the pK_a of 6.

5. Ca^{2+}, because selectivity for divalent cations is very high. Mg^{2+}, Cu^{2+} and Fe^{3+} may also be exchanged in trace amounts.
6. Water concentration of cuticles in Ca^{2+} form is lower, because the concentration of counter ions is only half that of Na^+, and Ca^{2+} ions bind much stronger to COO^- groups. This reduces osmotic pressure and electrostatic free energy.
7. I would measure $P_{\text{diffusion}}$ and P_{viscous} and $P_{\text{viscous}}/P_{\text{diffusion}} > 2$. Penetration of hydrated ions across an intact CM also indicates the presence of aqueous pores.
8. Because aqueous pores arise by swelling in presence of water, and specimens embedded in plastic resins contain no water.
9. The membrane is impermeable to the osmotic solute.
10. Using (4.16) we obtain $D = 1.21 \times 10^{-10}\,\text{m}^2\,\text{s}^{-1}$.
11. Three different models with three different options how waxes can be incorporated in the polymer matrix have been suggested. Model III A assumes that a thin layer of wax is deposited on the outer surface of the MX and forms the limiting barrier. Model III B assumes that all waxes are embedded in cutin and on the outer surface of the MX, leaving all aqueous pores unaffected. Model III C is a mixture of the two other models, assuming that the superficial wax layer covers a fraction but not all of the aqueous pores.
12. So far no convincing correlations between cuticular transpiration and thickness of the CM, thickness of the wax layer, and chemical composition of waxes have been detected.
13. Using (4.18) a permeance P_w of $1.88 \times 10^{-9}\,\text{m}\,\text{s}^{-1}$ can be estimated.
14. Using (4.19) a permeance P_w of $1.55 \times 10^{-10}\,\text{m}\,\text{s}^{-1}$ can be estimated for water.
15. This problem can be solved using (2.5). Using the thickness of the cuticle which is $3\,\mu\text{m}$, a D of $1.15 \times 10^{-15}\,\text{m}^2\,\text{s}^{-1}$ is calculated. Assuming that the transport-limiting barrier of the cuticle is formed by the layer of cuticular wax, which corresponds to about one tenth of the thickness of the cuticle, a D of $1.15 \times 10^{-17}\,\text{m}^2\,\text{s}^{-1}$ is obtained.
16. Using (4.20) it can be calculated that a monolayer composed primarily of C_{28} wax molecules has a thickness of 3.5 nm. A wax layer having $1.0\,\mu\text{g}\,\text{cm}^{-2}$ has a thickness of 11 nm. Thus, not more than three monolayers could be formed.

Chapter 5
Penetration of Ionic Solutes

In Chap. 4, we characterised the pathways for water in cuticles. Cutin is the major constituent of the polymer matrix. It is lipophilic and constitutes the lipophilic pathway. The polymer matrix contains polar polymers which sorb water and swell. This hydration water is continuous, and gives rise to aqueous pores which traverse the cuticle (Sect. 4.5). Waxes occur both sorbed in cutin and as epicuticular wax on the surface of the polymer matrix. Waxes associated with cutin greatly reduce permeability of the lipophilic pathway, and for this reason we have also used the term "waxy" pathway. Permeance of the waxy pathway is proportional to the partition coefficient that is to solubility of solutes in cutin and waxes. Polar solutes have very low partition coefficients, and for this reason permeance of the waxy pathway is very low but finite (Sect. 4.6).

With ionic solutes the situation differs, because under physiological conditions ionic groups are surrounded by water molecules. This hydration water is bound very strongly by ion–dipole interactions which renders them essentially insoluble in oils, fats, cutin and waxes. For this reason, hydrated ions cannot access the waxy pathway. Penetration of inorganic and organic ions is limited to the aqueous pathway (Schönherr 2006). Negative and positive ions must penetrate in equal numbers to maintain electroneutrality. Each cation must be accompanied by equivalent amounts of anions. For instance, Ca^{2+} must be accompanied by two Cl^{-} ions. This is true as long there is no drop of electric potential across the cuticle, which is always the case under natural conditions. Thus, the appropriate term is salt or electrolyte permeability, not ion permeability.

Whenever salt or electrolyte penetration is observed in the field or in the laboratory, this is definitive evidence for the presence of aqueous pores in the cuticles of leaves and stems investigated. Strugger (1939) was one of the first to demonstrate presence of aqueous pores in plant cuticles. Agriculturalists and horticulturalists interested in foliar nutrition have studied salt permeation into leaves or isolated CM (cf. Yamada et al. 1964; McFairlane and Berry 1974; Tuckey 1970; Schönherr 2000, 2001; Schlegel and Schönherr 2002; Schreiber 2005). Many agricultural chemicals are ionic (bentazon, glyphosate, paraquat) and penetrate into the foliage only when aqueous pores occur in cuticles of these leaves. Penetration of glyphosate was studied to better understand cuticular penetration and to improve efficacy of these foliar sprays (Schönherr 2002; Schönherr and Schreiber 2004a).

5.1 Localisation of Aqueous Pores in Cuticles

Strugger (1939) used the cationic dye berberine sulphate (BS) to trace the flow of water in the xylem and the leaf apoplast. BS is a salt consisting of two berberine cations and one sulphate group. The formula weight of this salt is 768.8 $g\,mol^{-1}$. BS shows only weak fluorescence when dissolved in water, but when bound to the cell wall very intense fluorescence is observed. BS is a fairly strong base, and ionic interaction between positively charged berberine and negatively charged carboxyl groups of cutin and pectins is probably involved.

When the cut stems of small branches of *Helxine soleirolii* were placed in aqueous BS (1 $g\,l^{-1}$), the ascent in the xylem vessels was rapid and took only a few minutes, especially when stomata were open and transpiration was taking place. The dye spread from the veins to interveinal regions of the leaves, and accumulated in cuticular ledges of guard cells, in the basal cells of trichomes and in anticlinal walls. Accumulation of BS in the upper epidermis, which lacks stomata, proceeded more slowly, but the pattern of distribution in epidermal cells and trichomes was similar, while fluorescence was not quite as intense. When transpiration was inhibited by submerging leaves in water, the extrafascicular spread of BS was eliminated. Fluorescence developed more slowly when stomata were closed, and uptake was faster with younger leaves. Strugger (1939) suggested that BS accumulated at sites where cuticular and stomatal transpiration occurred and that fluorescence intensity was proportional to the amount of BS deposited at these sites. The possibility that the density of carboxyl groups was higher at the sites of exit of water was not considered as a possible factor involved in intensity of fluorescence.

Additional experiments were conducted by Strugger (1939) to test if sites exhibiting fluorescence were porous and permeable to BS. Cut stems were placed in BS solution, and after about 30 min of transpiration in light all leaves were stained with BS. A solution consisting of 50 $g\,l^{-1}$ gelatine, 0.9 $mol\,l^{-1}$ glucose and 0.1 $mol\,l^{-1}$ potassium rhodanide (KCNS) was prepared and cooled, and immediately before its solidification some leaves were covered with this mixture, which then quickly solidified. In this way a strong osmotic gradient was generated, which induced water flux into the gelatine. With KCNS, BS forms insoluble crystals having intense fluorescence. After just 10 min, fluorescing crystals of berberine rhodanide could be observed in the gelatine over cuticular ledges and at the bases of glandular trichomes. With young leaves, some fluorescent crystals were seen over anticlinal walls. These results clearly demonstrate that cuticular ledges of guard cells and basal cells of glandular trichomes of *Helxine* have aqueous pores large enough to allow passage of ionic BS.

Similar experiments were conducted using modern equipment for fluorescence microscopy. Leaves were floated on BS solutions for 2–3 h. The dye penetrated the cuticle, and produced intense fluorescence in cuticular ledges and anticlinal walls.

In Fig. 5.1a the anticlinal ledges are in focus, and fluorescence can be seen clearly. Figure 5.1b shows the same specimen with focus on cuticular ledges. In all three species, glandular trichomes also fluoresced intensely (cf. Fig. 5.1f).

Schönherr (1969) and Schlegel et al. (2005) used silver nitrate as an ionic tracer for localising aqueous pores. Drops of silver nitrate solutions were placed for 1 h on the surface of *Phaseolus vulgaris* and *Vicia faba* leaves, and treated areas of epidermis were viewed with the bright field microscope. Black silver precipitates were observed in anticlinal cell walls, in trichomes and in cuticular ledges (Fig. 5.2). This indicates again that these were sites of preferential penetration of $AgNO_3$. Using energy dispersive X-ray analysis (EDX) and *Vicia* leaves, it was shown that these precipitates were AgCl. This is strong evidence that silver nitrate penetrated the cuticle and was precipitated as AgCl in the apoplast of the leaf. These investigations show that ionic compounds penetrate the lipophilic cuticle. However, penetration was not uniform, and specific sites exist where electrolytes penetrate.

Leaves treated with Gilson fixative containing mercuric chloride ($HgCl_2$) exhibited crystalline precipitates of mercurous chloride in anticlinal walls and guard cells. Post-treatment with potassium iodide reduces these precipitates, and metallic mercury can be seen as black dots in the outer epidermal wall (Schönherr and Bukovac 1970a, b). These precipitates are arranged in rows over anticlinal walls, as in onion leaves (Fig. 5.3a). Adaxial leaf surfaces of *Convallaria* leaf (Fig. 5.3b) exhibit a more random distribution of mercury precipitates, but over veins rows of precipitates over anticlinal walls are seen. Dense precipitates are also evident in guard cells. Lightly brushing the leaf surface or rinsing it briefly with chloroform destroyed the typical pattern. After brushing, new rows of precipitates appeared along the tracks of the bristles. When cuticles were isolated enzymatically from onion leaves or onion bulb scales and mounted on gelatine or agar containing ascorbic acid as reducing agent, the precipitate pattern was the same as in epidermal strips. Clearly, precipitates are formed in the cell wall or in the agar at sites where the cuticle is selectively permeable to $HgCl_2$, provided the matrices contain a reducing agent which is needed for the formation of insoluble HgCl (Schönherr and Bukovac 1970a).

Aqueous $HgCl_2$ solutions have an extremely low electrical conductivity, as mercury and chlorine are bound by divalent rather than ionic bonds. $HgCl_2$ dissolves in water, alcohols, ether, benzene and other organic solvents (Falbe and Regitz 1995). Non-ionic $HgCl_2$ is not surrounded by a hydration shell. It is lipid soluble and does not depend on aqueous pores, as do berberine sulphate or silver nitrate. Yet the distribution pattern in the cell wall of berberine sulphate (Fig. 5.1) mercury (Fig. 5.2) and silver (Fig. 5.3) is very similar. This raises the question as to the nature of sites in cuticles selectively permeable to $HgCl_2$. The Gilson solution is very acidic (*p*H 1.0), and carboxyl groups or phenolic hydroxyl groups are certainly not ionised at this *p*H value, while amino groups are. Polypeptides occur in cuticles, albeit only in small amounts (Schönherr and Bukovac 1973; Schönherr and Huber 1977). Possibly, Hg precipitates mark sites with non-ionic functional groups such as hydroxyl, aldehyde, phenolic hydroxyl, and ether or ester bonds. Carboxyl groups would be non-ionised at these acidic *p*H values, but they may also be present along the $HgCl_2$ diffusion paths.

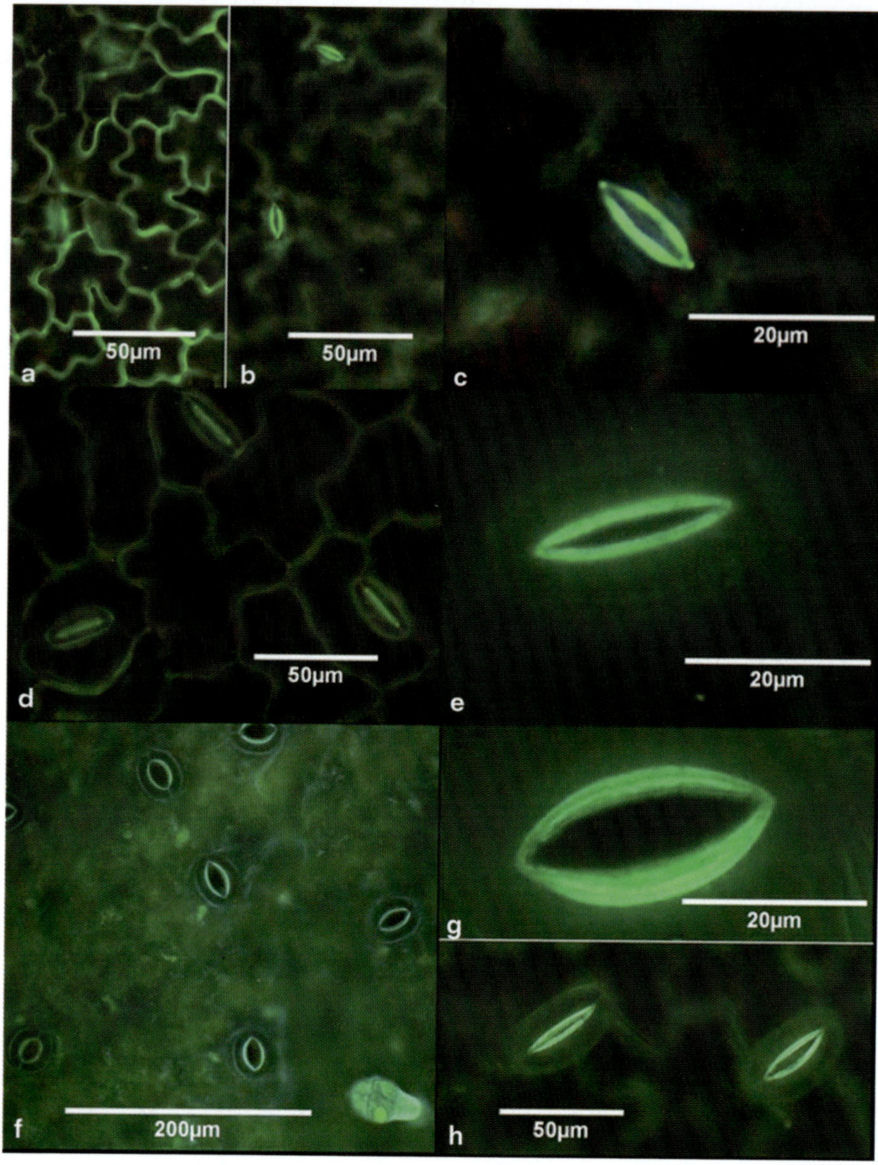

Fig. 5.1 Fluorescence micrographs of stomatous leaf surfaces of *Helxine soleirolii* (**a**–**c**), *Phaseolus vulgaris* (**d**, **e**) and *Vicia faba* (**f**–**h**) treated with berberine chloride ($1\,g\,l^{-1}$) for 2–3 h in light. The photographs were taken using a Zeiss Axioplan 2 microscope equipped with a Zeiss AxioCam digital camera and AxioVision 3.1 software (Zeiss, Germany). The Zeiss filter set No. 02 (excitation: G 365 nm; beam splitter FT 395 nm, excitation: LP 420 nm) was used. (Taken from Schönherr 2006)

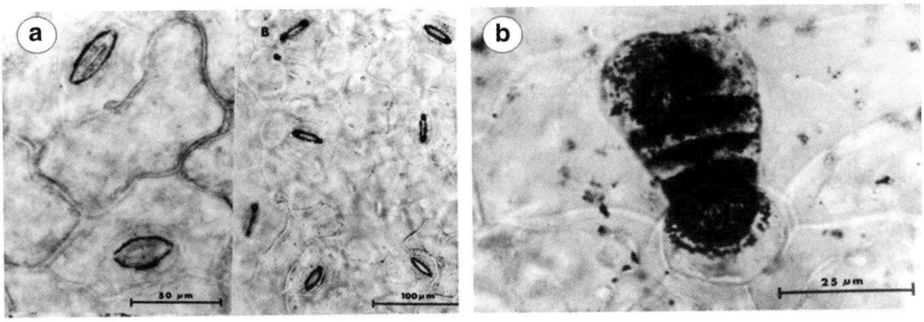

Fig. 5.2 Bright field micrographs of leaf surfaces of *Phaseolus vulgaris* after treatment with silver nitrate ($AgNO_3$) solution for 1 h. Sites of a strong silver precipitation are anticlinal cell walls, cuticular ledges and glandular trichomes. Data from Schönherr (1969)

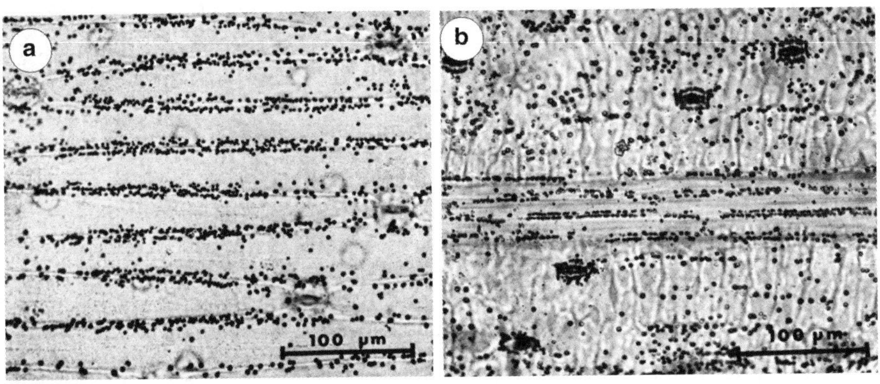

Fig. 5.3 Surface view of mercury precipitates in *Allium* (**a**) and *Convallaria* (**b**) leaf epidermal walls. (Taken from Schönherr and Bukovac 1970a)

Copious amounts of mercury precipitates can be found in anticlinal walls, while over periclinal walls mercury is rarely found. The effect of removing waxes on distribution pattern indicates that waxy domains are impermeable to $HgCl_2$, even though it dissolves in benzene. Perhaps crystalline wax domains are impermeable to $HgCl_2$, while amorphous waxes are not. This would imply that precipitates form in the cell wall wherever the cuticle is free of crystalline waxes. This apparently applies also to silver precipitates and distribution of berberine chloride (Figs. 5.1 and 5.2).

5.2 Experimental Methods

Permeability of cuticles to polar non-electrolytes (Sect. 6.2; Fig. 6.23) and electrolytes is very low and difficult to measure accurately. Quantitative measurements require maximum driving force—that is, concentrated solutions. This can be realised by applying small droplets to the outer surface of CM or to leaf discs. After evaporation of water, the solid salt residue is exposed to high humidity such that salts deliquesce. The point of deliquescence (POD) is defined as the humidity at which solid salt, saturated salt solution and relative humidity of the mixture are in equilibrium. POD depends only on temperature, but fortunately it varies little at physiological temperatures. For instance, humidity over a saturated solution of $CaCl_2 \times 6\ H_2O$ at 10°C, 15°C, 20°C, 25°C and 30°C is 38%, 35%, 32%, 29% and 26% (Kolthoff et al. 1969). If humidity is below POD the salt loses all its water and solidifies, and penetration ceases. With humidity higher than POD the salt takes up water and deliquesces, and penetration proceeds from a concentrated salt solution. Hence, accurate control of temperature and humidity of the salt residue are essential when studying cuticular penetration of salts.

Cuticular penetration of salts may be measured using isolated CM and the simulation of foliar penetration technique (SOFP) which is described in Sect. 6.4. CM are inserted into chambers shown in Fig. 5.4a. A finite dose of the radio-labelled ionic solute is placed as a small drop (5 μl) on the outer surface of the cuticle and allowed to dry. After drying, time-dependent penetration of the radio-labelled compound occurring from this finite dose on the cuticle surface is measured at controlled temperature and humidity by repeatedly changing the receiver and counting radioactivity. Non-linear rates of uptake are observed when relative amounts of penetration are plotted vs time. As an example, the penetration of ^{14}C-labelled isopropylamine salt of glyphosate (IPA glyphosate) across poplar CM *(Populus canescens)* at different humidities is shown (Fig. 5.5). Rates of penetration of the salt (slopes) are highest at the beginning and level off with time (Fig. 5.5a). This non-linearity is a consequence of the fact that the amount of IPA glyphosate in the deposit on the cuticle surface is decreasing in proportion to the amount which penetrated the cuticle. Hence, driving force decreases with time (Sect. 2.5). SOFP has great advantages, because the same set of CM can be used repeatedly with different humidities or solutes. Also, the time course can be followed for each CM individually, because penetration is followed by changing the receiver and assaying it for salt which has penetrated. By reference to Fig. 5.5a, it is clear that determining amount penetrated only at a single time interval is likely to lead to erroneous conclusions.

The effect of guard cells and glandular trichomes cannot be studied using isolated CM, because of the stomatal pores between guard cells. These holes would be filled with donor solutions and serve as parallel pathways for ionic solutes. In these cases, leaf discs must be used (Fig. 5.4b). Leaf disks are punched out and are inserted into the apparatus. The lower part is a hollow cylinder made of Teflon. It serves as a spacer to prevent contact of the leaf disk with water in the Petri dish. In this way, leaching of radio-labelled solutes from the leaf disk is prevented. The lid with an orifice in its centre is secured with silicon grease on the leaf disk, which prevents water loss from the cut edges of the disc. Ten of these penetration units are placed in a Petri dish lined with moist filter paper. A 5 μL droplet of donor solution is pipetted onto the leaf surface. After evaporation of the water, the Petri dish is closed and placed in a growth chamber either in light or in the dark. Humidity inside the Petri dish is 100% and it cannot be varied. This type of experiment has an

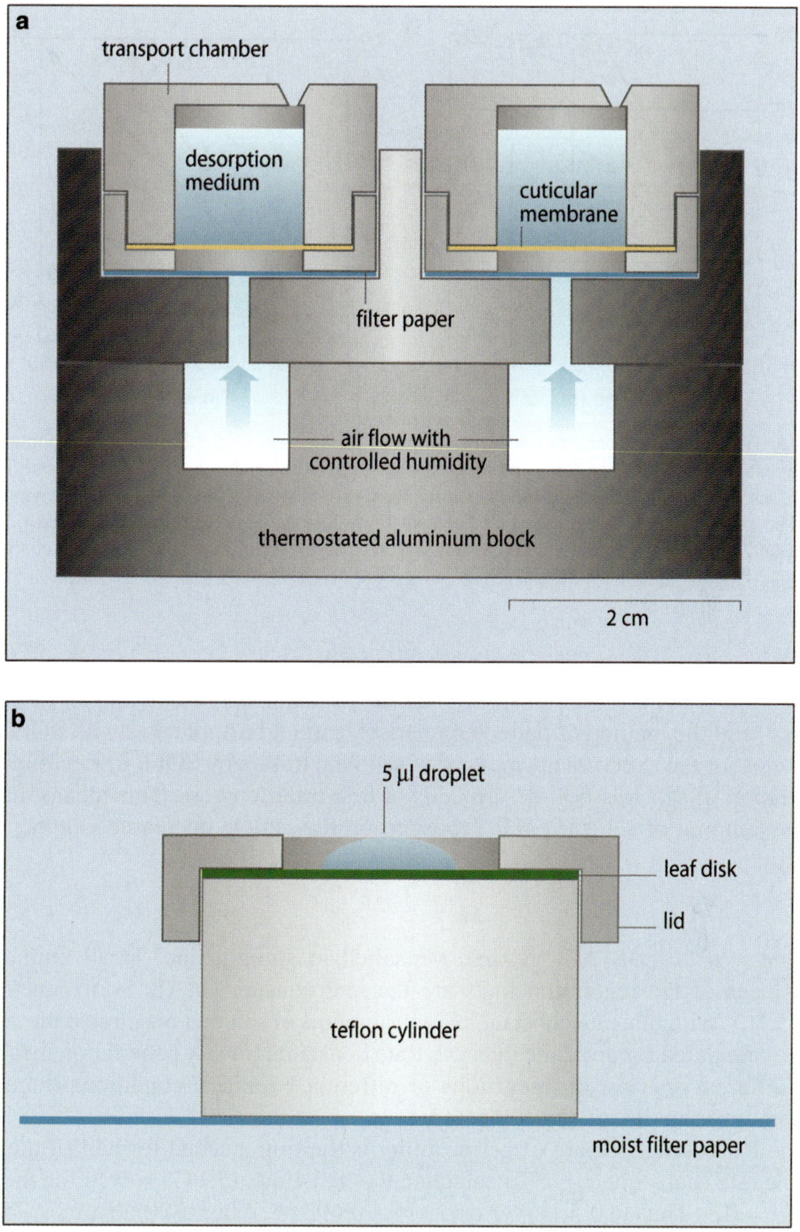

Fig. 5.4 **(a)** Transport chambers used for studying penetration of salts and other electrolytes across isolated CM. The chambers are positioned in wells of a thermostated aluminium block. Air of constant humidity is blown on the salt residues (*arrows*) at about ten air changes per second. Temperature and humidity can be varied (taken from Schönherr 2006). **(b)** Penetration unit used to study salt penetration into leaf disks. The leaf disk is positioned on a short hollow Teflon tube and covered with a stainless steel lid. Teflon and lid are coated with silicon grease. The units are placed on moist filter paper in Petri dishes. Humidity is always 100%

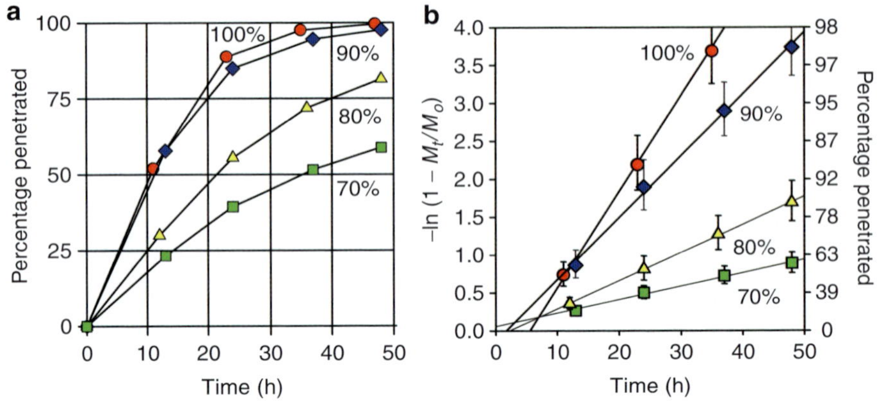

Fig. 5.5 Time course of penetration of the isopropylamine salt of glyphosate across poplar CM. **(a)** Percentage penetrated vs time. **(b)** Natural logarithm of the fraction of IPA-glyphosate left on the surface of the CM is plotted against time. Humidity (%) over the residue is shown on each plot. The same set of CM was used at different humidities. Donor solutions contained $0.2\,\mathrm{g\,l^{-1}}$ Glucopon 215 CSUP as wetting agent. Water served as receiver solution. Bars are 95% confidence intervals. (Redrawn from Schönherr 2002)

additional disadvantage. The leaf disk itself is the receiver, and repeated sampling is not possible. For each time interval a new set of leaf disks (usually 50 or 100) must be used, and the method of paired comparison cannot be applied as with SOFP.

In most of the experiments measuring salt penetration by SOFP it was found that penetration of salt was best described by a first-order process. This means that the relative amount of salt (M_t/M_0) deposited on the cuticle decreased exponentially with time:

$$\frac{M_t}{M_0} = 1 - \mathrm{e}^{-kt}. \tag{5.1}$$

Plotting $-\ln\,(1 - M_t/M_0)$ vs time t resulted in straight lines at all humidities. The slopes of the regression lines are the rate constants k (h^{-1}) of penetration (Fig. 5.5b). With this rate constant known, amounts of salt that penetrated the cuticle can be calculated for any time interval. Rate constants for salt penetration measured with different species, different ions or different boundary conditions (humidity, temperature) can directly be compared.

Another useful measure of permeability is the time needed for half of the dose to penetrate (half-time $t_{1/2}$). To calculate the half-time, (5.1) is solved for the case $M_t/M_0 = 0.5$. Thus $\ln 0.5 = -kt$ or $t_{1/2} = -0.693/k$. After two half-times, 75% of the salt has penetrated and after three half-times 87.5% has penetrated. For instance, the slopes at 70% and 100% humidity (Fig. 5.5b) are 0.056 and $0.127\,\mathrm{h}^{-1}$ and half-times are 12.4 and 5.5 h respectively. For 87.5% of the dose to penetrate, 37.2 or 16.5 h are required. Penetration at 100% humidity was 2.25 times faster than at 70%.

5.3 Cuticular Penetration of Electrolytes

5.3.1 Effect of Wetting Agents

Leaf surfaces are difficult to wet by water, and good contact of donor solutions with the waxy surface of the CM is essential. Rates of salt penetration could be significantly increased by adding small amounts of surfactants, which decrease surface tension (Fig. 5.6). Plantacare 1200P, APG 325 and Glucopon 215 CSUP are alkyl polyglucosides (Henkel, Düsseldorf, Germany). Average degree of polymerisation of glucose is about 1.4–1.5, and fatty alcohols linked to the sugar moieties have 8–14 carbon atoms. These surfactants have a good ecotoxicological profile, and phytotoxicity has not been observed.

At a concentration of $0.2\,g\,l^{-1}$, surface tension was reduced from 72 to 30–32 $mN\,m^{-1}$, and rate constants were increased by a factor of 12. There was no difference among the three surfactants. The wetting agent Glucopon 215 CSUP at a concentration of $0.2\,g\,l^{-1}$ was used in all subsequent experiments with CM and leaf disks.

Similar experiments were conducted using poplar CM and the isopropylamine (IPA) salt of glyphosate. In the control experiment, $0.2\,g\,l^{-1}$ of Glucopon 215 CSUP was added as wetting agent, and the rate constant was about $0.1\,h^{-1}$. If accelerator

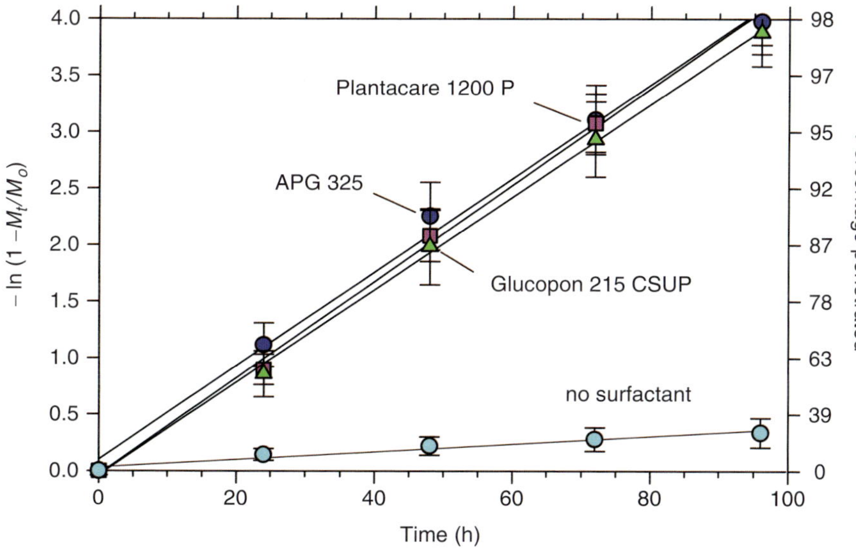

Fig. 5.6 Effects of wetting agents ($0.2\,g\,l^{-1}$) on rates of penetration of $CaCl_2$ across astomatous isolated CM of *Pyrus communis* leaves. A 5-μl drop containing $6\,g\,l^{-1}$ of $^{45}CaCl_2$ was added to the surface of the CM, and after drying, penetration was measured at 20°C and 90% humidity. Error bars are 95% confidence intervals. The same set of 100 CM was used as control (no surfactant) and with surfactants. (Redrawn from Schönherr 2001)

adjuvants (Genapol C-100, Ethomeen T25 or diethylsuberate) at rates of $4\,g\,l^{-1}$ were added to the IPA-glyphosate/Glucopon 215 CSUP, donor surface tension was not decreased further, but rate constants were about $0.125\,h^{-1}$. This slight increase was attributed to increased spreading of the droplet on the CM, which increased the area of CM in contact with the donor droplet. Accelerator adjuvants penetrate into wax and increase its fluidity. This increases permeability of CM of lipophilic solutes by orders of magnitude (Chap. 7). Such an effect was not observed with electrolytes, and this is good evidence that ionic solutes diffuse in aqueous pores and not in the waxy pathway.

5.3.2 Penetration of Calcium and Potassium Salts

For cuticular penetration of salts to occur, humidity of the salt must be above the point of deliquescence (POD). Each salt is characterised by a specific POD (Greenspan 1977). With all salts shown in Fig. 5.7, penetration graphs were linear, and rate constants were plotted against humidity. Rate constants measured with the various salts depended on the anions, which were not radio-labelled. The anions affected rate constants, because neither $^{45}Ca^{2+}$ nor $^{86}Rb^{+}$ can penetrate alone. The system must remain electrically neutral, and the fast ion pulls the slower one along. The entire salt penetrates, not individual ions.

Highest rate constants were observed at 100% humidity, and with inorganic calcium salts k increased from 50% to 100% humidity by a factor of almost

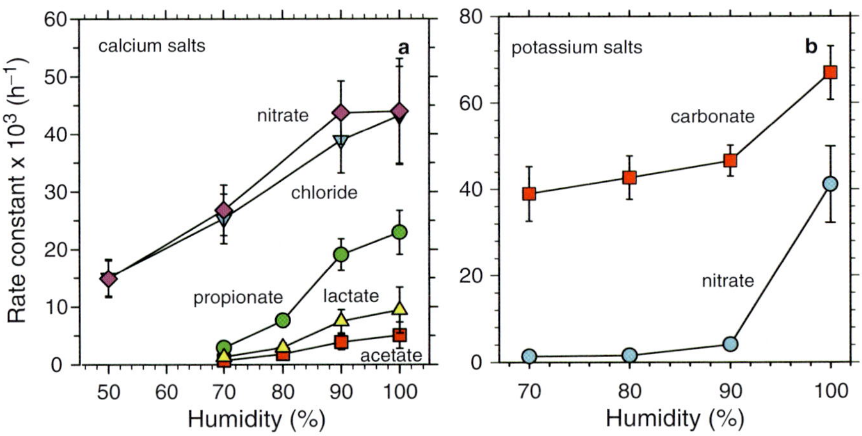

Fig. 5.7 Rate constants of penetration of calcium (**a**) and potassium salts (**b**) across astomatous isolated pear leaf CM (*Pyrus communis*). A 5-μl drop of $5\,g\,l^{-1}$ of the salt solution spiked with $^{45}Ca^{2+}$ calcium salts or $^{86}Rb^{-}$ potassium salts as tracers containing $0.2\,g\,l^{-1}$ Glucopon 215 CSUP as wetting agent was added to the outer surface of the CM. Temperature was 20°C and humidity over the salt residues was varied. Error bars are 95% confidence intervals. (Redrawn from Schönherr 2001, 2006)

three (Fig. 5.7a). POD of calcium chloride and calcium nitrate are 32% and 50%, respectively (Appendix B). Rate constants with the organic calcium salts were lower, and with acetate and lactate very little penetrated at 70% or 80% humidity. POD of these salts are much higher, and amounted to 95% (propionate), 95% (lactate) and 100% (acetate) respectively. At 80% humidity and below, all organic salt residues on the CM had a whitish appearance, indicating that they had crystallised. All these three salts should not have penetrated at 90% humidity and below, but significant rates were in fact measured. This is due to stagnant air layers directly over the CM. Since water constantly penetrates the CM from the receiver, this leads to thin water films between salt and CM. The POD of potassium nitrate (KNO_3) is 94% (Appendix B). As a consequence, KNO_3 penetration across the cuticle ceases at humidities below the 94% because the salt crystallises. However, potassium carbonate [$K_2(CO_3)_2$] having a POD of 45% penetrated at all humidities investigated (Fig. 5.7b).

Humidity has a dual function in cuticular penetration of salts. Humidity must be higher than POD for the salt to deliquesce. However, even above the POD penetration rates of the salts increased with increasing humidity (Figs. 5.5 and 5.7). The driving force for cuticular penetration is the donor concentration (Chap. 2). Inspection of residues shows that salt solutions are present on CM if humidity is above POD, and the salt is crystalline when humidity is below POD. However, the salt concentration is not known in these experiments, and for this reason $-\ln\,(1 - M_t/M_0)$ was plotted. Linearity shows that the salt fraction left on the CM decreased exponentially with time, and we accept this as evidence that the salt concentration also decreased exponentially with time. At high humidity the salt concentration is probably lower than at low humidity, and this implies that driving force is lower at high humidity. Nevertheless, rate constants increase with humidity, indicating permeability of CM did not depend on concentration. Concentrations decreased exponentially with time, and this did not depend on the magnitude of the initial amount or concentration (M_0 or C_0). Schönherr (2000) studied the effect of initial $CaCl_2$ in the donor droplets. At concentrations of 2, 4, 6 and $10\,g\,l^{-1}$ and 90% humidity, penetration plots were linear and rate constants were in fact independent of initial donor concentration and amounts of $CaCl_2$ deposited initially. At a concentration of $1\,g\,l^{-1}$, the rate constant decreased when more than 60% of the salt had penetrated. This was attributed to an inhomogeneous distribution of the salt residue on the cuticle.

The effect of humidity on rate constants indicates a change in permeability of CM. In Sects. 4.5 and 4.6 it was shown that water permeability and water content of CM increased with increasing humidity. This leads to an increase of the number of aqueous pores available for cuticular penetration of salts.

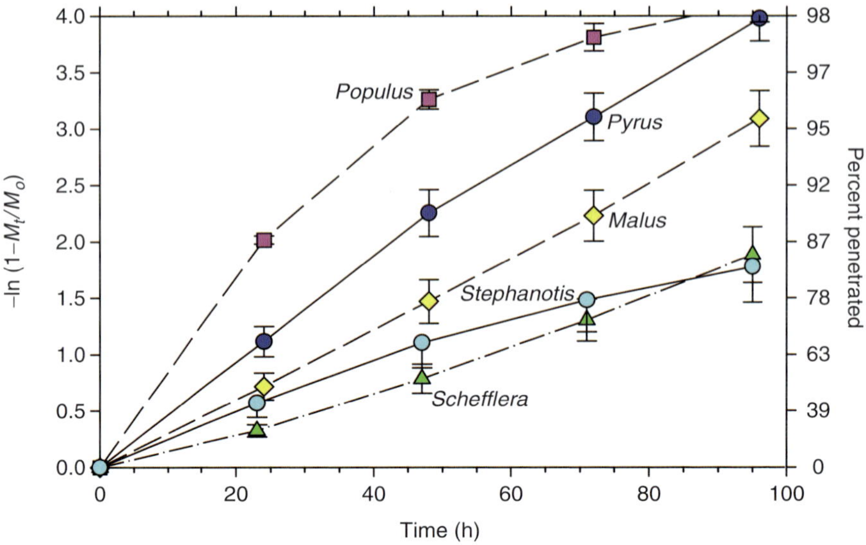

Fig. 5.8 Penetration of $CaCl_2$ across astomatous CM isolated from *Populus canescens*, *Pyrus communis*, *Malus domestica*, *Stephanotis floribunda* and *Schefflera actinophylla* leaves. A 5-µl donor droplet containing $5\,g\,l^{-1}$ $^{45}CaCl_2$ and 0.2 Glucopon 215 CSUP as wetting agent was applied to the outer surface of the CM and after droplet drying, penetration of $CaCl_2$ was measured at 20°C and 90% humidity. Error bars are 95% confidence intervals. (Redrawn from Schönherr 2000)

5.3.3 Rate Constants Measured with Leaf CM from Different Species

Rate constants for $CaCl_2$ penetration differed considerably among species, and not all penetration plots were linear to the end (Fig. 5.8), but linearity was maintained until about 80% of the dose had penetrated. *Populus* CM (grey poplar) had the highest rate constants ($8.4 \times 10^{-2}\,h^{-1}$), whereas lowest k ($1.4 \times 10^{-2}\,h^{-1}$) were measured with *Schefflera*. This corresponds to a factor of 6 between the two species, with lowest and highest permeability for $CaCl_2$. This shows that aqueous pores occur in CM of all of these species tested, but they may differ in number and/or size. These rate constants measured with $CaCl_2$ may be compared with water permeance (P_w) in Table 4.8. P_w of poplar CM ($26.8 \times 10^{-10}\,m\,s^{-1}$) was 5.7 times higher than P_w of *Stephanotis* CM ($4.7 \times 10^{-10}\,m\,s^{-1}$). For rate constants of $CaCl_2$ penetration, this ratio is only 2.9. This difference might be related to the fact that water also diffuses across the waxy pathway, which is excluded for $CaCl_2$. Too much should not be made of this comparison, however, because different lots of CM were used for estimating P_w and k for $CaCl_2$.

Another factor should be pointed out which is not visible in Fig. 5.8. Reproducible data and reasonably short error bars are obtained only with very large sample sizes of 50–100 CM. Rate constants measured with individual CM vary greatly, and in each population cuticles occurred which were practically impermeable. There are indications that frequency of aqueous pores in cuticles decreases during leaf expansion and maturation. Permeability to an Fe-chelate of *Vitis vinifera* and *Prunus persica* leaves decreased dramatically during leaf expansion (Schlegel et al. 2006). The upper astomatous leaf cuticle of poplar leaves can be isolated only from leaves which are just fully expanded. With these leaves, relatively high rate constants were measured for $CaCl_2$, while permeability of the upper cuticle of older leaves is almost zero (unpublished data). These two facts indicate a considerable dynamic in development of aqueous pores and their closure in older leaves.

5.3.4 Size Selectivity of Aqueous Pores

Citrus MX membranes were almost impermeable to raffinose (Fig. 4.9), having a molecular radius of 0.65 nm. The pentahydrate of raffinose has a molecular weight of 595 g mol^{-1}. Estimates of pore radii of other species are not available. Cuticles over anticlinal walls and in cuticular ledges are permeable to berberine sulphate, which has a molecular weight of 769 g mol^{-1}. In Sect. 4.5 we argued that diffusion of solutes in aqueous pores is increasingly hindered as solute size approaches pore size. Size selectivity of aqueous pores in grey poplar CM (Schönherr and Schreiber 2004b) and *Vicia faba* leaf disks (Schlegel et al. 2005) has been studied using calcium salts ranging in molecular weights from 111 to 755 g mol^{-1}.

Anhydrous molecular weight was used in the original literature, since the hydration numbers of these salts were not known. Size selectivity of the lipophilic or waxy pathway has been characterised using equivalent molar volumes (V_X) calculated according to McGowan and Mellors (1986). When size selectivity of aqueous pores and of the waxy pathway are to be compared, the same variable for size should be used. V_X of calcium salts was estimated based on a plot V_X vs molecular weight (Fig. 5.9). The training set was composed of aliphatic and aromatic organic molecules containing C, O and H atoms. With this set, V_X was smaller than the molecular weight by a factor of 0.895. This factor was used to estimate V_X of calcium salts (squares). A direct calculation of V_X was not possible since the equivalent molar volume of Ca^{2+} was not available.

Rate constants of penetration decreased with increasing molecular weight of the Ca^{2+} salts (Fig. 5.10), showing again coupling of cation and anion fluxes. Each divalent calcium ion must be accompanied by two monovalent anions. Aqueous pores in all cuticles were large enough to accommodate Ca-lactobionate having a molecular weight of 755 g mol^{-1}. A good linear correlation was obtained when log k was plotted against molecular weights (solid lines) or molar volumes (dashed lines). With *Populus* CM, rate constants decreased from 0.199 to 0.0285 h^{-1} when molecular weight increased from 100 to 500 g mol^{-1}. Rate constants measured with *Vicia* leaf disks were higher in light than in the dark, but slopes were similar. Broad bean leaves have stomata and glandular trichomes on both leaf surfaces, while poplar adaxial CM are astomatous and at the time of isolation unicellular trichomes have already all been shed. The cuticle over guard cells and glandular trichomes is probably

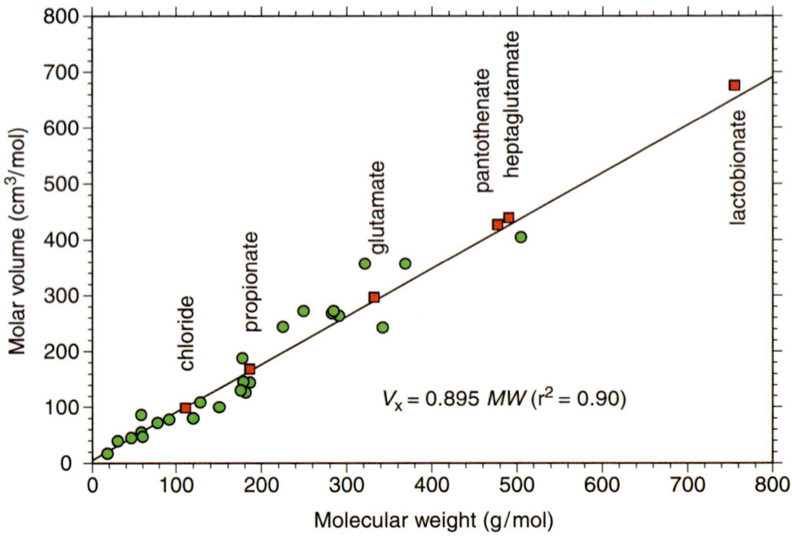

Fig. 5.9 Molar volumes (V_x) of a set of aliphatic and aromatic organic compounds plotted against their molecular weights (*green circles*). The slope (0.895) was used to estimate V_x of calcium salts (*red squares*) used in Fig. 5.10

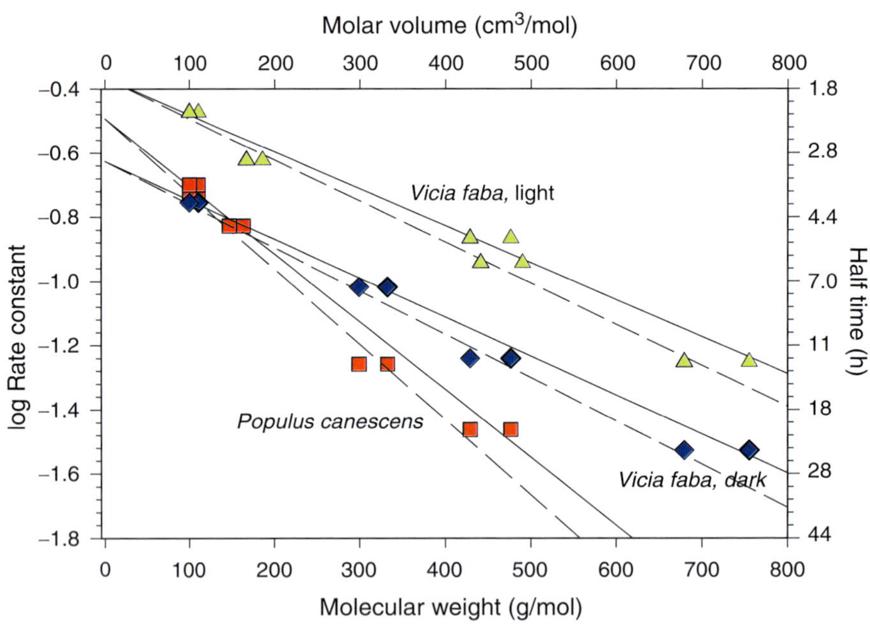

Fig. 5.10 Effect of molecular weight (*solid lines*) or molar volumes (*dashed lines*) of salts on rate constants and half-times of penetration of Ca salts into adaxial epidermis of *Vicia faba* leaf disks and across CM of *Populus canescens* at 20°C and 100% humidity. (Redrawn from data of Schönherr and Schreiber 2004a, b and Schlegel et al. 2005)

Table 5.1 Parameters of the regression lines shown in Fig. 5.10 and rate constant (k) calculated for salts having molar volumes of 100 or 500 $cm^3\,mol^{-1}$

Species	$\log k_0 (h^{-1})$	$\sigma' (mol\,cm^{-3})$	$k(100)\ (h^{-1})$	$k(500)\ (h^{-1})$	$\frac{k(100)}{k(500)}$
Vicia light (MW)	−0.37	-1.15×10^{-3}	–	–	–
Vicia light (V_x)	−0.37	-1.28×10^{-3}	0.32	0.098	3.3
Vicia dark (MW)	−0.63	-1.21×10^{-3}	–	–	–
Vicia dark (V_x)	−0.63	-1.35×10^{-3}	0.17	0.050	3.4
Populus (MW)	−0.49	-2.11×10^{-3}	–	–	–
Populus (V_x)	−0.49	-2.34×10^{-3}	0.19	0.022	8.7

$\log k_0$ is the y-intercept, and σ' corresponds to the slope of the regression lines in Fig. 5.10

responsible for the difference in rate constants between broad bean and poplar. This is in accord with preferential penetration of berberine chloride at these sites (Fig. 5.1).

Slopes and y-intercepts of the straight lines which quantify selectivity of poplar and broad bean cuticles (Fig. 5.10) have physical meaning. The y-intercepts are the log k_0 values for a hypothetical compound of zero molecular weight or molar volume (cf. Sect. 6.3.2, subsection: "Variability of Solute Mobility with Size of Solutes"). The slopes (σ') of the lines characterise size selectivity. These parameters are summarised in Table 5.1.

These k_0-values characterise permeability of cuticles from different species and at different experimental conditions. With broad bean (dark) k_0 is $0.23\,h^{-1}$, while with poplar CM lacking stomata the k_0 was $0.32\,h^{-1}$. Since slopes also differ, the difference between species increases with molar volume of solutes (Fig. 5.10). Permeability of broad bean in light ($k_0 = 0.40\,h^{-1}$) was 1.7 times higher than in the dark. As was already pointed out, the difference indicates that cuticles over stomata are more permeable when stomata are open (cf. Fig. 6.13). As size selectivity was not affected by light, higher permeability appears to be due to an increased number of pores. Since k_0-values mark permeability of a salt having zero molar volume or molecular weights, they are the same no matter which size variable is used in calculation.

Slopes of the lines (σ') are negative, which implies that log k decreases by the factor $\sigma' \times V_x$. Size selectivity of aqueous pores was larger with poplar CM than with bean leaf cuticles. Size selectivity was not very pronounced, because increasing molar volume by a factor of 5 decreased rate constants by factors of only 3.3–8.6 (Table 5.1). Size selectivity of the lipophilic path (β') in cuticles has been studied [(6.21) and (6.23); Table 6.8] and was found to be the same with all species investigated. Average sized dependence β' is 9.5×10^{-3} mol cm^{-3}. Thus, size discrimination of lipophilic solutes in the waxy pathway is 4–7.4 times larger than discrimination of ionic solutes in aqueous pores.

Rate constants are proportional to permeance (2.26), and permeance depends on the diffusion coefficient, partition coefficient and on thickness of the membrane (2.18). For a given species, the thickness of the cuticle or the lengths of the aqueous pores are the same for all solutes. Partition coefficients should be around 1.0, since the salts are dissolved in the aqueous phases of the donor and the pore fluid. There might be a slight decrease with increasing molar volume, but there are no data available. Diffusion coefficients decrease as molecular weights or molar volumes increase (4.16), and this is more pronounced in narrow pores than in bulk liquid. Hence, size-dependent rate constants measured for the cuticles of different species mainly reflects differences in diffusion coefficients, thus they are mobility parameters. Rate constants observed for different plant species probably also depend on thickness of cuticles and path lengths.

5.3.5 Penetration of Organic Ions and Zwitter Ions

In addition to inorganic ions and their chelates, important for foliar nutrition, there are organic ions used as plant-protecting agents. One of the best known herbicides is glyphosate, which is sprayed on the foliage as an organic ion. Glyphosate is characterised by two acidic and one basic ionisable group (Schönherr and Schreiber 2004a, b), and for foliar application it is formulated as sodium, potassium, isopropyl amine or trimesium salt (Tomlin 1997). This affects water solubility and molecular weights of glyphosate. Betaine, a stress metabolite, and proline, an amino acid are both Zwitter ionic species which form inner salts. Putrescine is a biogenic polyamine having growth-regulating activity. Permeability of *Populus canescens* CM to all of these compounds increased with increasing humidity (Schönherr 2006) and depended on POD (Fig. 5.11). Putrescine has a POD around 77%, and at 70% humidity penetration was practically zero. The calcium salt of glyphosate needs 100% humidity to deliquesce, and rate constants were zero below 100% humidity. Thus, these organic ionic species respond to humidity similarly as KNO_3 (Fig. 5.7b). All other compounds had their POD below 70%, and their rate constants at 70% humidity amounted to about 15–35% of the maximum values measured at 100% humidity. The increase in rate constants with humidity if above POD again indicates that permeability of CM increases with humidity.

Data presented have practical significance. Size dependence of permeation in aqueous pores is smaller than in the lipophilic pathway. Unfortunately, for stomatous leaves we have data only for broad bean, and additional species need to be investigated. In the case that other important crop plants or weeds respond in a similar fashion, it would be advantageous to formulate weak electrolytes as salts, especially if they have a high molecular weight. However, if this strategy is followed, it should be remembered that calcium salts of carboxylic acids generally have very high POD, and if the anionic species is monovalent two anions are needed to maintain electroneutrality; this nearly doubles molecular weight. Glyphosate is a good case in point (Fig. 5.11). The rate constant for the Ca-salt was much lower than with the potassium salt, and since POD of Ca-glyphosate is 100% (Fig. 5.11) herbicidal activity of glyphosate may be reduced when sprays are prepared using hard water (Schönherr and Schreiber 2004a, b). In the experiments depicted in Fig. 5.12, $CaCl_2$ and Ca-gluconate were included as reference chemicals, and excellent linearity

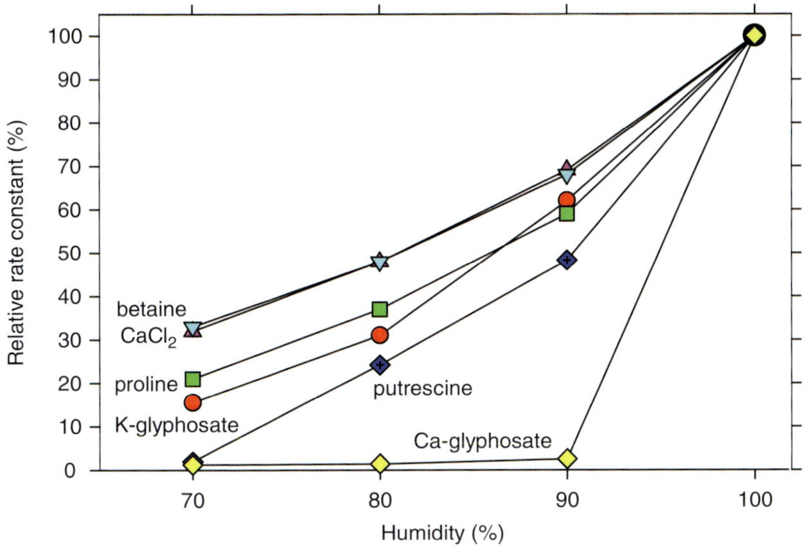

Fig. 5.11 Permeability of *Populus canescens* CM to selected organic ions and to $CaCl_2$ as affected by humidity and POD. Donor solutions contained $0.2\,g\,l^{-1}$ Glucopon 215 CSUP as wetting agent; solute concentration was $5\,g\,l^{-1}$ and temperature 20°C. Maximum rate constants at 100% humidity were $0.99\,h^{-1}$ (putrescine), $0.66\,h^{-1}$ ($CaCl_2$, proline and betaine) and $0.33\,h^{-1}$ (IPA-glyphosate). (Redrawn from Schönherr 2006, and Schönherr and Schreiber 2004a, b)

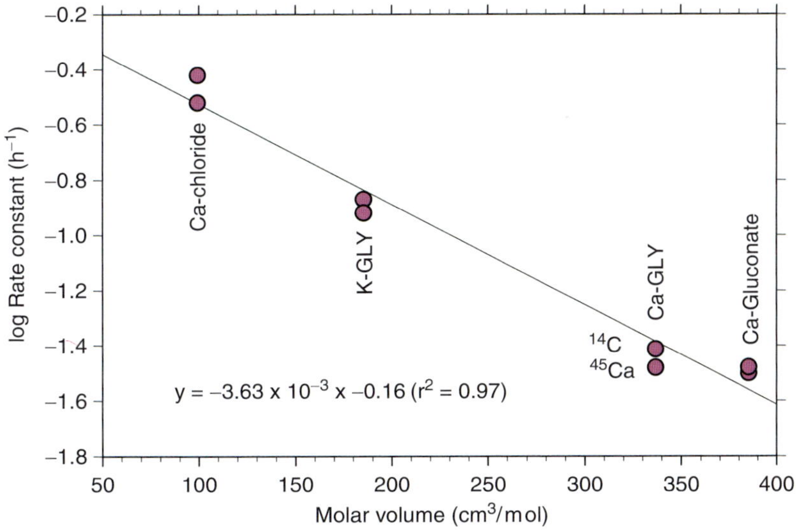

Fig. 5.12 Effect of molar volume of selected ionic compounds on rate constants of penetration across *Populus canescens* CM at 20°C and 100% humidity. Donor solutions contained $5\,g\,l^{-1}$ solutes and $0.2\,g\,l^{-1}$ Glucopon 215 CSUP as wetting agent. The same set of CM was used for all experiments, and duplicate determinations were made with all compounds. The calcium salt of glyphosate was labelled either as ^{45}Ca-glyphosate and or as Ca-^{14}C-glyphosate. The slope of the regression line is $3.63 \times 10^{-3}\,mol\,cm^{-3}$ (Schönherr and Schreiber 2004a)

was observed. The slope of the regression line is $-3.63 \times 10^{-3}\,\mathrm{mol\,cm^{-3}}$, and the y-intercept was -0.16. Both values are larger than the values for the lot of poplar CM shown in Table 5.1. For this reason the same lot of CM should be used to compare different ionic solutes and test dependence on molar volumes of rate constants. By reference to Fig. 5.12, it is clear that duplicate determinations using the same lot of CM give very similar results, and rate constants are the same within experimental error if the Ca-glyphosate is labelled with $^{45}Ca^{2+}$ or with ^{14}C-glyphosate.

Many growth regulators such as abscisic acid, indolacetic acid, 2,4-D, NAA and gibberellic acids are carboxylic acids. Their Ca^{2+} salts have molar volumes almost twice as high as their K^{+} salts, and their POD values are close to 100%. This is no problem in scientific experiments, where deionised water can be used. But in the field, sprays are prepared with water from wells or rivers, and if this water contains Ca^{2+} ions these weak acids form calcium salts and biological activity may be greatly reduced or lost. This problem is much greater than with Ca-glyphosate, because the above growth regulators are used at very low concentrations of 100–500 ppm.

5.4 Cuticular Penetration of Fe Chelates

Many important crops (peach, pear, *Citrus*) develop chlorotic leaves when grown on calcareous soils in which Fe^{3+} occurs as insoluble carbonate or hydroxide. Foliar sprays of ferric chelates are recommended as remedy. Chelates must be used because inorganic Fe^{3+} salts are stable only at pH 1, which is highly phytotoxic (Fernández and Ebert 2003, 2005). However, effects of such treatments are often poor and variable. Penetration of Fe chelates across *Populus* CM (Schönherr et al. 2005) and into leaf disks of different plant species (Schlegel et al. 2006) turned out surprisingly different compared to Ca^{2+} and K^{+} salts. Penetration of radioactive $^{59}FeCl_3$ could hardly be measured due to rapid formation of insoluble hydroxides on the leaf surface. Penetration of various ^{59}Fe chelates [FeEDTA, FeIDHA (Fe imidodisuccinic acid), FeEDDHA (Sequestren 138Fe), natrel (Fe ligninsulfonic acid), ferric citrate and ammonium ferric citrate] was measurable, but it immediately ceased at humidities below 100%. Compared to penetration of Ca salts, half-times of penetration of all Fe chelates were significantly lower and did not depend on molecular weights. This contrasts to results obtained with all other ionic compounds (Figs. 5.10 and 5.12) and lipophilic solutes (Sect. 6.3.2, subsection: "Variability of Solute Mobility with Size of Solutes"). With concentrations of FeIDHA and FeEDTA increasing ten-fold from 0.002 to $0.02\,\mathrm{mol\,l^{-1}}$, rates of penetration of both Fe chelates decreased three- to four-fold (Fig. 5.13). A similar decrease of about five-fold was observed in $^{45}CaCl_2$ penetration when non-radioactive FeEDTA chelates were added to $CaCl_2$ at 0.01 and $0.02\,\mathrm{mol\,l^{-1}}$ (Fig. 5.13). This decrease of Fe chelate penetration at increasing concentration, and the effect on Ca penetration, is surprising and difficult to explain. Within experimental error, the slopes in Fig. 5.13 are not different, and this indicates structural changes in the CM which do not depend on the ions investigated. Fe chelates apparently lead to deswelling or

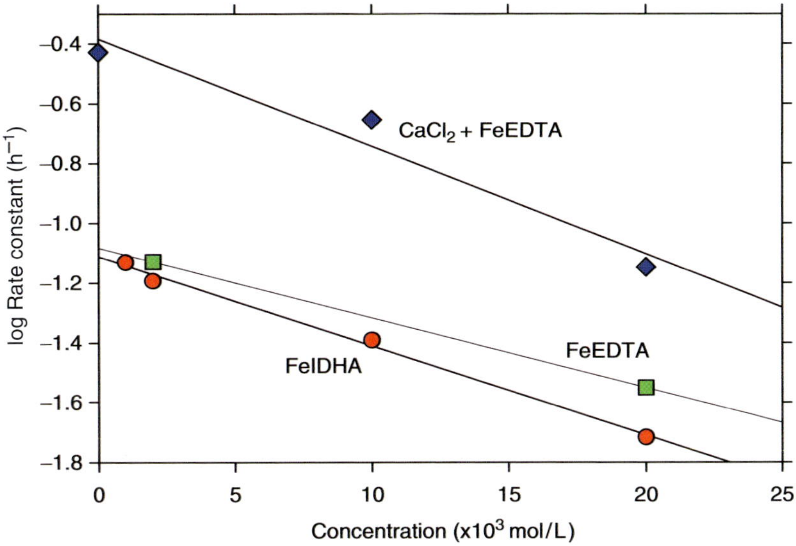

Fig. 5.13 Penetration at 100% humidity of ^{59}FeIDHA and ^{59}FeEDTA across *Populus* CM at concentrations increasing from 0.002 to 0.02 mol l^{-1}, and penetration of $^{45}CaCl_2$ in the presence of non-radioactive FeEDTA. Glucopon 215 CSUP was added to the donor solutions. Temperature was 20°C. The same set of CM was used for all experiments. (Redrawn from data of Schönherr et al. 2005)

blockage of aqueous pores in the *Populus* CM, serving as polar path of transport for both Fe chelates and Ca salts, and as a consequence rates of penetration decrease.

Leaf age greatly affected penetration of Fe chelates. Highest rates of penetration were observed with young unfurling leaves, whereas penetration into fully expanded leaves of different species was hardly measurable. These results show that young growing leaves should be sprayed, and humidity must be 100%. Hence, spraying in the evening is recommended when due to decreasing temperatures the dew point may be reached. Fe chelates are destroyed by UV radiation, which requires spraying after sun set (Schönherr et al. 2005).

Problems

1. Why is the cuticular permeability of polar compounds very low?
2. Why is the diffusion of ionic solutes in the lipophilic cutin and wax phase impossible?
3. Which are preferential sites in the leaf surface where aqueous pores are located?
4. What is the half-time of cuticular penetration $t_{1/2}$ of a Ca^{2+} salt having a rate constant k of $-0.0001\,s^{-1}$?
5. How does relative humidity affect the penetration of a salt across the cuticle?

6. How is cuticular penetration across grey poplar CM affected when the molecular volumes of Ca^{2+} salts increase from 100 to $500\,cm^3\,mol^{-1}$?
7. How does light affect cuticular penetration of Ca salts across leaf surfaces of *Vicia faba*?

Solutions

1. The partition coefficient of polar compounds (e.g., sugars) between the external aqueous phase and the lipophilic cutin and wax phase is very low. Consequently, according to (2.18) permeance P is very low.
2. Ionic solutes strongly bind hydration water, and this renders them insoluble in lipophilic phases like cutin and wax.
3. Trichomes, stomatal ledges and anticlinal cell walls are sites of the leaf surface where aqueous pores are preferentially located.
4. For this problem, (5.1) must be solved for $M_t/M_0 = 0.5$. The half-time of penetration $t_{1/2}$ is 6,931 s or 1.92 h.
5. Humidity affects cuticular penetration of salts in two different ways. It interacts both with the salt deposit on the cuticle surface and with the cuticular membrane. (a) Humidity must be higher than the POD of the salt, otherwise the salt crystallises and it becomes immobile. The salt dissolves only above the POD. (b) With increasing humidity, more water is sorbed to polar functional groups in the cutin matrix, and this increases the number of aqueous pores, which in turn leads to increased rates of salt penetration.
6. Based on the regression equation shown in Fig. 5.12, rate constants decrease from 0.30 to $0.011\,h^{-1}$, which is a factor of 27.
7. Size selectivity is not affected (Table 5.1), but rate constants of penetration are higher in light ($0.32\,h^{-1}$) than in the dark ($0.17\,h^{-1}$). This is attributed to an increased number of aqueous pores in stomatal ledges in light when stomata are open.

Chapter 6
Diffusion of Non-Electrolytes

6.1 Sorption in Cuticular Membranes, Polymer Matrix, Cutin and Waxes

Chemicals are characterised by their physicochemical properties such as molecular weight, volatility, ionisation constants, water solubility and solubility in lipids. *Agrochemicals* and environmental xenobiotics are often lipophilic, and these compounds are sorbed and accumulate in lipophilic compartments of the environment, such as the organic fraction in soil, the storage lipids of man, animals or plants. Rates of diffusion across cuticles and other membranes are proportional to the partition coefficient (2.17) of the compound, which is a measure for differential solubility in lipids and water (2.12). Thus, when analysing sorption in and penetration through cuticles, partition coefficients for the substances must be known or measured.

6.1.1 Definition and Determination of Partition Coefficients

The octanol/water partition coefficient (K_{ow}) of a compound is most commonly used. It relates the equilibrium concentrations in n-octanol and water (6.1). Since concentration is the ratio of amount (M) over mass (m), this equation can be written in terms of ratios of amounts and masses. Molal concentrations $(\mathrm{mol\ kg^{-1}})$ are used for both phases, because they do not depend on temperature. These units cancel, and partition coefficients are dimensionless. By convention, concentration in lipid phases are in the numerator:

$$K_{\mathrm{ow}} = \frac{C_{\mathrm{octanol}}(\mathrm{mol\ kg^{-1}})}{C_{\mathrm{water}}(\mathrm{mol\ kg^{-1}})} = \frac{\frac{M_{\mathrm{octanol}}}{m_{\mathrm{octanol}}}}{\frac{M_{\mathrm{water}}}{m_{\mathrm{water}}}} = \frac{M_{\mathrm{octanol}}}{M_{\mathrm{water}}} \times \frac{m_{\mathrm{water}}}{m_{\mathrm{octanol}}}. \tag{6.1}$$

K_{ow}-values have been measured and tabulated for a large number of substances (Hansch and Leo 1979; Leo et al. 1971; Sangster 1997). A partition coefficient of

1.0 indicates that solubility is the same in both phases. Compounds having partition coefficients >1 are lipophilic, and they are hydrophilic when $K_{ow} < 1$. Partition coefficients vary by several orders of magnitude, and to avoid exponents log values (e.g., log K_{ow}) are used.

Partition coefficients can also be determined for solid lipid phases such as fats, waxes or cuticles. Solutes sorbed in a solid constitute a solid solution (Fig. 2.6). Sorption sites in a solid are finite, and for this reason partition coefficients decrease with increasing concentration in the aqueous phase (Riederer and Schönherr 1986a).

Partition coefficients CM/water (K_{cw}) can easily be determined using radiolabelled solutes. Isolated CM are equilibrated with an aqueous solution of a radiolabelled compound at constant temperature. After equilibration, the amounts of radioactivity in water are determined by scintillation counting. The amounts of radioactivity in the CM can be calculated from the decrease in the concentration in water after equilibration. If this method is used, the drop in concentration should be large, especially when radiochemical purity is less than 99%. It is better to determine radioactivity in both phases after equilibration. Care should be taken to remove water films adhering to the cuticles by blotting with soft tissue paper. Cuticles are thin, often only about 3 μm thick or less. The inner surface of the cuticles is easy to wet, and water films cannot be avoided. If partition coefficients are >10, liquid films of the same mass as the cuticular materials introduce little error. The error can be estimated and corrected by weighing the CM after blotting and after air drying. With the weight of both phases (CM and water) known, K_{cw} can be calculated according to (2.12). Similarly, partition coefficients can also be measured with MX (K_{mxw}), cutin (K_{cuw}) and cuticular waxes (K_{ww}) as solid lipid phases.

Time needed to establish equilibrium depends on diffusion coefficients in cuticles and on membrane thickness. Equilibration usually takes only a few hours (Sect. 6.3, Figs. 6.13 and 6.14). However, with compounds carrying a carboxylic group (e.g., 2,4-D) and cuticles with epoxyfatty acids (Chap. 2), it was observed that partition coefficients slowly but constantly increased for many days (Riederer and Schönherr 1986b). This was due to formation of ester bonds between reactive epoxide groups and carboxyl groups of 2,4-D. Hence, partition coefficients increased with time and were overestimated. This problem can be avoided by washing the CM in 1.5 M HCl, which converts epoxy groups in vicinal hydroxyl groups.

6.1.2 Cuticle/Water Partition Coefficients K_{cw}

K_{cw} have been measured for 2,4-D and a series of cuticles isolated from ten different species (Riederer and Schönherr 1984). With leaf CM, partition coefficients ranged from 240 to 470, and with fruit CM the range was 424–579, depending on plant species. Mean K_{cw} of six leaf CM (*Clivia, Ficus, Citrus, Hedera, Pyrus* and *Olea*) was 316. The mean K_{cw} of four fruit cuticles (*Lycopersicon, Capsicum, Solanum* and *Cucumis*) was 476, which is about 50% higher. Based on this study, leaf CM from

Citrus and *Ficus* and tomato and pepper fruit CM cover the full range of partition coefficients observed with 2,4-D.

In the same study, K_{mxw} and K_{cuw} for 2,4-D were also measured. Both partition coefficients were larger than K_{cw}. These CM contain different amounts of wax (Table 1.1), which is either sorbed in the MX or deposited as epicuticular wax. K_{mxw} were 20–160% higher than K_{cw}, depending on species. There was a positive correlation between increase in partition coefficients on extraction of waxes and the weight fraction of waxes in cuticles. This indicates that additional sorption sites became available after wax extraction. K_{ww} are considerably lower than K_{cw} (see below), and this most likely contributed to the difference between K_{cw} and K_{mxw}. Variability of K_{cw} among plant species could also be caused by differences in the cutin fraction. With most species K_{cuw} values were higher than K_{mxw}, but variability between species did not disappear when polar polymers were hydrolysed and eliminated from the MX. This indicates that sorptive properties of cutin from different species are not the same. This is not too surprising, since in some species cutin is composed of two fractions (Sect. 1.2), ester cutin and non-ester cutin (cutan), and their sorptive properties probably differ.

Using cuticles isolated from *Lycopersicon* and *Capsicum* fruits and from *Ficus* and *Citrus* leaves, K_{cw} of eight different organic chemicals was measured (Kerler and Schönherr 1988a). Log K_{cw} values ranged from 2 to 8 and variability between species was small, compared to the large differences in log K_{cw} values between different compounds (Table 6.1). The mean log K_{cw} calculated from all values of these four species were very similar to log K_{ow} values (Table 6.1). These results show that plant cuticles are very efficient sorbers for lipophilic environmental chemicals, and non-volatile lipophilic compounds accumulate from the environment over time (Schönherr and Riederer 1989). K_{mxw} for these eight compounds and four plant species were all slightly higher than log K_{cw} values (Kerler and Schönherr 1988a).

Table 6.1 Cuticle/water partition coefficients (log K_{cw}) measured with the CM of *Citrus aurantium, Ficus elastica, Lycopersicon esculentum* and *Capsicum annuum* and eight substances (log K_{cw} and log K_{ow}). Data taken from Kerler and Schönherr (1988a)

Substance	log K_{cw} *Citrus*	log K_{cw} *Ficus*	log K_{cw} *Lycopersicon*	log K_{cw} *Capsicum*	log K_{cw} mean	log K_{ow}
4-NP	1.79	1.80	1.89	1.97	1.87	1.92
2,4-D	2.47	2.50	2.63	2.76	2.61	2.50
AT	2.15	2.16	2.12	2.19	2.16	2.64
2,4,5-T	3.13	3.11	3.19	3.21	3.16	3.40
PCP	4.42	4.55	4.57	4.66	4.56	4.07
HCB	5.70	5.74	5.83	5.80	5.77	5.47
PER	6.45	6.20	6.50	6.55	6.36	6.50
DEHP	7.22	7.28	7.32	7.48	7.34	7.86

4-Nitrophenol (4-NP), 2,4-dichlorophenoxyacetic acid (2,4-D), atrazine (AT), 2,4,5-trichlorophenoxyacetic acid (2,4,5-T), pentachlorophenol (PCP), hexachlorobenzene (HCB), perylene (PER) and diethylhexylphthalate (DEHP)

6.1.3 Wax/Water Partition Coefficients K_{ww}

Waxes greatly contribute to barrier properties of the CM (Sects. 4.6 and 6.3), and solubility of penetrants in cuticular wax is important when analysing transport properties of cuticles. Wax/water partition coefficients K_{ww} have been determined using reconstituted cuticular waxes extracted from leaves of *Hordeum vulgare* and from isolated cuticles of *Prunus laurocerasus, Ginkgo biloba* and *Juglans regia* (Burghardt et al. 1998; Schreiber and Schönherr 1992a; Kirsch et al. 1997). Cuticular wax was extracted using chloroform and was recrystallised on thin aluminium disks (Sect. 9.6). Wax amounts ($200 \pm 30\,\mu g$) were calculated from differences in weight of the aluminium disks before and after applying the wax. Wax samples were incubated in solutions of the radio-labelled compounds, and radioactivity in both phases was determined after equilibration.

K_{ww} values were always significantly lower than K_{cw}, by factors of 3–10 (Table 6.2). Cuticular waxes are mixtures of linear long-chain aliphatic molecules (Table 1.3) and in some species pentacyclic triterpenoids are abundant (Fig. 1.4). These waxes are solids at physiological temperatures, and about 20–50% of the waxes are crystalline at room temperature (Reynhardt and Riederer 1991, 1994). Crystals are impermeable and inaccessible for solutes. Only the amorphous wax fraction can sorb lipophilic solutes. Barley leaf wax is about 50% crystalline, and

Table 6.2 Wax/water partition coefficients (log K_{ww}) measured with cuticular wax isolated from *Hordeum vulgare, Prunus laurocerasus, Ginkgo biloba* and *Juglans regia*, mean log K_{ww} values and mean log K_{cw} values of ten compounds. Values in brackets indicate log K_{ww} values of *Hordeum* corrected for the crystallinity of 50%, assuming that the crystalline wax fraction does not contribute to sorption

Substance	log K_{ww} *Hordeum*	log K_{ww} *Prunus*	log K_{ww} *Ginkgo*	log K_{ww} *Juglans*	log K_{ww} mean	log K_{cw} mean
MET	0.54 (0.84)[a]	1.11[c]	–	–	0.82	1.48[c]
4-NP	–	1.15[c]	–	–	1.15	1.87[d]
BA	–	1.32[c]	1.35[c]	1.34[c]	1.34	1.71[c]
AT	–	1.45[c]	–	–	1.45	2.16[d]
SA	1.48 (1.78)[a]	1.66[c]	1.67[c]	1.68[c]	1.62	2.03[c]
TRI	1.51 (1.81)[b]	–	–	–	1.51	2.88[b]
2,4-D	1.66 (1.96)[b]	2.13[c]	2.14[c]	2.16[c]	2.02	2.61[d]
TB	2.81 (3.11)[a]	–	–	–	2.81	3.54[e]
BIT	3.02 (3.32)[b]	–	–	–	3.02	4.05[b]
PCP	3.55 (3.85)[b]	–	–	–	3.55	4.56[d]

Metribuzin (MET), 4-nitrophenol (4-NP), benzoic acid (BA), atrazine (AT), salicylic acid (SA), triadimenol (TRI), 2,4-dichlorophenoxyacetic acid (2,4-D), tebuconazole (TB), bitertanol (BIT) and pentachlorophenol (PCP).

[a] Burghardt et al. (1998)
[b] Schreiber and Schönherr (1992a)
[c] Kirsch et al. (1997)
[d] Kerler and Schönherr (1988a)
[e] Baur et al. (1996b)

the K_{ww} values should be doubled (Table 6.2) before they are compared to K_{cw} or K_{ow}. After this adjustment, K_{ww} values are still much smaller than K_{cw}. Thus, even amorphous barley wax sorbs much less than cuticles. Crystalline wax fractions for the other species are not known, but they are likely to be smaller than 50%.

6.1.4 Concentration Dependence of Partition Coefficients

At low concentrations in the micromolar and millimolar range, partition coefficients showed a small but significant decrease with increasing aqueous concentrations of 4-nitrophenol (Riederer and Schönherr 1986a). Thus, for practical reasons, as long as low concentrations are used it should be sufficient to determine partition coefficients at one single concentration. However, at high concentrations in the millimolar and molar range, sorption capacity of the CM can become limiting (Sect. 8.1). It was estimated for 4-nitrophenol (4-NP) and isolated cuticles of *Lycopersicon* and *Ficus* that about 21% of the volume fraction of the CM are available for the sorption of molecules like 4-NP (Riederer and Schönherr 1986a). If this is true for all sorbates, it would imply that only 20 weight percent cuticular waxes can be accommodated in cutin as embedded waxes.

6.1.5 Prediction of Partition Coefficients

K_{ow} values are frequently used in modelling the environmental fate of chemicals. Since K_{ow} values for all environmental chemicals are not available and the determination is time-consuming, there have been many efforts to predict K_{ow} from basic physicochemical properties such as solubility, fragment group contributions, linear solvation energy relationships and others (Sangster 1997) Similar concepts have been adopted to predict K_{cw} values.

Since log K_{cw} and log K_{ow} are similar (Table 6.1), attempts were made to predict cuticle/water partition coefficients from octanol/water partition coefficients. This is possible with fairly good accuracy (Schönherr and Riederer 1989):

$$\log K_{cw} = 0.057 + 0.970 \times \log K_{ow} \quad (r^2 = 0.97). \qquad (6.2)$$

This equation is based on 13 different chemicals with log K_{cw} values varying over eight orders of magnitude. Using (6.2), cuticle/water partition coefficients in the range of 10^2 and 10^8 can be estimated for low solute concentrations in water.

Organic solutes with no or few polar groups (i.e., –OH, –COOH, –CHO, $-NO_2$, $-NH_2$) are hydrophobic and have a low water solubility. Large K_{ow} values are mainly caused by low water solubility, rather than high solubility in octanol (Sangster 1997). For this reason K_{ow} can be predicted from water solubility. This alternative approach offers the advantage that log K_{cw} values can be calculated if log K_{ow}

values are not available, or if values appear unreliable. For a set of 13 compounds and water solubilities (S_{water}) ranging from 10^{-1} to 10^{-11} mol l^{-1} (6.3) has been established (Schönherr and Riederer 1989).

$$\log K_{cw} = 1.118 - 0.596 \times \log S_{water} \quad (r^2 = 0.96). \qquad (6.3)$$

The coefficient of determination (r^2) is only slightly lower than that of (6.2), and both equations are valuable tools for the prediction of log K_{cw} values. However, K_{cw} of compounds with infinite water solubility (e.g., methanol and ethanol) cannot be predicted from (6.3).

Molecular connectivity indices have been used to predict log K_{cw} values (Sabljic et al. 1990). Molecular connectivity indices are exclusively derived from the structure of the molecules (type and number of atoms and bonds), and they do not represent experimental values. The advantage of this approach is the fact that molecular connectivity indices are not subject to experimental errors, which are always associated with experimental values.

When log K_{ww} values are plotted against mean log K_{cw} (Table 6.2), a reasonable correlation is obtained (Fig. 6.1). Equation (6.4) shows that sorption in wax is lower by a factor of about 6.8 $(10^{0.83})$ than sorption in the CM

$$\log K_{ww} = -0.29 + 0.83 \times \log K_{cw} \quad (r^2 = 0.93). \qquad (6.4)$$

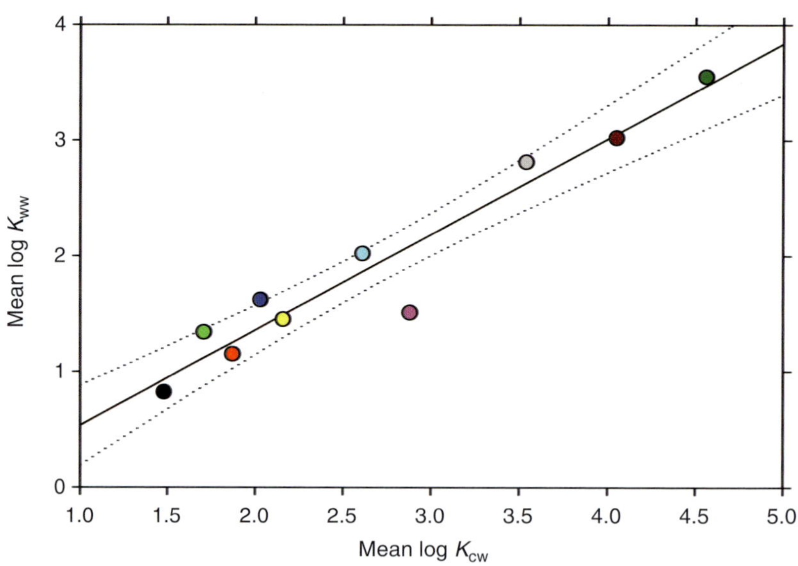

Fig. 6.1 Plot of the logarithm of mean wax/water partition coefficients (mean log K_{ww}) measured with isolated wax of four species and ten compounds as a function of the logarithm of mean cuticle/water partition coefficients (mean log K_{cw}). *Dotted lines* represent the 95% confidence interval of the regression. Data are taken from Table 6.2

6.1.6 Problems Related to the Measurement of Partition Coefficients

6.1.6.1 Solutes with Ionisable Acidic and Basic Groups

When weak acids or bases are ionised, their water solubility increases by orders of magnitude, because ions are surrounded by water molecules (hydration water). At the same time, lipid solubility becomes practically zero (Riederer and Schönherr 1984; Burghardt et al. 2005). Partition coefficients shown in Tables 6.1 and 6.2 always refer to the non-dissociated molecules. With weak acids and bases, partition coefficients must be determined using buffered aqueous solutions, and only the non-ionised fraction of the compound in the external solution is used for calculating the partition coefficients. The degree of ionisation α of an acid can be calculated from (6.5a) (Fujita et al. 1964)

$$\alpha = \frac{1}{(10^{p\mathrm{K_a}-p\mathrm{H}})+1}. \tag{6.5a}$$

The degree of ionisation α of a base can be calculated from (6.5b)

$$\alpha = \frac{1}{(10^{p\mathrm{H}-p\mathrm{K_b}})+1}. \tag{6.5b}$$

6.1.6.2 Hydrophobic Solutes with Extremely Low Water Solubility

The methods sketched above for measuring K_{cw} are easy and straightforward as long as compounds have partition coefficients between 10 and 10,000. With extremely lipophilic solutes, serious experimental problems can arise due to insufficient radiochemical purity and extremely low water solubility. These solutes quickly disappear from the aqueous phase, even when saturated solutions are used. Very large volumes of water may be used, or solid solute particles may be added to the water. These alternatives are awkward, and the problem can be circumvented by dissolving very lipophilic molecules in a mixture of water and acetone, methanol or ethanol at different ratios. This increases the solubility of the lipophilic compound in the solution and decreases partition coefficients. Partition coefficients are determined at different water/solvent ratios, and K_{cw} is obtained by extrapolating to 100% water. This approach has been successfully applied using methanol/water mixtures ranging from 80:20 to 50:50 and the extremely lipophilic compounds diethylhexylphthalate (DEHP), hexachlorobenzene (HCB) and perylene (Kerler and Schönherr 1988a). Their log K_{cw} values ranged from 5.7 to 7.7. Methanol added to the water did not interfere with sorption of the lipophilic molecule in the CM, and the K_{cw} values calculated were in good agreement with K_{ow} values taken from the literature. It appears that the method works well with cuticles, but it cannot be used to determine K_{ow} because methanol is soluble in octanol.

6.1.6.3 Polar Solutes with Extremely High Water Solubility

Solutes having a very high water solubility (i.e., urea, glucose and sucrose) dissolve in extremely small amounts in octanol, cuticles or cutin. These polar compounds are highly soluble in water, and the concentration in the aqueous phase will hardly decrease during equilibration even with very large mass cuticle/water ratios. Amounts sorbed in the CM must be measured by separating cuticles from the aqueous phase and determining the amounts of polar solutes associated with the cuticle. Small droplets sticking to the waxy surface of the CM, or aqueous films spread over the inner surface, present huge problems and prevent the obtaining of valid partition coefficients.

This problem can be illustrated by a simple model calculation. At an external aqueous solution of the polar compound of $1\,\mu g\,g^{-1}$, and a partition coefficient of 0.01, the solute concentration in the CM calculated from (2.12) is $0.01\,\mu g\,g^{-1}$. This amount of solute is contained in 10 mg of solution. With an average weight of $300\,\mu g\,cm^{-2}$ typical for many leaf CM (Table 4.8), one gram of cuticle has a total area of $0.33\,m^2$. Let us assume that after equilibration and blotting, a thin water film of $0.1\,\mu m$ thickness remains on the inner side of the CM and the waxy outer surface is dry and free of water. This water film would have a mass of 33 mg water, which contains $0.033\,\mu g$ solute. Since only $0.01\,\mu g$ solute are sorbed in the CM, the amount contained in the water film would be 3.3 times higher. The total amount of solute associated with 1 g CM is $0.043\,\mu g$, and dividing this by concentration of the solution ($1\,\mu g\,g^{-1}$) we obtain a partition coefficient of 0.043 rather than the real one of only 0.01. This example clearly shows that it is practically impossible to accurately determine partition coefficients of very polar solutes. Apart from the tedium involved in isolating 1 g of thin cuticles, the assumption of a very thin water film of only $0.1\,\mu m$ is very optimistic. In our experience, wet cuticles contained 30–50 weight percent water after thorough blotting with tissue paper, while the amount of sorbed water at 100% humidity is less than 10% by weight (Chamel et al. 1991).

Popp et al. (2005) published partition coefficients measured for various sugars and ivy CM. Predicted log K_{ow} values ranged from -1.52 (erythrose) to -7.36 (maltotriose). Sugar concentration in water was $50\,g\,kg^{-1}$ (Popp, personal communication), and experimental K_{mxw} value for erythrose was 99 and the values for glucose, maltose and maltotriose were between zero and 10. They were not related to predicted log K_{ow}. The authors suggested that the sugars were not sorbed in cutin but in aqueous pores traversing the CM and MX membranes. This appears reasonable, since cuticles are heterogeneous membranes composed of lipids and polar polymers (Chaps. 1 and 4). Cutin and polar polymers sorb up to $80\,g\,kg^{-1}$ water when exposed to 100% humidity (Chamel et al. 1991). Only about $20\,g\,kg^{-1}$ are sorbed in cutin. Water content in aqueous pores would amount to not more than $60\,g\,kg^{-1}$ (Fig. 4.6). Popp et al. (2005) measured a total water content of ivy cuticles of $30\,g\,kg^{-1}$. If it is assumed that these aqueous pores can be accessed by the sugars, 1.5–3 g sugar would be in 1 kg cuticle and the partition coefficient would be 0.03 or 0.06. These estimates are maximum values, and they are about two orders of magnitude smaller than the values published by Popp et al. (2005). Liquid films on the surfaces of the cuticles containing dissolved sugars are most likely the cause for this difference.

The above two model calculations leave little doubt that partition coefficients for highly water soluble compounds and very thin cuticles cannot be determined with sufficient accuracy, and they should be looked at with suspicion.

6.2 Steady State Penetration

Non-electrolytes are neutral solutes which carry no electrical charges. Weak acids and bases can be treated as non-electrolytes if the pH is adjusted such that dissociation is suppressed. For instance, with weak acids 99% of the molecules are undissociated when the pH is 2 units below the $p\mathrm{K_a}$. In this section, we restrict our attention to solutes which are sufficiently water-soluble and not too volatile, such that working with open donor and receptor compartments is possible.

Solute permeability of cuticles can be characterised by permeance. Combining (2.3) and (2.18), we obtain

$$J = P\left(C_{\mathrm{donor}} - C_{\mathrm{receiver}}\right) = \frac{KD}{\ell}\left(C_{\mathrm{donor}} - C_{\mathrm{receiver}}\right). \tag{6.6}$$

Permeance (P) is calculated by dividing the steady state flux (J) of a solute by the driving force, which is the difference of the solute concentration between donor and receiver. In the steady state the concentration in the receiver can be maintained negligibly small, and driving force is simply the donor concentration. P can be determined using isolated CM, leaf disks or detached leaves. We shall demonstrate application of (6.6) using examples taken from the literature.

6.2.1 *Permeance of Isolated Cuticular Membranes*

Riederer and Schönherr (1985) measured 2,4-D permeability of CM isolated enzymatically from astomatous leaf surfaces and fruits, and these data are suitable to demonstrate variability of P among plant species. 2,4-D is a weak acid ($p\mathrm{K_a}$ is 2.77) and it was ^{14}C-labelled. Donor and receiver solutions were buffered at pH 2, and donor concentrations ranged from 0.2 to $1 \times 10^{-3}\,\mathrm{mol\,l^{-1}}$. At a pH of 2.0, 85.5% of the molecules are non-ionised (Sect. 6.1). Plotting the amount diffused into the receiver vs time, linear plots are obtained and their slopes represent the flow rate (F in $\mathrm{mol\,h^{-1}}$). Dividing F by membrane area $(1\,\mathrm{cm^2})$ and concentration of non-dissociated 2,4-D in the donor gives the permeance (P). Non-dissociated 2,4-D is lipophilic, as the partition coefficients CM/water(K_{cwr}) range from 240 to 579, depending on species (Riederer and Schönherr 1984). It follows that 2,4-D is sorbed in cuticular lipids, while 2,4-D concentration in the polar cuticular polymer phase is negligible.

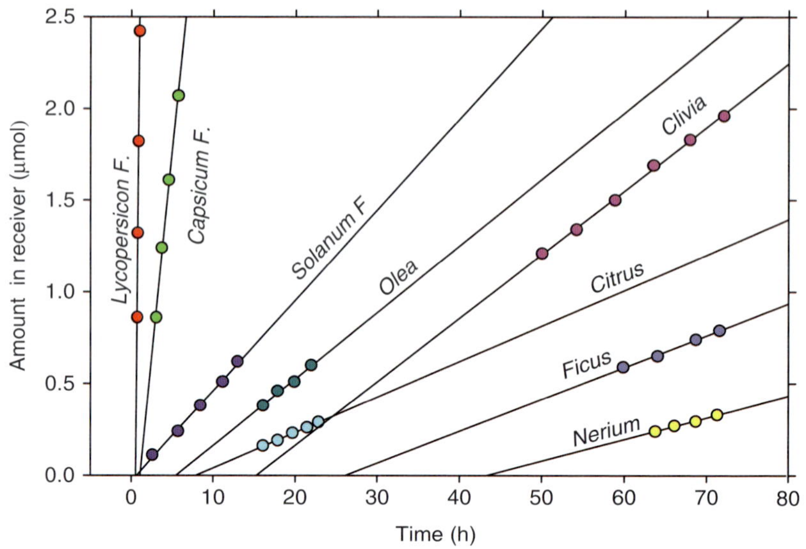

Fig. 6.2 Time course of 2,4-D diffusion across CM of various plant species at 25 °C. The steady state flow was extrapolated to the time axis, which yields the extrapolated hold-up time (t_e). Fruit CM were obtained from tomato, pepper and egg plants; all others were isolated from astomatous adaxial leaves. (Redrawn from Riederer and Schönherr 1985)

Steady state flow rates and the hold-up times (t_e) varied greatly among species. The amount diffused was greatest with fruit CM, and t_e was short. With leaf CM, slopes are considerably smaller and hold-up times are longer (Fig. 6.2). Permeance calculated from these data range from 2.72×10^{-8} (pepper) to 1×10^{-10} m s^{-1} (*Ficus*), that is, among species they differed by a factor of 272. With P known, the steady state fluxes caused by a given concentration can be calculated. For instance, if donor concentration is 1×10^{-3} mol l^{-1}, the steady state flux in 24 h would be 2.35×10^{-3} mol m^{-2} s^{-1} and 8.64×10^{-6} mol m^{-2} s^{-1} across CM of pepper fruit and *Ficus* leaf, respectively. At the same driving force and time interval, 272 times more 2,4-D penetrates into a pepper fruit than in a *Ficus* leaf. It should be realised that this represents a rather small number of species, and that with more species included variability might likely be larger.

In the laboratory, growth regulators and other biologically active materials can be applied to leaves under controlled conditions, and the steady state can be maintained for a long time. Applying the same substance at constant concentration for the same time period, the dose delivered to leaves, stems or fruits of different species, genotypes or mutants differs if permeance is not the same. If P are not known, this type of experiment is likely to lead to wrong conclusions in dose–response experiments.

By reference to Fig. 6.2 it is clear that penetration experiments with only one time point are not a good practice. Permeance can still be calculated using (6.6), but hold-up time is not known and must be assumed to be zero. With *Citrus* CM, 28 µmol 2,4-D penetrated in 24 h and extrapolated hold-up time was 7.9 h. Had only one sample been taken after 24 h, the calculated flow would have been 1.17 µmol h^{-1}. However, time available for steady state penetration was only 16.1 h (24–7.9 h), and the proper slope is 1.74 µmol h^{-1}. Neglecting the hold-up time results in an erroneous estimate of flow, which is too small by a factor of 0.67. In the vast majority of published data, foliar penetration was measured only after one time interval.

In the field, chemicals are applied by spraying plants with small droplets. They dry up quickly, concentration in the droplets increases, 2,4-D probably crystallises and wetting of cuticles may be a problem. These practical problems reduce steady state penetration to a very short time, and we shall address these and other practical problems and their solutions later (Sects. 6.3 and 6.4).

Inspecting Fig. 6.2 immediately raises the question as to what might have caused the differences among species. Permeance is a mixed quantity and proportional to the diffusion and partition coefficients and inversely proportional to membrane thickness (6.6). From these quantities only ℓ and P are known, and D can be obtained from the hold-up time (2.5). The partition coefficient K can be calculated as $P \times \ell/D$ (6.6). P, D and K_{calc} are summarised in Table 6.3.

Thickness of CM ranged from 2.6 to 10.7 µm, while permeances differed by a factor of 272, and there is no correlation between P and $1/\ell$ as suggested by (6.6). *Citrus* CM is the thinnest in this collection, and its P is similar to permeances of the thick CM of *Clivia, Ficus* and *Nerium*. Tomato and pepper fruit CM are also very thick, but their permeances are more than 200 times higher than that of the rubber leaf CM.

Diffusion coefficients ranged from 6.1×10^{-15} (*Lycopersicon*) to $5.4 \times 10^{-17}\,m^2\,s^{-1}$ (*Citrus*), and D values of fruit CM are about an order of magnitude higher than in leaf CM. The lowest D was measured with the thinnest CM (*Citrus*). Most strikingly, calculated partition coefficients (K_{calc}) are much smaller than partition coefficients determined directly (K_{det}) in sorption experiments (Table 6.3).

Table 6.3 Permeance (P), diffusion (D) and partition coefficients (K_{calc}) obtained with cuticular membranes from the species shown in Fig. 6.2. For comparison, partition coefficients determined directly (K_{det}) by sorption experiment are included (Riederer and Schönherr 1985)

Species	ℓ(µm)	$P(m\,s^{-1})$	$D(m^2\,s^{-1})$	K_{calc}	K_{det}	$P(MX)/P(CM)$
Capsicum F	9.2	2.72×10^{-8}	5.20×10^{-15}	48	579	46
Lycopersicon F	8.1	2.55×10^{-8}	6.08×10^{-15}	34	428	29
Solanum F	6.3	3.36×10^{-9}	7.97×10^{-15}	2.7	424	1,557
Olea	6.2	2.67×10^{-9}	2.63×10^{-16}	63	469	2,442
Citrus	2.6	2.80×10^{-10}	5.40×10^{-17}	14	300	1,767
Clivia	8.9	1.30×10^{-10}	2.22×10^{-16}	5.2	240	3,876
Ficus	9.8	1.00×10^{-10}	2.28×10^{-16}	4.3	315	9,192
Nerium	10.7	1.80×10^{-10}	1.44×10^{-16}	13	300	656

D and K_{calc} were obtained using average values from 6–13 CM for ℓ and P respectively. In the original publication, calculations were performed for individual CM and averaged. For this reasons, above values differ somewhat from those of the original publication. K_{det} (6th column) taken from Riederer and Schönherr (1984)

Clearly, diffusion in CM of various species is not quantitatively accounted for by (6.6), which is valid only for homogeneous membranes. Hence, these CM are not homogeneous. Transport cannot be explained based on single values of D and K for the entire CM, and thickness is not related to P and D as indicated in (6.6).

For water we had come to a similar conclusion, but with water the situation is much more complicated (Chap. 4). Water can move along two parallel pathways, and viscous flow is involved as well as diffusion. 2,4-D is lipophilic, and its concentration in cutin and waxes (Sect. 6.1) is much higher than in water. Furthermore, volume fractions of polar cuticular polymers are much smaller than volume fractions of cutin and waxes (Sect. 2.1). Taken together this implies that transport of 2,4-D across CM occurs solely in membrane lipids, and aqueous pores (if they are present) should not be involved. Heterogeneity shown in Table 6.3 means that cutin and waxes do not form a homogeneously mixed barrier, and they are not uniformly distributed in CM.

The role of waxes in heterogeneity can be estimated by comparing data for CM and MX Riederer and Schönherr (1985). Extracting waxes increased P of fruit CM 29- to 1557-fold, while leaf MX membranes were 656–9,192 times more permeable than CM (Table 6.4). Clearly, waxes greatly reduce permeance, even though their weight fractions are small (Table 1.1). Part of this increase in P upon extraction can be attributed to an increase in diffusion coefficients. D (averaged over all lipid phases) measured with CM of fruits were $4.0 \times 10^{-15}\,m^2\,s^{-1}$ while with leaf CM average D was $1.7 \times 10^{-16}\,m^2\,s^{-1}$ (Riederer and Schönherr 1985). Extracting cuticular waxes increased D and eliminated the differences in D between fruits and leaves and the average D is now $1.3 \times 10^{-14}\,m^2\,s^{-1}$ (Riederer and Schönherr 1985).

Focussing on partition coefficients (Table 6.4) it is clear that extracting waxes greatly increased solubility of 2,4-D in MX. The only lipid phase in the MX is cutin, and depending on species it amounts to 70% (fruits) or 76% (leaves) of the MX. Dividing K_{calc} by the cutin faction in MX (calculated from data given in Table 1.1) increases partition coefficients (corrected K_{calc}). Calculations are based

Table 6.4 The effect of extraction of waxes on 2,4-D permeance; partition coefficients calculated (K_{calc}) for the MX from the $P\ell/D$ ratio. Partition coefficients obtained in a sorption experiment (K_{det}) are shown for comparison

Species	$P(MX)/P(CM)$	K_{calc} for CM	K_{calc} for MX	K_{calc} correct.	K_{det} for MX
Capsicum fruit	46	48	670	989	768
Lycopersicon fruit	29	34	470	635	612
Solanum fruit	1,557	2.7	350	522	755
Olea leaf	2,442	63	700	1,000	990
Citrus leaf	1,767	14	430	558	435
Clivia leaf	3,876	5.2	160	200	307
Ficus leaf	9,192	4.3	180	240	485
Nerium leaf	656	13	240	324	648

Data taken from Riederer and Schönherr (1984, 1985). Corrected K_{calc} were obtained by dividing K_{calc} by the weight fraction of cutin in MX taken from Table 1.1

on average values of P and D measured or calculated for individual MX membranes. Accounting for considerable variability among individual membranes, the agreement between corrected K_{calc} and with K_{det} is good to fair. Good agreement obtained with MX from fruits and leaves of *Citrus* and *Olea* indicates that 2,4-D diffused in cutin and cutin is fairly homogeneous throughout the MX membranes. With *Clivia, Ficus* and *Nerium* MX, corrected K_{calc} are still considerably lower than K_{det}. Cuticles of these three species have two different types of cutins, i.e., ester cutin near the epidermal cell wall and cutan at the outer surfaces of the CM (cf. Sects. 1.2 and 1.4). These types of cutin differ in polarity and structure, and it appears that D in cutan is lower than in ester cutin. This could account for low K_{calc} seen in Table 6.4. Hence these three species still have a heterogeneous MX, but heterogeneity is much less than in CM.

Having established that permeance of 2,4-D varies widely among species, we now turn to the role of partition coefficients as determinants of P. Equation (6.6) states that P is proportional to the partition coefficient, and Kerler and Schönherr (1988b) measured permeance of *Citrus* CM using a selection of important agricultural and environmental chemicals (Fig. 6.3). Solutes 1–5 are weak electrolytes, and donor solutions were appropriately buffered to assure high and constant

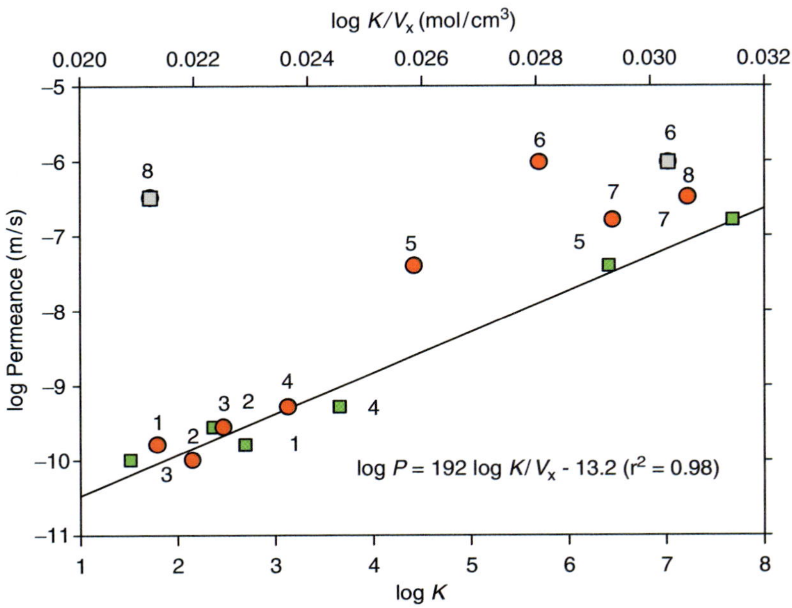

Fig. 6.3 Logarithms of permeance P of *Citrus aurantium* CM measured at 25 °C as a function of log K_{cw} of solutes (*red circles*). Log P plotted vs log K_{cw}/V_x is shown as *green squares*. Numbers refer to the test compounds used: 4-nitrophenol (1), 2,4-D (2), atrazine (3), 2,4,5-trichlorophenoxyacetic acid (4), pentachlorophenol (5), hexachlorobenzene (6), perylene (7), diethylhexyl phthalate (8). The linear regression equation applies to the squares, with the exception of compounds 6 and 8 *(grey squares)*. Data were replotted or recalculated from data given by Kerler and Schönherr (1988a, b). V_x is the equivalent molar volume of solutes in cm^3 mol^{-1}

concentrations of non-ionised molecules. The partition coefficients CM/water for the non-ionised species varied from 83 (4-nitrophenol) to 7.24×10^7 (diethylhexyl phthalate), which is a factor of almost one million. It is not trivial to measure extremely high partition coefficients (Sect. 6.1) and permeances. With hexachlorobenzene (HCB), perylene and diethylhexyl phthalate (DEHP), Kerler and Schönherr (1988b) measured steady state diffusion using solid residues of the solutes in the donor to prevent rapid depletion in the donor solutions.

When log P was plotted vs log K_{cw}, a positive dependence could be seen (red circles). From 4-nitrophenol to DEHP, permeance increased by a factor of nearly 2,000. This is considerably less than the difference in partition coefficients. One of the reasons for this is the fact that molecular weights of the compounds were not constant but ranged from 139 (4-nitrophenol) to 390 g mol^{-1} (DEHP), and diffusion coefficients in CM and waxes greatly depend on size of solutes. This will be dealt with quantitatively in Sect. 6.3. There are various ways to estimate the volumes of molecules. With liquids it is not difficult to determine the volume of one mole of substance, but these figures greatly depend on temperature, and it is not possible to obtain good estimates for solids. We have been using the method of Abraham and McGowan (1987) to characterise size of solutes. Molar volumes (in cm^3 mol^{-1}) are obtained by adding the volumes of the atoms and correcting for the influence of bonds on molar volumes. Volumes of all relevant atoms have been tabulated, and the molar volumes (V_x) calculated are characteristic volumes at zero Kelvin. For this reason characteristic volumes are smaller than molecular weights, but for the purpose of comparing permeances or diffusion coefficients of compounds having different molecular weights, V_x turned out to be well-suited.

Dividing log K_{cw} by V_x corrects for differences in size of solutes, and a good correlation with log P_{cw}/V_x is obtained (green squares in Fig. 6.3) if data for DEHP and HCB were disregarded. The fact that permeances for DEHP and HCB are much higher than predicted by the regression line is probably related to their plasticising activity (Chap. 7). Plasticisers render solid polymers more flexible and increase solute mobility (Buchholz 2006). DEHP is a typical plasticiser used commercially in synthetic polymers. Some ethoxylated alcohols are very effective plasticisers (Schönherr 1993a, b). Pentafluorophenol is also a very effective plasticiser (Schönherr and Baur 1996). Plasticising activities of PCP and HCB seem not to have been studied, but in view of structural similarity with pentafluorophenol it is likely that they are plasticisers as well. The effect of plasticisers is concentration-dependent, but concentration of solutes in the CM cannot be calculated, as donor concentrations were not given in the original publication. Their concentration in the CM probably differed, and their intrinsic activity probably as well. It should be realised that plasticiser activity of solutes would have gone unnoticed had penetration been measured using only one time interval, and results would have been expressed as percent penetration in an arbitrary time interval.

Prediction of permeances of *Citrus* CM at 25 °C to other solutes is possible using the equation

$$\log P = 192 \frac{\log K_{cw}}{V_x} - 13.2. \tag{6.7}$$

Equation (6.7) returns permeances (in m s^{-1}) of *Citrus* CM for lipophilic solutes which are inert, that is solutes which have no plasticiser activity. The dimension of V_X is $cm^3\,mol^{-1}$, and with the solutes used it ranged from 95 (4-nitrophenol) to 340 (DEHP). The equation is valid over a very wide range of partition coefficients. Since K_{cw} of neutral solutes varies much more than molar volumes of solutes, it is more important than size in determining permeance. Since log K_{cw} can be calculated from octanol/water partition coefficients (Sect. 6.1) which have been tabulated or can be calculated using various methods, prediction of permeance of *Citrus* CM is possible with good precision without having to conduct experiments.

Equation (6.7) should not be used with compounds that are better soluble in water than in cutin $(K < 1)$. Such polar neutral solutes (i.e., urea, glycerol, glucose) diffuse in the water of the polar cuticular polymer fraction. There are no reliable data on permeances of small neutral polar solutes, but based on the arguments of Sect. 4.4 it is not likely that valid estimates of using partition coefficients and (6.7) can be obtained. Likewise, prediction of permeances of ionic solutes (inorganic salts or organic electrolytes such as glyphosate) is not possible (Chap. 5).

The steady state flux (J) of a solute across cuticles is proportional to permeance and driving force, that is, solute concentration of the donor (6.6). Fluxes of different compounds are proportional to K_{cw} provided V_X and driving force are the same. In this case, very lipophilic solutes penetrate faster than slightly lipophilic ones. However, rates of penetration must not be confused with the amounts that penetrate. Amounts are limited by the dose contained in the donor $(C_{donor} \times V_{donor})$, and given sufficient time the total dose will penetrate (see Sect. 6.3).

It is of practical relevance that C_{donor} cannot be varied arbitrarily, because aqueous solubility (S_{water}) of very lipophilic solutes can be extremely low. In fact, high partition coefficients are not the result of high lipid solubility but rather of low water solubility, and log K_{cw} is proportional to $-\log S_{water}$. With some simplifying assumptions, Schönherr and Riederer (1989) derived the (6.3) which is repeated here for convenience:

$$\log K_{CM/water} = 1.118 - 0.596 \times \log S_{water} (r^2 = 0.96). \tag{6.8}$$

This equation can be used to calculate K_{cw} from the aqueous solubility $(mol\,l^{-1})$. The data base includes solutes shown in Fig. 6.3, and aqueous solubilities ranged from 0.17 to $2.4 \times 10^{-11}\,mol\,l^{-1}$.

6.2.2 Steady State Penetration into Detached Leaves: The Submersion Technique

In the experiments discussed above, concentrations in donor and receiver were monitored and permeance was calculated using (6.6). Concentration of the donor was constant, while concentration of the receiver was kept close to zero by constantly changing the solutions in the receiver compartment. When studying foliar penetration, it is not possible to manipulate the receiver. Donor concentrations can be adjusted and held constant, but the influx is measured by analysing the entire leaf after blotting or rinsing off the donor solution. The entire leaf serves as receiver, and it is not known a priori where the solute is located. Depending on water and lipid solubility it will be distributed in aqueous and/or lipid compartments, including cuticular wax and cutin. With intact leaves, permeances can be estimated from time-course experiments, and using a desorption technique sorption compartments can be characterised. We will demonstrate this with barley leaves and conifer needles. Slender leaves as typical for conifer needles and some monocot species can be submerged in large volumes of donor solutions in glass test tubes. They are maintained at constant temperature, and are lightly agitated for mixing to minimise unstirred boundary layers.

6.2.2.1 Penetration into Cut Edges

Whenever leaves or needles are removed from the plants a wound is generated, and solutes can penetrate into the wound. This necessitates corrections for solute penetration into the wound. Wounds may be closed by dipping into paraffin wax, grease, glue or other materials. In these cases, it must be established that the seal is effective and that the materials do not sorb solutes. With barley leaves and conifer needles we have used a different approach. The cut edge was dipped into a shallow (1 mm) aqueous solution of ^{14}C-labelled triadimenol ($K_{cw} = 760$), the vessels were closed to establish 100% humidity, and they were incubated in the dark to minimise transpiration. After 6 or 24 h the leaves were removed from the bath, the base was blotted dry, and leaves were dissected in 1 mm-wide strips. These strips were combusted, $^{14}CO_2$ was trapped and radioactivity was determined by scintillation counting. If M_0 is the radioactivity associated with the first mm of leaf and M_x is the amount contained in segments above, the concentration gradient along the leaf is proportional to M_x/M_0. With increasing distance from the solution, radioactivity (that is, M_x/M_0) decreased, and in the segments 9 (6 h incubation) to 15 from the cut edge (24 h incubation) practically no radioactive triadimenol had arrived (Fig. 6.4).

This approach is superior to trying to seal the wound, as it gives a result that can be generalised to other non-volatile compounds which are not readily metabolised. Cyan symbols represent M_x/M_0 values calculated from (6.9)

$$\frac{M_x}{M_0} = 1 - \operatorname{erf}\left(\frac{x}{2\sqrt{Dt}}\right), \tag{6.9}$$

where x is the distance from the cut edge and t is the duration of incubation. The error functions of the term in brackets are tabulated (Crank 1975, Table 2.1, p. 375). With x and t given, various D values must be tried which result in the best fit to the data. With triadimenol, the best fit was obtained with a diffusion coefficient of $2 \times 10^{-10}\,m^2\,s^{-1}$. This is a value typical for diffusion in water at 25 °C. Hence, triadimenol diffused up the leaf in an aqueous compartment, and viscous flow by

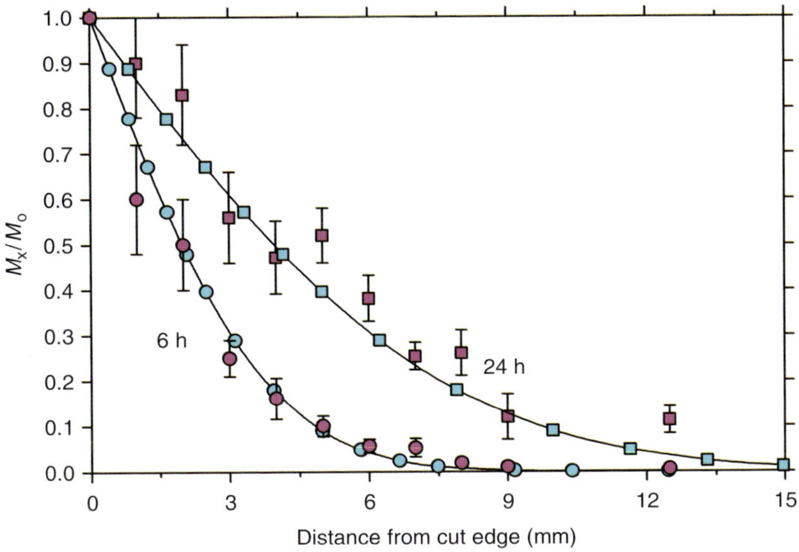

Fig. 6.4 Distribution of ^{14}C-triadimenol in barley leaves which had been dipped with their cut edges 1 mm deep for 6 or 24 h into radio-labelled aqueous triadimenol. *Pink symbols* are experimental values, while *cyan symbols* were calculated using a D of $2 \times 10^{-10}\,m^2\,s^{-1}$ *Vertical bars* are 95% confidence intervals. Redrawn from Schreiber and Schönherr (1992a)

transpiration was absent. D in water decreases with increasing molecular weight (MW), but since it is proportional to $\sqrt{MW}$, the effect of MW is small for most solutes of interest. Thus, results depicted in Fig. 6.4 will be similar, with many compounds diffusing in the apoplast of barley leaves. If leaves are submerged in solutions for up to 24 h, discarding the lower15 mm portion of the leaf eliminates practically all radioactivity which entered through the cut edge, und the remaining radioactivity in the leaf has penetrated through the cuticle. This method has been used with conifer needles and other lipophilic compounds, and distribution of radioactivity along the needle was the same except for the first 3 mm segment near the cut edge. This segment was discarded prior to combusting the needles (Schreiber and Schönherr 1992b, 1993a).

6.2.2.2 Cuticular Penetration

Cuticular penetration into barley leaves and conifer needles was studied by submerging them in large volumes of water or buffer containing radiolabelled solutes. Depending on partition coefficient, the mass of the donor solution should be about 20–50 times larger than the mass of the leaves. This will keep donor concentrations constant, but this should be checked by taking samples at the end of the experiments. During incubation the glass test tubes were closed with screw caps lined with aluminium foil, and they were lightly agitated for mixing. Incubation was in the dark, and temperature was 25 °C. It is important to establish this temperature of the donor before closing the vessel. If cold solutions are warmed up in a closed vessel, the pressure increases above ambient pressure, and this may cause stomatal infiltration (Schönherr and Bukovac 1972a). When studying very lipophilic chemicals it is essential to use glass tubes, as they may be sorbed to and in the walls of plastic vessels.

Barley leaves and conifer needles are covered profusely with microcrystalline surface waxes which render them very difficult to wet by aqueous solutions as long as no surfactant is added. The water is in contact only with the tips of the surface wax crystallites. When leaves are withdrawn from the donor solution, they are not covered by water films and appear dry. After incubation the leaves were lightly blotted with soft tissue paper, combusted and radioactivity was measured by scintillation counting. Results obtained with pentachlorophenol (PCP) are shown in Fig. 6.5. Donor concentration was 10^{-7} mol l^{-1} and the *p*H of the donor was 3.0, that is, 98% of the PCP was non-ionised. Penetration is expressed in moles penetrated per cm^2 of leaf surface, which is the sum of upper and lower surfaces. Amounts of chemical are calculated by dividing radioactivity (Bq) in the leaf by specific radioactivity (Bq mol^{-1}).

The plot marked "measured" (Fig. 6.5) refers to amount penetrated per cm^2 leaf surface. Permeance is calculated by dividing this slope by leaf area and donor concentration. Permeance is 1.3×10^{-7} m s^{-1}, which is larger by a factor of 3.25 than permeance of *Citrus* CM (Fig. 6.3). The plot is straight, but it intersects the *y*-axis at about 12×10^{-12} mol cm^{-2}. This may appear strange, since with CM plots amount

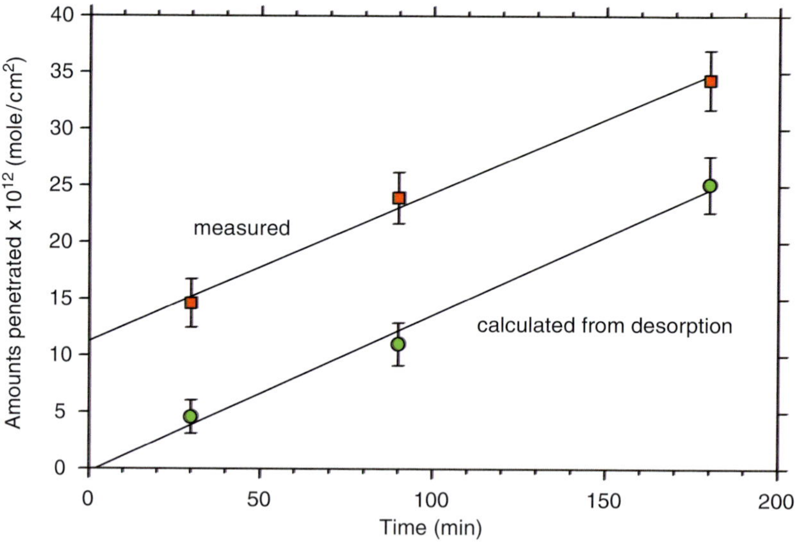

Fig. 6.5 Foliar penetration at 25 °C of PCP into barley leaves. Amounts penetrated increased linearly with time, but measured penetration (*red symbols*) was biphasic and the regression line intersects the ordinate at a positive value

penetrated vs time has a positive intersection with the time axis (extrapolated hold-up time seen in Fig. 6.2), because it takes some time before the first solute molecules appear in the receiver and penetration becomes steady. When penetration across CM is measured, concentration of the receiver is monitored and sorption in the CM goes unnoticed. During penetration into leaves, solutes are first sorbed in epicuticular waxes and in cuticles before they reach the tissue (apoplast and the symplast). Penetration of PCP into barley leaves was biphasic. During the first 30 min it was much faster than later on. The rapid penetration represents soprtion to the leaf surface and diffusion into epicuticular wax, while slow penetration marks cuticular penetration into the leaf apoplast. Compartmental analysis substantiates this conclusion.

6.2.2.3 Compartmental Analysis

From leaves preloaded with PCP in an experiment as shown in Fig. 6.5, the PCP was desorbed by submerging the leaves in borax buffer of *p*H 9. At this *p*H all PCP is ionised, because the pK_a is 4.73. PCP sorbed in wax and in cutin is non-ionised, and the concentration of non-ionised PCP in borax buffer is zero. Hence, there is a large driving force favouring efflux of PCP from the wax. Initially, desorption is very rapid and large amounts of PCP are desorbed during the first 5 min (Fig. 6.6). As time progresses, desorption curves approach a plateau which is lower the longer preloading lasted.

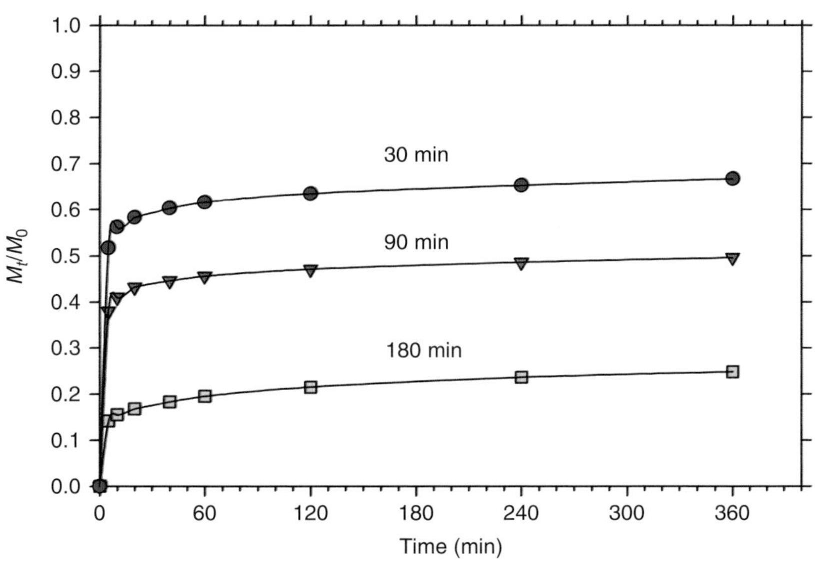

Fig. 6.6 Desorption of PCP from barley leaves preloaded with PCP for 30, 90 or 180 min. During preloading, PCP penetrated from the donor solution at *p*H 3 into leaves. After being briefly blotted between soft tissue paper, the leaves were submerged in borax buffer at *p*H 9. This changed the direction of the PCP concentration difference, and PCP diffused out of the leaves. (Redrawn from Schreiber and Schönherr 1992a)

Table 6.5 Distribution of PCP in the compartments (CPT) of barley leaves as affected by duration of preloading. CPT_1 and CPT_2 were obtained by non-linear regression analysis (6.10). CPT_3 is the fraction of PCP unaccounted for by CPT_2 and CPT_1

Preloading for	CPT_1	CPT_2	CPT_3
30 min	0.56	0.10	0.34
90 min	0.41	0.08	0.51
180 min	0.15	0.10	0.75

Desorption kinetics were analysed by non-linear regression. The best fit ($r^2 > 0.99$) was obtained with a model consisting of two compartments (CPT_1 and CPT_2) and two rate constants (k_1 and k_2):

$$\frac{M_t}{M_0} = CPT_1\left(1 - e^{-k_1 t}\right) + CPT_2\left(1 - e^{-k_2 t}\right). \tag{6.10}$$

Rate constants were independent of duration of preloading, and amounted to 8.5×10^{-3} and $3.7 \times 10^{-4}\,\mathrm{s}^{-1}$ for k_1 and k_2 respectively. Rate constants were defined by (2.6). From the rate constant, the half-time ($t_{1/2}$) required for 50% sorption or desorption ($M_t/M_0 = 0.5$) can be calculated as $t_{1/2} = \ln 0.5/k$. Hence, $t_{1/2}$ is 81.5 and 1,873 s for compartments 1 and 2 respectively. It takes about five half-times before the compartments are empty. In the case of barley leaves and PCP this amounted to 6.8 and 156 min for compartments 1 and 2, respectively.

The compartment sizes in (6.10) represent fractions. The absolute size of a compartment in Bq is obtained by dividing radioactivity in the CPT by total radioactivity in the leaf. The fractions of PCP contained in the compartments are given in Table 6.5.

Sizes of CPT_1 and CPT_3 depended on duration of preloading, while CPT_2 was constant and 8–10% of total PCP was sorbed in this compartment. CPT_1 decreased and CPT_3 increased with duration of preloading. PCP in CPT_1 and CPT_2 was reversibly sorbed, while PCP in CPT_3 increased with time and could not be recovered from the leaves. This indicates that this fraction of PCP was in the leaf tissue in dissociated form and was held there by an ion trap mechanism. *p*H in the apoplast and symplast are around 5–6, and this favours dissociation.

If the absolute amounts contained irreversibly in CPT_3 are plotted vs time, a straight line is obtained (Fig. 6.5, plot "calculated from desorption") which intersects the origin and has the same slope as the plot "amount penetrated" vs time. Permeance calculated from desorption experiments is $1.4 \times 10^{-7}\,\mathrm{m\,s}^{-1}$, and is identical within experimental error to permeance obtained from the slope of a penetration experiment.

The parallel displacement is caused by PCP reversibly sorbed on the leaf surface, in epicuticular wax and in the cuticle underneath (cutin and intracuticular wax). The donor solution is in contact only with the tips of the epicuticular wax crystallites. Hence, epicuticular waxes are filled first and serve as intermediate phase between aqueous donor and cuticle. PCP sorbed in wax continues to diffuse into the cuticle and eventually into the leaf tissue. When PCP is desorbed from leaves, the efflux takes place at first from the epicuticular wax and later from the remainder of the cuticle. Comparing the two rate constants, it is seen that k_1 is 23 times larger than k_2, and for this reason diffusion across epicuticular wax was not rate-limiting. Diffusion across cutin and embedded waxes was the rate-limiting step in foliar penetration of PCP.

The y-intercept seen in Fig. 6.5 has practical implications. Permeance can only be calculated if penetration is measured for more than one time interval. From the slope, rate of penetration and permeance can be calculated. If penetration is measured using only one time interval, sorption in wax and cuticle is overlooked and permeance estimated is too high. For instance, after 90 min penetration amounted to 24×10^{-12} and the y-intercept was 11.3×10^{-12} mol cm^{-2} (Fig. 6.5). Subtracting the y-intercept, the flux is 1.3×10^{-13} mol cm^{-2} min^{-1}, while flux calculated from the amount penetrated would have been 2.7×10^{-13} mol cm^{-2} min^{-1}. Permeance calculated without correcting for the y-intercept would be 2.1 times too high. Furthermore, a desorption experiment provides more information, as rate constants and compartment sizes can be estimated.

Half-time of desorption for the first compartment, that is, for desorption from epicuticular waxes is only 82 s. Frequently, adhering donor is rinsed off using water, buffer, or aqueous acetone. In the experiment shown in Fig. 6.6, this would have reduced the magnitude of the y-intercept but it would not have eliminated it. Extended rinsing time will remove some more PCP from the cuticle, but complete elimination is unlikely because the half-time for the second compartment (the cuticle) is 31.2 min. If leaves are washed or rinsed after the penetration experiment, a variable and unknown fraction of solute sorbed in the cuticle is considered to have penetrated, even though it had not yet reached the leaf tissue. The error is larger with short experiments. A controlled desorption experiment after blotting leaves to remove adhering donor solution is clearly the better approach, since it provides information about number of compartments, compartment sizes and rate constants by which they drain.

Using the approach outlined above, penetration of other lipophilic solutes into barley leaves was studied (Fig. 6.7). Permeance increased with partition coefficients.

The approach used with barley leaves was successfully applied to study penetration of PCP and other lipophilic chemicals into conifer needles (Schreiber and Schönherr 1992b). With conifer needles amounts penetrated vs time were also biphasic, but magnitudes of y-intercepts differed greatly among species (Fig. 6.8). Rates of penetration (slopes) have the dimension amount PCP penetrated per min and mm needle length. The projected areas of needles were 5 mm^2 mm^{-1} with the *Abies* species and 3 mm^2 mm^{-1} with all others. Dividing the slopes of the plots by projected needle area (A_{pro}) and donor concentration yields the permeance in m s^{-1} if SI units were used in calculations (Table 6.4). P differed greatly among the species.

Desorption analysis as shown for barley leaves in Fig. 6.6 resulted again in two compartments having different rate constants. Rate constants of desorption of PCP from the first compartment (k_1) ranged from 8 to 11×10^{-3} s^{-1}, that is, PCP was

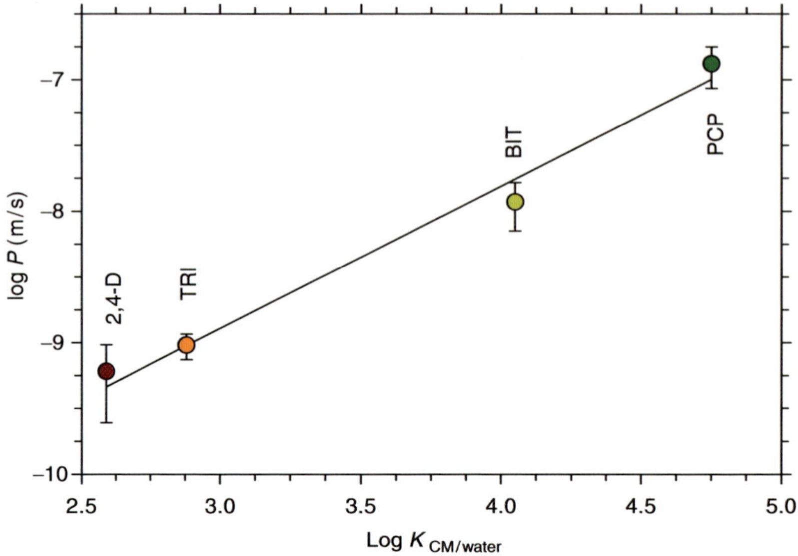

Fig. 6.7 Penetration into barley leaves at 25°C of 2,4-D, triadimenol (TRI), bitertanol (BIT) and pentachlorophenol (PCP). Permeance was calculated from rates of penetration. Vertical bars represent 95% confidence intervals. (Data taken from Schreiber and Schönherr 1992a)

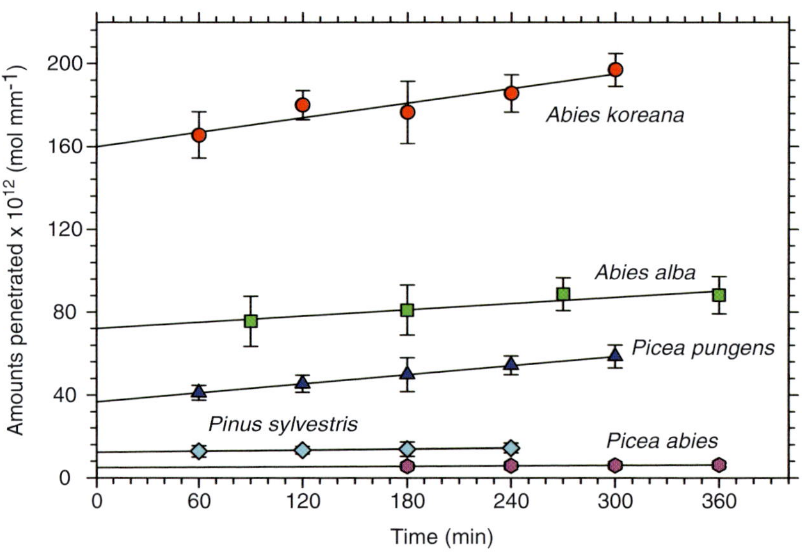

Fig. 6.8 Penetration of PCP into conifer needles at 25°C. Penetration was biphasic. Rates of penetration (*slopes*) and *y*-intercepts differed among plant species. Vertical bars represent 95% confidence intervals. (Redrawn from data of Schreiber and Schönherr 1992b)

desorbed from epicuticular waxes of conifers at similar rates to those from barley leaf wax. Rate constants of desorption (k_2) from the second compartment (i.e., the cuticle underneath the epicuticular wax) differed somewhat depending on species. They were a little higher with *Picea abies* and *Taxus baccata* ($2.4 \times 10^{-4}\,s^{-1}$) than with *Abies koreana* and *Abies alba* ($1.5 \times 10^{-4}\,s^{-1}$). These rate constants are also similar to k_2 measured with barley leaves ($1.4 \times 10^{-4}\,s^{-1}$). As with barley leaves, the size of the second compartment (CPT_2) did not depend on time of loading and was similar for all conifers. About 7% of PCP in needles was sorbed in compartment 2, that is in the cuticle. Fractions of PCP in compartment 1 decreased and compartment 3 increased with time, as was the case with barley leaves.

Permeances obtained from penetration were not significantly different from permeances obtained from desorption experiments. When rates of penetration were plotted against amount sorbed in compartments 1 and 2, a straight line was obtained. Hence, rates of penetration into compartment 3 (mesophyll) were proportional to amounts of chemicals sorbed in waxes and cuticles (Fig. 6.9). The amounts of chemicals sorbed are proportional to mass of wax, mass of cutin and partition coefficients. The mass of epicuticular wax per mm needle length was determined by dipping the needles for 1–2 s into chloroform. This method can be questioned (Sect. 2.2) but it is likely that most epicuticular waxes were dissolved without extracting too much embedded wax. Since compartments 1 and 2 were added (Fig. 6.9), the problem is not crucial. Amounts of surface wax varied among species (Table 6.6) and ranged from 1.56 to 7.1 μg mm^{-1} needle length (Schreiber and Schönherr 1993b). *Abies koreana* and *A. alba* had the highest amount of surface wax, and this is the reason

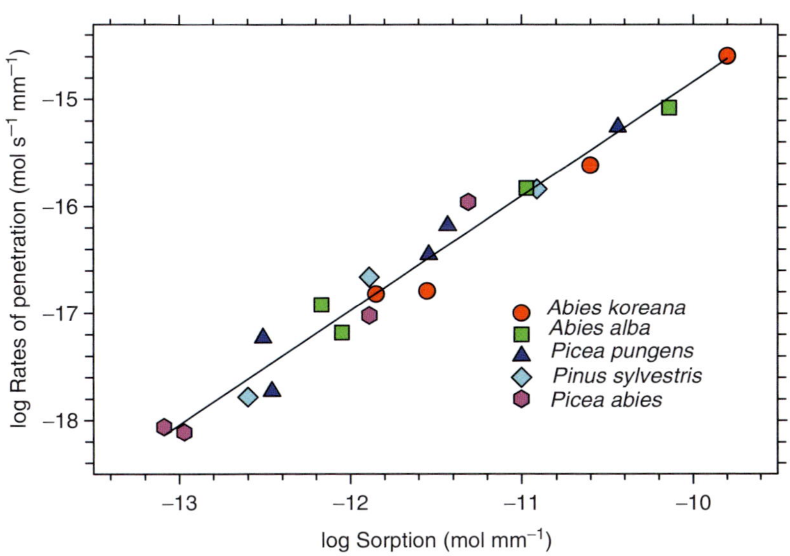

Fig. 6.9 Dependence of rate of penetration of 2,4-D, triadimenol, bitertanol, lindane and PCP into conifer needles on amounts of chemicals sorbed reversibly in compartments 1 and 2. (Redrawn from data of Schreiber and Schönherr 1992b)

Table 6.6 Amounts of surface wax, projected and specific surface areas and permeances of conifer needles from *Abies koreana, Abies alba, Picea pungens and Pinus sylvestris* (Schreiber and Schönherr 1992b)

Species	Surf. wax ($\mu g\ mm^{-1}$)	$A_{projected}$ ($mm^2\ mm^{-1}$)	$A_{specific}$ ($mm^2\ mm^{-1}$)	Solute	$P_{specific}$ ($m\ s^{-1}$)	$P_{projected}$ ($m\ s^{-1}$)
A. koreana	7.10	5	683	2,4-D	2.71×10^{-11}	3.70×10^{-9}
A. alba	2.50	5	254	2,4-D	3.11×10^{-11}	1.58×10^{-9}
P. pungens	–	3	65.7	2,4-D	2.24×10^{-11}	4.91×10^{-10}
P. sylvestris	–	3	41.3	2,4-D	3.78×10^{-11}	5.22×10^{-10}
P. abies	1.56	3	29.4	2,4-D	3.51×10^{-11}	3.43×10^{-10}
A. koreana	7.10	5	683	Triadim	2.54×10^{-11}	3.47×10^{-9}
A. alba	2.50	5	254	Triadim	5.58×10^{-11}	2.83×10^{-9}
P. pungens	–	3	65.7	Triadim	7.00×10^{-11}	1.53×10^{-9}
P. abies	1.56	3	29.4	Triadim	4.00×10^{-11}	3.92×10^{-9}
P. pungens	–	3	65.7	Lindane	4.23×10^{-10}	9.26×10^{-9}
P. sylvestris	–	3	41.3	Lindane	5.01×10^{-10}	6.91×10^{-9}
A. koreana	7.10	5	683	Bitertanol	4.23×10^{-10}	5.78×10^{-8}
A. alba	2.50	5	254	Bitertanol	6.84×10^{-10}	3.47×10^{-8}
P. pungens	–	3	65.7	Bitertanol	7.95×10^{-10}	1.74×10^{-8}
P. abies	1.56	3	29.4	Bitertanol	4.36×10^{-10}	4.27×10^{-9}
A. koreana	7.10	5	683	PCP	4.20×10^{-9}	5.74×10^{-7}
A. alba	2.50	5	254	PCP	3.90×10^{-9}	1.98×10^{-7}
P. pungens	–	3	65.7	PCP	6.63×10^{-9}	1.45×10^{-7}
P. sylvestris	–	3	41.3	PCP	3.30×10^{-9}	4.55×10^{-8}
P. abies	1.56	3	29.4	PCP	5.00×10^{-9}	4.90×10^{-8}

why both amounts penetrated and *y*-intercepts were highest (Fig. 6.8). The other chemicals had smaller partition coefficients than PCP, and for this reason sorption is smaller, and rates of penetration as well (Fig. 6.8).

This correlation between rates of penetration and sorption in surface wax and cuticles is not really surprising, since amounts sorbed and rates of penetration are proportional to concentration in the donor (6.6). After loading compartments 1 and 2 from the aqueous solution, the concentration in waxes and cuticles is the driving force for loading the mesophyll, at least as long as concentration of non-ionised PCP in the mesophyll remains small and insignificant and external concentration does not change.

6.2.2.4 Projected and Specific Surface Area

Whenever permeances are calculated, rates of penetration are divided by area and donor concentration (6.6). Glaucous leaves have large amounts of epicuticular wax, which can greatly increase surface area. With regard to foliar penetration, the question arises as to how much of this surface wax area is in contact with the aqueous donor solution. This in turn depends on wetting, and in absence of surfactants it seems plausible that only the tips of the wax rodlets typical for barley and conifers are in contact with water. Hence, the real contact area may be smaller or larger than the projected area. This problem was generally ignored, and the projected surface area was used as reference.

The standard method for estimating porosity and surface area of solids is sorption of N_2. This gas penetrates into porous solids, and knowing the area of a N_2 molecule the total internal and external surface is estimated from monolayer sorption. The amount in the monolayer is obtained by measuring N_2 sorption at various pressures (Gregg and Sing 1982).

Schreiber and Schönherr (1992b) attempted to measure the amount of PCP sorbed in a monolayer on the surface of epicuticular wax crystallites. PCP was dissolved in an aqueous buffer to ensure a high concentration of non-ionised species. The amount of PCP associated with needles was studied at various PCP concentrations, and data were analysed as BET isotherms. PCP is a planar lipophilic solute, and its area when laying flat on the surface of wax is known. From these isotherms the PCP in the monolayer was obtained, and assuming all to be sorbed superficially the area of this monolayer was calculated. The specific surface areas so obtained are shown in Table 6.6. Depending on species they vary between 29 and 683 $mm^2\,mm^{-1}$, which means that the specific surface area is larger by factors of 10–137 than the projected needle surface.

Permeances calculated using these specific areas ($P_{specific}$) eliminated all differences between plant species, and permeance depended only on type of solute via its cuticle/water partition coefficient. However, large differences in P are evident when it is calculated based on projected needle area. With 2,4-D, $P_{projected}$ was largest with *Abies koreana* and smallest with *Picea abies*, and the difference amounts to a factor of 10.8 (Table 6.6). Only permeances based on projected leaf area can be compared to permeances calculated for isolated CM or with barley leaves. With barley leaves, P for the same solutes ranged from 1.1×10^{-7} (PCP) to $5 \times 10^{-10}\,m\,s^{-1}$ (2,4-D) (Fig. 6.7). Similar $P_{projected}$ were obtained with *Picea pungens* (Table 6.6). Permeances measured with isolated CM can also be compared. 2,4-D permeability of *Citrus* CM was $2.8 \times 10^{-10}\,m\,s^{-1}$ (Table 6.3), and this is similar to 2,4-D permeability of *Picea abies* (Table 6.6).

Y-intercepts were obtained by extrapolating steady state penetration of PCP (Figs. 6.5 and 6.8). Specific surface area of needles, as calculated from sorption isotherms of PCP, are based on the assumption that all PCP molecules are located in a monolayer on the surface of wax crystallites (Schreiber and Schönherr 1992b). The two measurements are completely independent but they are correlated (Fig. 6.10). The slope is 0.82, showing that $A_{specific}$ was underestimated somewhat, but the correlation between the two parameters is convincing evidence that initial sorption of PCP molecules as surface monolayers and concomitant penetration into surface waxes cannot be separated. Model calculations based on amounts of epicuticular wax of the needles and wax/water partition coefficients show that 25–50% of all PCP, estimated from the positive y-intercepts (Fig. 6.8), were located inside the epicuticular wax and not on the waxy surface as assumed. This conclusion is supported by the fact that $A_{specific}$ increased with the amount of surface wax. Specific surface area estimated from the BET isotherms was larger with species having

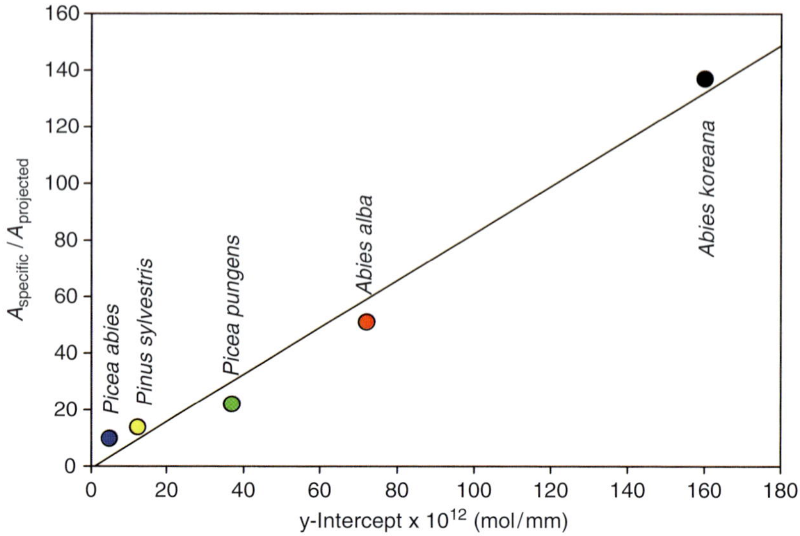

Fig. 6.10 Correlation between *y*-intercept measured by a penetration experiment (Fig. 6.8) with ratio $A_{specific}/A_{projected}$ estimated from sorption of PCP in conifer needles. The slope is 0.82. (Data taken from Schreiber and Schönherr 1992b)

more epicuticular wax (Table 6.6). Thus, specific surface areas estimated from BET isotherms are overestimates and not very precise.

Monolayer formation and sorption in waxes are related, and the sum of both constitutes the driving force of cuticular penetration. The *y*-intercept measured by steady state experiments (Figs. 6.5 and 6.9) characterise the sizes of the two compartments in which lipophilic chemicals are reversibly sorbed. Their sizes are proportional to the amount of wax and cutin and to partition coefficients of solutes (Table 6.5). CPT_1 and CPT_2 are intermediate compartments from which solutes penetrate into apoplast and symplast. Rates of penetration into CPT_3 are proportional to the sizes of CPT_1 and CPT_2. The absolute amounts of solutes sorbed reversibly in CPT_1 and CPT_2 are proportional to the solute concentration in the donor. This follows from the definition of the partition coefficient. For this reason, permeance is also proportional to the *y*-intercept.

6.2.2.5 Evaluation of Compartmental Analysis

Data generated using barley leaves and conifer needles demonstrate that steady state penetration in combination with compartmental analysis is a powerful tool for measuring and interpreting data on foliar penetration. Measurements are simple, and can be made with detached leaves if not too large. Isolating cuticles is not required, and leaves with stomata can be used. When stomata open in the light, infiltration of the intercellular space is possible, but this can be avoided when donor solutions have a relatively high surface tension. With leaves of *Zebrina* it was shown that infiltration of stomata will not occur if surface tension is 35 mN m^{-1} or higher (Schönherr and Bukovac 1972a). This even permits using low concentrations of surfactants to improve wetting of leaf surfaces. There is no need to worry that infiltration of stomata occurs but is not noticed. Infiltration can be detected with the bare eye, because dark spots will be seen in incident light which look bright in transmitted light. This phenomenon is due to a local change in refractive index when intercellular air spaces are filled with water.

Diffusion of solutes into the wound caused by cutting off the leaf at the petiole is no problem, because it can be quantified and corrected for easily (Fig. 6.4). The donor solution must be agitated to ensure mixing. If donor solutions are at ambient pressure (vessels open), there is no need to worry about pressure forcing liquid into open stomata. When working with barley leaves and conifer needles this was not a problem, even though test tubes were tightly closed.

So far the methods have been used only with lipophilic solutes. There is no reason why it should not work with polar non-electrolytes or with ions. With polar solutes the sizes of CPT_1 and CPT_2 are probably very small, and the *y*-intercept is close to zero. Permeance can be calculated from the slope of the penetration or the desorption graphs which are superimposed if sorption in wax and cuticles is insignificant (cf. Fig. 6.13). With leaves that are easily wetted, a surface film of donor can be estimated by desorption, and all donor solution will be washed off with the first change of desorption medium. This offers the possibility to measure permeability of delicate leaves such as *Arabidopsis*.

The method works well as long the leaf surface is not densely populated by microorganisms, as was observed when working with older conifer needles sampled from forest trees (Schreiber and Schönherr 1992c). Plants grown in growth chambers or greenhouses usually have clean surfaces. In any case, it is good practice to check for surface contaminations.

6.2.3 Steady State Penetration into Leaf Disks Using the Well Technique

Broadleaved plants can have very large leaves which are not suitable for the submersion technique. In these cases, penetration can be measured using a droplet method. Small droplets of donor solutions are placed on the upper or lower surfaces of leaves attached to or dissected from plants. This approach is much closer to the situation after spray application to the foliage, and this is often considered an advantage. However, experiments of this kind are beset with severe problems, as penetration proceeds under uncontrolled conditions. For instance, concentrations and *p*H of donor solutions change during droplet drying, temperature of the donor differs from leaf and surrounding air, contact area between donor and leaf is difficult to estimate precisely and may vary with time, and the solutes may solidify or crystallise. These problems are almost as bad in growth chambers than in the field.

If researchers manage to distinguish between solutes in and on the leaves, the best result of such experiment is fractional penetration during some arbitrary time after droplet application.

Even in this case it is generally not realised that the velocity of penetration depends on size of droplets, more precisely on the ratio droplet volume (V_{droplet}) over area of contact (A_{contact}) between droplet and leaf surface. The situation can be demonstrated assuming a hemispherical droplet positioned on a leaf. This means that the leaf is difficult to wet and the contact angle is 90°. It is assumed that V_{droplet}, A_{contact} P and C_{donor} are constant and do not vary with time. Hence, the droplet must not dry up. Our starting point is (2.25), which is repeated here with appropriate subscripts:

$$\frac{-P \times A_{\text{contact}} \times t}{V_{\text{droplet}}} = \ln \frac{C_{\text{donor}}}{C_0}. \tag{6.11}$$

C_0 is the initial donor concentration ($t = 0$) and C_{donor} is the concentration at any later time. If penetration occurs, C_0 decreases with time and we want to calculate the time needed for 50% of the dose to penetrate into the leaf, that is $C_{\text{don}}/C_0 = 0.5$ or $\ln C_{\text{don}}/C_0 = 0.693$. Rearranging (6.11), we see that the half-time

$$t_{1/2} = \frac{0.693}{P} \times \frac{V_{\text{droplet}}}{A_{\text{contact}}} \tag{6.12}$$

depends on the volume of the droplet and the contact area. For a hemispherical droplet $V_{\text{droplet}}/A_{\text{contact}} = (2/3) \times r_{\text{droplet}}$. We have calculated half-times for frequent values of permeances of cuticles and droplet sizes produced by conventional spraying equipment (Fig. 6.11). Such spherical droplets have mean diameters ranging from 100 to 500 μm, which corresponds to volumes of 0.5–65 nl.

When droplet radii increase from 33 to 133 μm, half-times increase by a factor of 1,000; and depending on permeance, half-times were in the range of minutes to 280 h. If permeance is very high (10^{-7} m s^{-1}) it might be possible to maintain $V_{\text{droplet}}/A_{\text{contact}}$ fairly constant, but with lower P this is impossible. Contact angles on leaves vary greatly, and they depend on surface tension of the donor solutions. Both factors greatly affect half-times because they affect $V_{\text{droplet}}/A_{\text{contact}}$. Better wetting leads to smaller $V_{\text{droplet}}/A_{\text{contact}}$, even with constant droplet volumes, and this greatly reduces half times.

These purely physical considerations have consequences for spray applications. Loss of agrochemicals by rain and volatilisation can be minimised by using a larger number of small droplets. There is a limit to this strategy because very small droplets can be lost by drift. However, for rapid penetration it is a good strategy to deliver a constant dose with more droplets of small size, or use higher concentrations instead of low concentrations and large droplets.

Apart from these practical aspects, it should be clear that experiments with small droplets are extremely difficult to analyse, and misinterpretations of cause and effect are unavoidable. However, such experiments can be meaningful if fractional

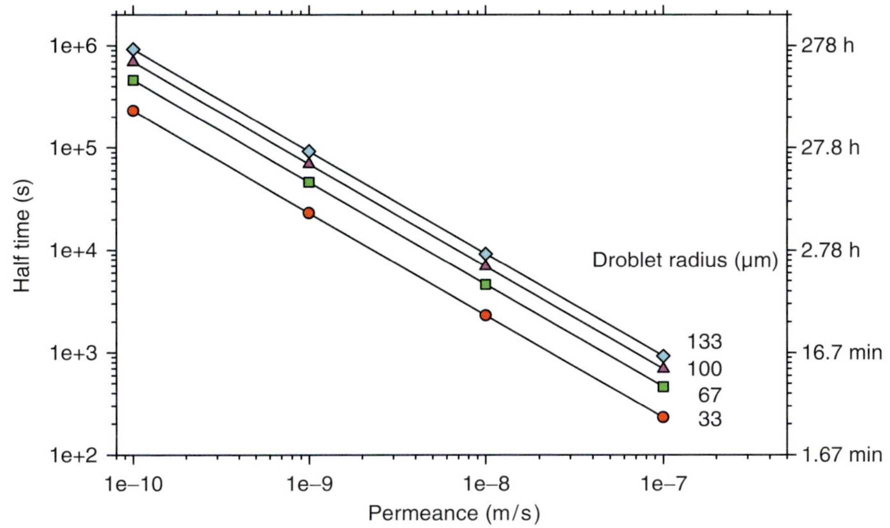

Fig. 6.11 Half-times for solute penetration from hemispherical droplets of different size as a function of permeance

penetration of a constant dose is aimed at, and we demonstrate this in Chap. 5 and Sects. 6.3.1–6.3.4.

In an attempt to circumvent problems associated with droplet experiments, glass wells have been glued to leaf surfaces using silicone rubber (Fig. 6.12). Relatively large volumes of up to 1 ml donor can be pipetted into these wells, and both contact area and donor volume can be kept constant. Problems may arise if the glue is phototoxic or when the solute is sorbed in the glue. Schönherr (1969) and Schönherr and Bukovac (1978) have used this approach for studying foliar penetration of succinic acid-2,2-dimethyl hydrazide (Alar), which is a zwitterion. Small glass tubes (10 mm diameter and 7 mm height) were attached to leaf discs using silicon rubber and a non-toxic catalyst. Silicon rubber provides a good seal even over veins, and surfactant solutions did not leak out. At the end of the experiment the rubber remained attached to the glass and the leaf disk could be peeled easily and did not interfere with subsequent processing of leaf disks (autoradiography and counting radioactivity). Rates of penetration were constant, as penetration plots were linear with all treatments. Plots intersect the origin, that is, there was no measurable hold-up time and no positive *y*-intercepts due to sorption in wax and cutin. From the slopes and the donor concentration, permeance can be calculated. Permeance was very low, depending on treatment. It ranged from 5×10^{-11} to 25×10^{-11} m s^{-1} (Fig. 6.13).

Permeability of the lower leaf surface was higher than that of the upper one (Fig. 6.13). Light increased permeability of both leaf surfaces, and the wetting agent Tween 20 (polyoxylethylene sorbitane monolaurate) increased rates of penetration. This wetting agent did not cause stomatal infiltration, and its effect on rates of penetration was related to increased contact areas between donor and leaf. The light

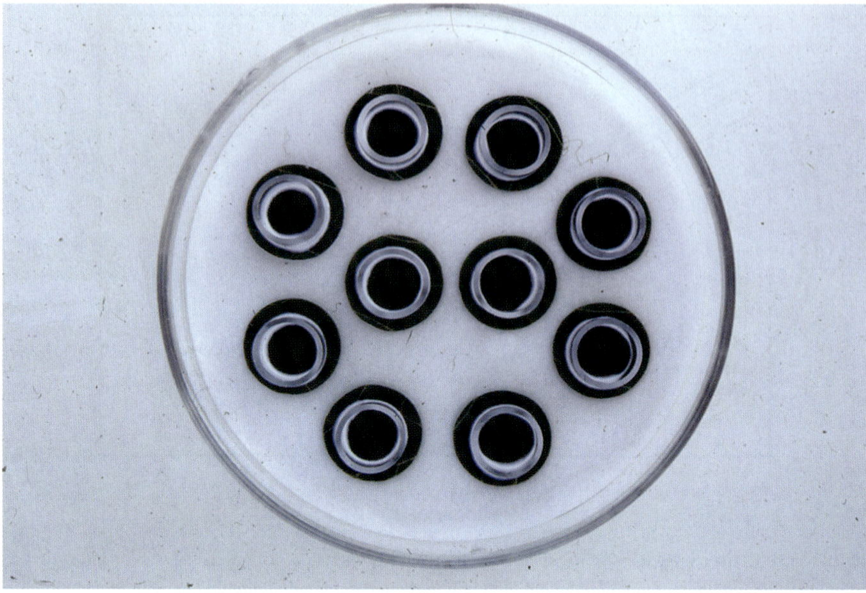

Fig. 6.12 Penetration units consisting of bean leaf disks (17 mm in diameter) and glass tubes (10 mm in diameter) attached with silicon rubber. The units were positioned on moist filter paper in Petri dishes which permitted studying the effect of light on rates of penetration (taken from Schönherr 1969)

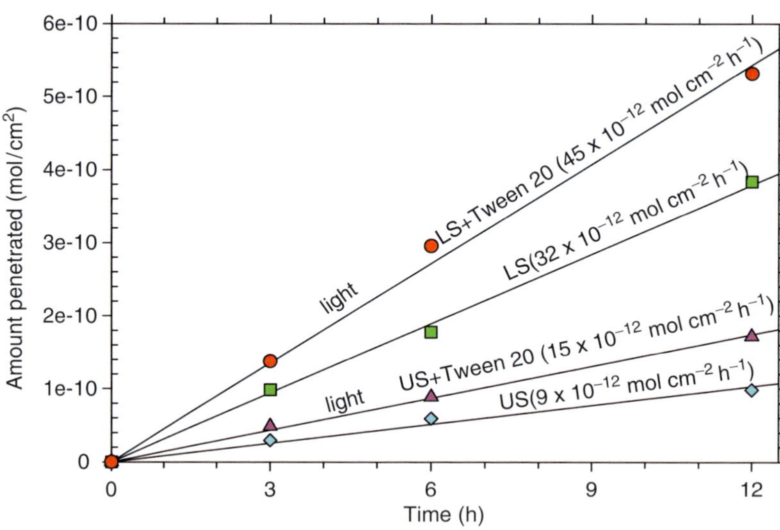

Fig. 6.13 Penetration of succinic acid-2,2-dimethylhydrazide (Alar) into primary leaves of kidney bean at 25 °C. Donor concentration was 5×10^{-4} mol l^{-1} and was buffered with citrate–phosphate buffer at *p*H 5. Upper (US) and lower (LS) leaf surfaces are marked on the plots. Fluorescent light (5.25 mW m^{-2}) was used when indicated and Tween 20 was added to the donor at 0.1%. Rates of penetration are given in parentheses on the plots. (Data taken from Schönherr and Bukovac 1978)

effect on rates was shown to be caused by stomatal opening which increased permeability of cuticular ledges (Schönherr and Bukovac 1978). Alar is an electrolyte, and at *p*H 5 it is neutral because negative and positive charges are present in equal amounts (Schönherr and Bukovac 1972b). The role of stomata in foliar penetration of ionic compounds is treated comprehensively in Chap. 5.

Alar penetration into bean leaves using the leaf disk method with attached wells is a good example to demonstrate the merits of the method. The main advantages are the facts that permeability of large leaves can be measured, and it is possible to test if permeability of lower and upper leaf surfaces differ. With submerged leaves this is not possible. Kirsch et al. (1997) have used this method to compare solute permeability of isolated CM with non-isolated cuticles (Sect. 6.5). It is a somewhat laborious undertaking, since sampling is destructive and when rates of penetration (that is the time course of penetration) are studied, different leaf disks must be used for each time interval. This increases variability, and a large number of leaf disks (25 and more) must be used for representative sampling and for testing linearity. We have shown above why this is absolutely essential.

Autoradiography makes it possible to study distribution of radioactive labels, provided the leaves are freeze-dried quickly to avoid redistribution and metabolism. Autoradiographs of selected bean leaf disks show that in the presence of Tween 20 more succinic acid-2,2-dimethylhydrazide penetrated and radio-label was much more uniform (Fig. 6.14). Sometimes the label spread along the veins and reached the cut edges. In later experiments with $CaCl_2$, this was avoided by placing the leaf disks on stainless steel washers rather than directly on moist filter paper (Fig. 5.4b).

Permeances for non-ionised 2,4-D, salicylic acid and benzoic acid were measured using the upper, astomatous leaf surfaces or CM obtained from *Prunus laurocerasus, Ginkgo biloba* and *Juglans regia* leaves (Kirsch et al. 1997). With leaf disks and CM, penetration plots were linear and permeances could be calculated from slopes. Permeances measured with leaf disks and CM did not differ significantly with all three species and compounds. Clearly, enzymatic isolation of cuticles did not affect permeability of cuticles.

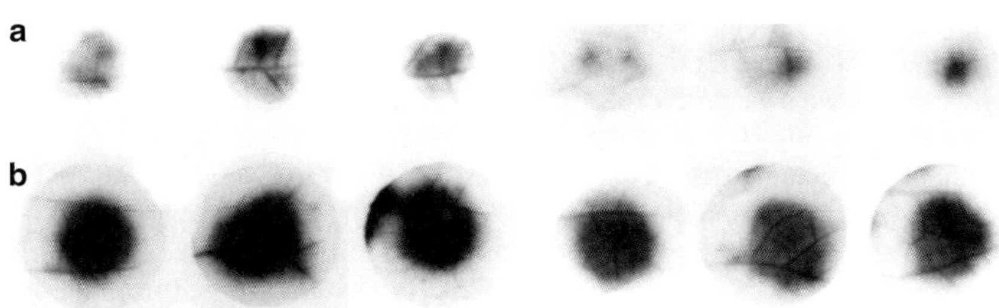

Fig. 6.14 Autoradiographs of bean leaf disks after penetration of ^{14}C labelled succinic acid-2,2-dimethylhydrazide (Alar) in light for 12 h. Penetration without (**a**) and with 0.1% Tween 20 (**b**) as wetting agent

6.3 Diffusion with Changing Donor Concentrations: The Transient State

Permeance (P) is a composite property (2.18). It is proportional to solubility (K) and mobility (D) of water and solutes in membrane, and inversely proportional to thickness of the membrane (ℓ). In homogeneous membranes, solute permeability is characterised using D and K because they are easily measured. Cuticles are highly asymmetrical (Sect. 1.4), and both water concentration and mobility vary with position (Chap. 4). Water transport occurs in aqueous pores, cutin and waxes, and their mutual arrangements determine water permeability. The question arises whether this also applies to solutes. In Sect. 6.2 we used permeances to characterise solute permeability, which can be measured easily in the steady state using various methods. Unfortunately, since P is a mixed quantity it is not possible to explain why permeances differ among cuticles from different species. In an attempt to obtain a better understanding of solute permeability in cuticles, we have developed methods to estimate solute mobility in CM, MX and in waxes.

6.3.1 Simultaneous Bilateral Desorption

Diffusion coefficients can be estimated from sorption and desorption experiments (Sect. 2.6). We have used this approach to measure mobility of lipophilic 2,4-D in CM and MX membranes. Astomatous CM and MX membranes obtained from leaves and fruits were submerged in aqueous buffer (*p*H 3) containing ^{14}C-labelled 2,4-D. After equilibration they were removed from the buffer, blotted dry, flattened on a piece of Teflon and air-dried. Dry membranes were inserted between the two half-cells of a transport apparatus made of stainless steel (Fig. 9.4). This made it possible to desorb 2,4-D separately from the outer and inner surfaces of the membranes. Borax buffer (*p*H 9.18), in which 2,4-D is fully ionised, was used as desorption medium. By changing desorption media quickly and repeatedly, the amounts of 2,4-D desorbed at various times (M_t) were studied.

Desorption was very rapid, and 90% of the 2,4-D initially contained in the CM and MX membranes were desorbed in 4 h (Fig. 6.15a). With a homogeneous membrane, 50% of solute should be desorbed from either side and the two desorption plots should be symmetrical. This was clearly not the case, neither with CM and MX. Only very small amounts (5%) were desorbed from the morphological outer surfaces of the *Citrus* CM, while 17% could be desorbed from the outer surface of the MX membranes. The largest fraction of 2,4-D was desorbed from the inner surface, as in 240 min 86% (CM) and 71% (MX) left the membranes through the morphological inner surfaces. Clearly, these membranes are highly asymmetrical, and this asymmetry can be characterised by comparing the initial rates of desorption from the inner and outer surfaces respectively. During the first 3 min, 163 times more 2,4-D was desorbed from the inner surface of the CM than from the outer. With

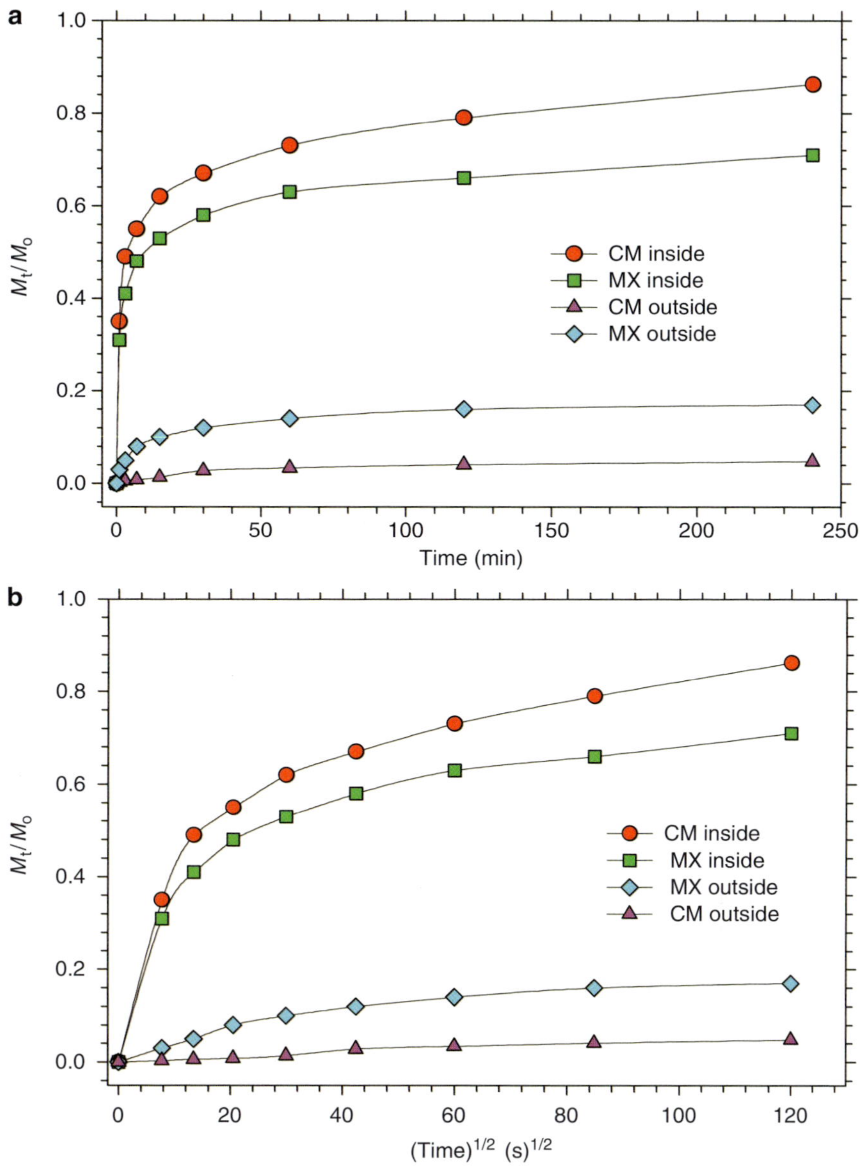

Fig. 6.15 Simultaneous bilateral desorption of 2,4-D from CM and MX membranes of *Citrus aurantium* leaves. The membranes had been preloaded with ^{14}C-2,4-D, and were desorbed at 25°C with borax buffer having a *p*H of 9.18. M_0 is the amount of 2,4-D initially contained in the membrane and M_t is the amount desorbed at time t. In **(a)** M_t/M_0 is plotted vs time, while in **(b)** it was plotted vs the square root of time. (Redrawn from Schönherr and Riederer 1988)

MX membranes this asymmetry factor was only 10, because desorption from the outer surface was higher. This is still a high asymmetry, since with a homogeneous membrane the factor should have been unity.

This asymmetrical desorption pattern indicates that CM and MX membranes are composed of at least two compartments. The bulk of the 2,4-D was contained in the inner volume element, and it was desorbed through the inner surface. We call this inner domain of the cuticle the sorption compartment (soco), and morphologically it is identical with the cuticular layer(s) seen in TEM (Sect. 1.4). The outer domain across which only very small amounts of 2,4-D were desorbed is the cuticle proper, and we refer to it as limiting skin or limiting layer (Fig. 6.16). Volumes and thicknesses of these two layers cannot be deduced from desorption plots. As in 4 h only 5% 2,4-D diffused across the outer surface of the CM, diffusion coefficients in the limiting skin must have been very low and much lower than in the sorption compartment. The data do not reveal if it is a wax layer on top of the cuticle or if the barrier consists of waxes embedded in the outer fraction of the MX. A combination of both is possible as well. In any event waxes are involved, since extracting them reduced asymmetry greatly (Fig. 6.15).

When M_t/M_0 (desorption through the inner surface) was plotted against the square root of time (Fig. 6.15b), plots were not linear up to M_t/M_0 equal to 0.5, as would be expected with homogeneous membranes (Fig. 2.10b), but diffusion coefficients can still be estimated from the initial slope using (2.35). Since both CM and MX are heterogeneous, these D-values are some kind of average characterising diffusion of 2,4-D in the sorption compartment (cuticular layer). For these calculations,

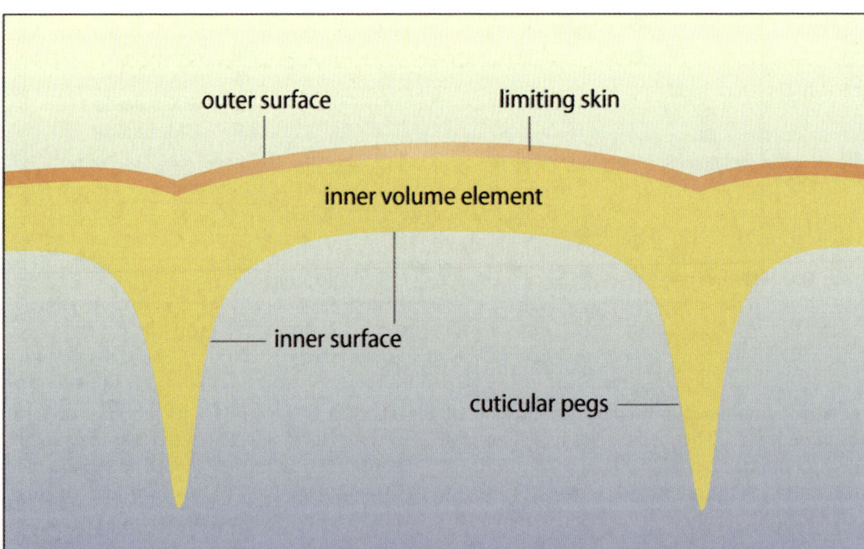

Fig. 6.16 Schematic drawing of a cross section of a cuticle showing the thin limiting skin and the thick sorption compartment (not to scale). Modified from Bauer and Schönherr (1992)

the thickness (ℓ) of the compartments which drained through the inner surface of the membranes must be estimated. Using the weight average thickness (2.6 μm), which amounts to neglecting thickness of the limiting skin, mean diffusion coefficients for desorption from the *Citrus* CM and MX membranes are $3 \times 10^{-15}\,m^2\,s^{-1}$ and $2 \times 10^{-15}\,m^2\,s^{-1}$ respectively. Since thickness enters as the square (2.35), the difference is not significant. Thickness of the cuticle proper ranges from 0.05 to 0.5 μm (Jeffree 2006), and if it is assumed that in *Citrus* 10% of the mass represents limiting skin, diffusion coefficients for CM and MX are 2.2×10^{-15} (CM) and $1.7 \times 10^{-15}\,m^2\,s^{-1}$ (MX). These D values are a little smaller since ℓ is only 90% of the total thickness of the cuticles. Averaging these four values, we obtain a mean D of $2 \times 10^{-15}\,m^2\,s^{-1}$ for desorption of 2,4-D from the sorption compartment through the inner surface of *Citrus* CM and MX.

Desorption plots obtained with fruit (*Lycopersicon esculentum, Capsicum annuum*) and *Ficus decora* leaf CM resemble those shown in Fig. 6.15. Again, in 6 h only 2–3% of the total amount of 2,4-D was desorbed through the outer surfaces of these CM. With MX membranes, efflux of 2,4-D across the outer surface amounted to 12–26%. Extraction of waxes increased efflux through the outer surface of the MX significantly, but asymmetry was not eliminated. Diffusion coefficients calculated from the first 3 min desorption intervals are $7 \times 10^{-15}\,m^2\,s^{-1}$ (*Ficus*), $4 \times 10^{-14}\,m^2\,s^{-1}$ (*Capsicum*) and $5 \times 10^{-14}\,m^2\,s^{-1}$ (*Lycopersicon*) respectively. The D values for fruit CM are larger than for leaf CM, but this may be related to extensive cutinisation of anticlinal walls. This leads to overestimation of cuticle thickness by factors of about 2 (Riederer and Schönherr 1985). If calculations are repeated using half of the gravimetric thicknesses, diffusion coefficients become $9 \times 10^{15}\,m^2\,s^{-1}$ (*Capsicum*) and $1.4 \times 10^{-14}\,m^2\,s^{-1}$ (*Lycopersicon*), which is still a little higher than the value estimated for *Citrus* and *Ficus* leaf CM.

Asymmetry as revealed by these desorption experiments is remarkable. It is excellent evidence that the CM and MX membranes have a limiting barrier at their outer surfaces in which solute mobility is much lower than in the sorption compartment. With all fruit and leaf CM, efflux through these limiting skins amounted to only 2–3% of the total 2,4-D. Asymmetry was smaller with MX membranes but it was still pronounced, showing that waxes greatly contributed to barrier properties of the limiting skin, but diffusion coefficients in cutin itself differ at the outer surface and in the sorption compartment.

Extraction of waxes had little effect on diffusion coefficients in the sorption compartments. This is astounding, since these sorption compartments probably contain embedded waxes. At least for *Ficus* cuticles, strong birefringence in the cuticular layer has been demonstrated (Sitte and Rennier 1963), yet extraction of waxes had little effect on diffusion coefficients in the sorption compartments (Schönherr and Riederer 1988). Only waxes in the limiting skin reduce solute mobility, but D cannot be calculated from simultaneous bilateral desorption, simply because solutes escape nearly quantitatively through the inner surface of the membranes.

6.3.2 Unilateral Desorption from the Outer Surface

Solute mobility in the limiting skin can be estimated from unilateral desorption from the outer surface. This method eliminates solute loss through the inner surface because no desorption medium is in contact with it. Some changes in apparatus were made to simplify handling and to adopt it to neutral solutes. The cuticles are inserted in an apparatus shown in Fig. 6.17. The morphological inner surface is exposed in the orifice of the lid. ^{14}C-labelled test compounds are dissolved in water or buffer and applied as 200 µL droplet to the centre of the inner surface of the CM. During evaporation of water the lipophilic solutes quantitatively penetrate into the sorption compartment (soco). As seen in Fig. 6.15, desorption is a very rapid process as half of the 2,4-D was desorbed in about a minute. Sorption is the reverse process but it proceeds with the same velocity. As soon as the water had evaporated, the opening of the lid was closed with transparent sticky tape (Tesafilm). This ensures 100% humidity in the air over the inner surface of the CM, and it prevents a radioactive spill in case a membrane breaks.

Desorption from the outer surface is initiated by pipetting the desorption medium into the chamber. The chambers are placed with the cuticle facing down into wells of a thermostated aluminium block which is rocked slightly to ensure mixing. As a desorption medium, a buffer in which the solutes are ionised can be used with weak acids or bases. With neutral solutes, a phospholipid suspension ($10\,\mathrm{g\,l^{-1}}$) is suitable. It is prepared by sonicating soybean lecithin in hot water (60 °C). Small vesicles are

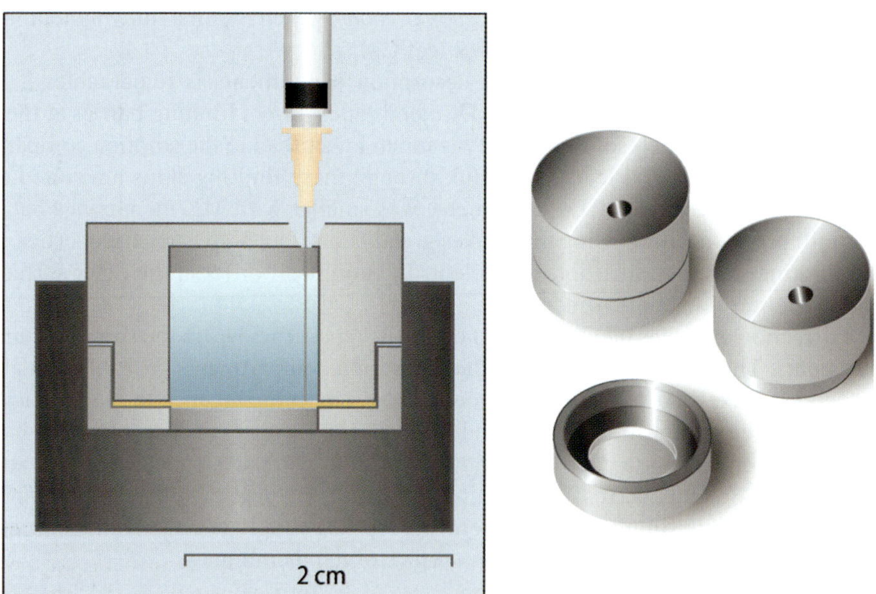

Fig. 6.17 Apparatus used for UDOS (unilateral desorption from the outer surface) experiments

formed, and lipophilic solutes are sorbed in these PLS vesicles. This maintains the solute concentration in the water surrounding the vesicles at practically zero, and it ensures good wetting of the waxy cuticle surfaces, because surface tension is lower than in water. The desorption medium is periodically withdrawn quantitatively and replaced by fresh medium. At the end of the experiment the cuticle exposed in the orifice of the lid is cut out, and residual radioactivity in the CM is extracted with scintillation cocktail. Radioactivity in desorption media and cuticles is determined with a scintillation counter.

Radioactivity in the desorption media at time t is M_t and the sum of the radioactivity in desorption media and cuticle is M_0, which was routinely compared to the amount applied; recovery was always 100%. M_t/M_0 is the solute fraction desorbed, and $(1 - M_t/M_0)$ is the solute fraction remaining in the CM. Plotting the natural logarithm of $(1 - M_t/M_0)$ vs time always resulted in straight lines (Fig. 6.18). The slopes of the plots are the rate constants (k^*) as defined by the equation

$$-\ln\left(1 - M_t/M_0\right) = k^* t. \tag{6.13}$$

These rate constants are related to permeance (P) as shown in (2.26), which for convenience is repeated here

$$\frac{\ln\left(C_{\text{donor}}/C_0\right)}{t} = \frac{-PA}{V_{\text{donor}}} = k^* \tag{6.14}$$

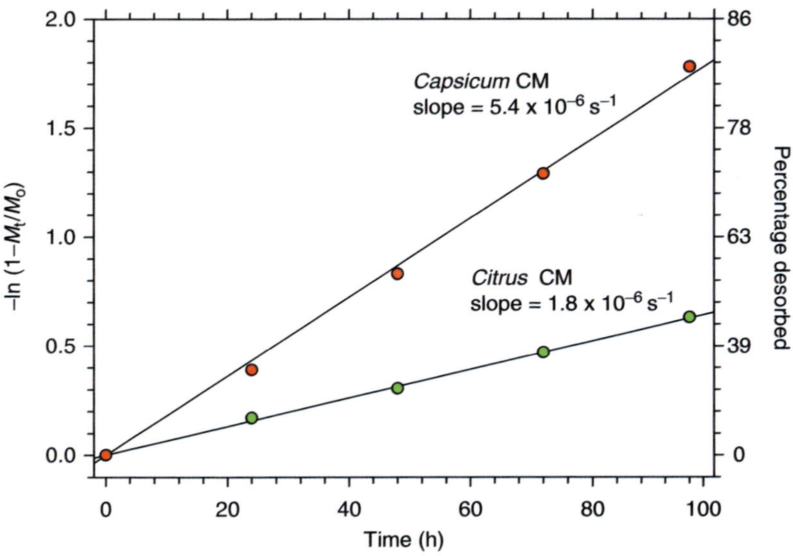

Fig. 6.18 Unilateral desorption of pentachlorophenol from the outer surface of *Citrus* and *Capsicum* CM. Slopes of the plots are the rate constants (k^*). (Redrawn from Bauer and Schönherr 1992)

with the exception that amounts (M) are used instead of donor concentrations (C_{donor}). Equation (6.14) states that the donor concentration decreases exponentially with time, while in UDOS the fraction of solute contained in the sorption compartment $(1 - M_t/M_0)$ decreases exponentially with time. It is convenient to use amounts rather than concentrations, because the volume of the sorption compartment is not known precisely (Fig. 6.16). Using amounts instead of concentrations introduces no error, since $C_{donor} = M/V_{donor}$ and the volume of the sorption compartment is constant during the experiment. At the beginning of the experiment, the solutes are dissolved in the lipids (cutin and wax) of the sorption compartment, and an aqueous donor phase is absent. Hence, the driving force in UDOS is not the concentration of an aqueous donor but the concentration in the sorption compartment. The ratio of this two concentrations is the partition coefficient

$$K_{soco/water} = \frac{C_{soco}}{C_{water}} \tag{6.15}$$

and this is taken care of by marking this type of permeance with an asterisk (P^*). With these definitions, (6.14) becomes

$$\frac{\ln(M_t/M_0)}{t} = \frac{-P^* A_{soco}}{V_{soco}} = \frac{-P^*}{\ell_{soco}} = k^*. \tag{6.16}$$

This permeance (P^*) is related to the permeance obtained from a steady state experiment, using the solute concentration of the aqueous donor by the partition coefficient:

$$P^* = \frac{P}{K_{soco/water}}. \tag{6.17}$$

Thus, with two independent measurements it is possible to fully describe solute diffusion across cuticles. From UDOS k^* and P^* are obtained, and P may be derived from a steady state experiment. K_{cw} may be determined in a sorption experiment, or it can be calculated from (6.17).

This was tested by at first measuring permeance of pepper fruit CM to 2,4-D in a steady state experiment followed by an UDOS experiment. The same four CM were used in both experiments, which eliminated variability among individual CM. Diffusion of 2,4-D was measured from the inner to the outer side of the CM. K and P were measured in a steady state experiment. Partition coefficients were obtained from the mass balance of 2,4-D (amount in the CM = total amount added to the donor – amount in the receiver) and the mass of the CM exposed to the donor solution. P was calculated from the steady state flux and the concentration difference between donor and receiver solutions. At the end of the experiment, all 2,4-D was removed from the CM by extensively washing donor and receiver chambers with borax buffer, followed by washing with water to eliminate borax.

With the receiver side empty, fresh donor was added to the compartment in contact with the inner side of the CM. The system was equilibrated for 3 h. Due to sorption of 2,4-D in the sorption compartment, the donor concentration decreased rapidly, and after equilibrium had been established the partition coefficient was

Table 6.7 Comparison of partition coefficients and permeances determined at 25 °C in the steady state and with UDOS

CM	Steady state			UDOS		$\frac{P^*_{UDOS}}{P^*_{steady\ state}}$
	$K_{cuticle/buffer}$	$P(\text{m s}^{-1})$	$P^*(\text{m s}^{-1})$	$K_{soco/buffer}$	$P^*(\text{m s}^{-1})$	
1	416	1.05×10^{-8}	2.52×10^{-11}	400	1.75×10^{-11}	0.69
2	396	1.49×10^{-8}	3.76×10^{-11}	386	3.02×10^{-11}	0.86
3	326	4.76×10^{-8}	14.60×10^{-11}	325	12.60×10^{-11}	0.86
4	398	11.20×10^{-8}	28.10×10^{-11}	389	24.50×10^{-11}	0.87

Taken from Bauer and Schönherr (1992). Donor solutions had a *p*H of 3.0

calculated from the mass of the CM, the volume of the donor solution, the decrease in donor concentration, and the equilibrium concentration of the donor. The donor solution was removed, and borax buffer was added as receiver and k^* was measured in an UDOS experiment. P^* was calculated from (6.16) using the weight average thickness (ℓ_{CM}) of the CM, rather than the unknown thickness of the sorption compartment (ℓ_{soco}).

Partition coefficients were virtually identical in the two types of experiments (Table 6.7), except that $K_{soco/buffer}$ tended to be slightly lower, possibly because the limiting skin was not yet in equilibrium after 3 h of loading. Steady state permeance (P) varied among CM by a factor of about 10, which is typical for CM of most species. P^* values calculated for steady state experiments varied by the same factor because $K_{cuticle/buffer}$ varied between CM only slightly. Agreement between P^* obtained from rate constants determined in UDOS experiments (6.16) and those calculated from steady state data (6.17) is very good, considering that both permeances varied among CM by a factor of more than 10. This shows that both types of experiment provide comparable data, and P can be calculated from rate constants (k^*) measured using UDOS.

P^* calculated as $\ell_{CM} \times k^*$ (UDOS) is consistently smaller than P^* calculated from steady state data (Table 6.7). The factor averaged over all four CM is 0.805, and this cannot be attributed to differences in partition coefficients, which amount for only 2% on average. The limiting skin has an unknown but finite thickness, hence $\ell_{soco} \times k^*$ should be smaller than $\ell_{CM} \times k^*$. However, P^* calculated from rate constants were smaller than P^* calculated from steady state permeance P and partition coefficients. It appears that ℓ_{CM} systematically underestimates the real path length.

6.3.2.1 Estimating Solute Mobility from Rate Constants

Steady state permeance (P) is proportional to the partition coefficient (2.18). Permeance of CM is determined by the thickness of the limiting skin, while K reflects solubility in the sorption compartment. These definitions result in

$$P = \frac{D_{\text{limiting skin}} \times K_{\text{soco/water}}}{\ell_{\text{limiting skin}}}. \tag{6.18}$$

P^* calculated from rate constants is independent of K, because in UDOS the driving force is the solute concentration in the sorption compartment, not the concentration of an aqueous donor (6.14) and (6.16). Combining (6.18) with (6.17) and solving for D we obtain

$$D_{\text{limiting skin}} = P^*_{\text{limiting skin}} \times \ell_{\text{limiting skin}} \tag{6.19}$$

and substituting $k^* \times \ell_{\text{soco}}$ for P^* (6.16) we obtain

$$D_{\text{limiting skin}} = k^* \times \ell_{\text{soco}} \times \ell_{\text{limiting skin}}. \tag{6.20}$$

which shows that the rate constant characterises solute mobility in the limiting skin. D could be calculated from the rate constant if the two thicknesses were known. They are not known, however, but for the sake of argument reasonable assumptions can be made.

For an order of magnitude estimate of D it suffices to assume that the limiting skin has the same thickness of the cuticle proper seen in TEM. Jeffree (2006) has summarised the available data and even though these are not always well-defined he estimated that in most species the CP has a thickness ranging from 0.05 to 0.5 μm. To obtain thickness of the sorption compartment [equated to the cuticular layer(s)], these values can be subtracted from total thickness. As already pointed out, weight average thickness overestimates thickness of cuticles over periclinal walls due to anticlinal pegs, and in fruits extensive cutinisation of cell walls in a multiple epidermis can occur (Schönherr and Riederer 1988). Since we have no alternative we have to live with this situation, and since we only aim at an order of magnitude estimate these assumptions can be tolerated.

Let total thickness be 11 and 2.5 μm for pepper fruit and *Citrus* leaf CM respectively. Their limiting skins are taken to be 0.5 (*Capsicum*) and 0.25 μm (*Citrus*). With these thicknesses and the rate constants shown in Fig. 6.18, PCP diffusion coefficients in *Capsicum* and *Citrus* CM calculated from (6.20) are 2.8×10^{-17} and $1 \times 10^{-18}\,\text{m}^2\,\text{s}^{-1}$ respectively.

In Fig. 6.19, a model calculation demonstrates the effect of thickness of the limiting skin on magnitude of diffusion coefficients, when total thickness (3 μm) of the CM is kept constant. With increasing thickness of the limiting skin, D increases. When thickness of the limiting skin increases from 0.1 to 0.5 μm, D increases by a factor of 7.75. This shows that an order of magnitude estimate of D is possible, even when the thickness of the limiting skin is not precisely known.

In most instances there is no need for calculating D, because rate constants can be used as measures of solute mobility. A case in point is when plant species or solutes are to be compared, or when effect of temperature and accelerators on solute mobility must be quantified. In determining rate constants, no assumptions regarding thicknesses of limiting skin and sorption compartment must be made.

There is one assumption implicit in calculating rate constants using UDOS. In all CM the bulk of the CM is the sorption compartment, in which the solutes are

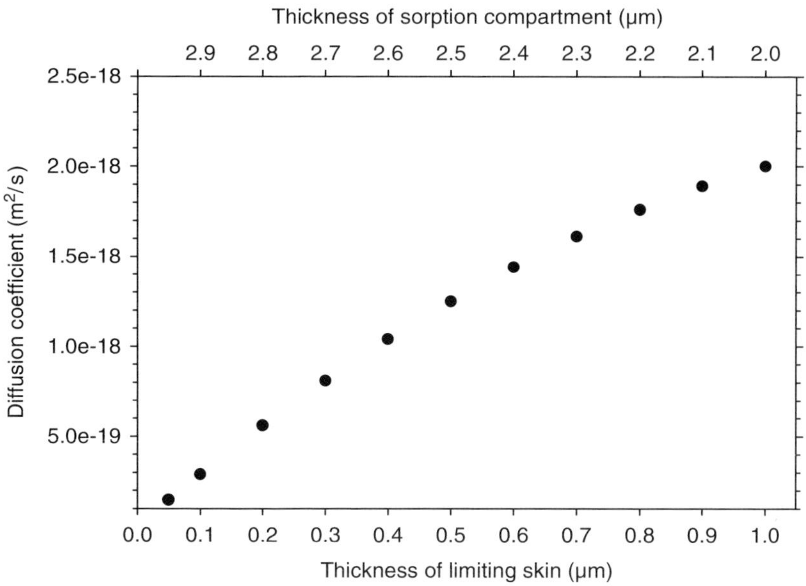

Fig. 6.19 Effects of thicknesses of cuticular compartments on diffusion coefficient. Total thickness of cuticle is 3 μm, and rate constant was taken to be $1 \times 10^{-6}\,s^{-1}$

dissolved initially. It is implicitly assumed that desorption from the outer surface is not limited by diffusion in anticlinal pegs and other portions of the sorption compartment remote from the limiting skin. This can be taken for granted as long as diffusion coefficients in the limiting skin is 50–100 times lower than in the sorption compartment. Whenever this is the case, desorption plots (Fig. 6.18) are linear, and in this case there is no need to worry.

Using simultaneous bilateral desorption, the diffusion coefficient of 2,4-D in the sorption compartment was estimated to be 10^{-14} to $10^{-15}\,m^2\,s^{-1}$ (Fig. 6.15). D values estimated from Fig. 6.18 are 10^{-18} to $10^{-19}\,m^2\,s^{-1}$, which is lower by more than three orders of magnitude. We can also estimate D in the sorption compartment using unilateral desorption. If PCP is desorbed from the inner surface of the CM a nonlinear plot is obtained with an initial slope of $0.0017\,s^{-1}$ (Fig. 6.20). This is larger by a factor of 944 than measured for desorption from the outer surface (Fig. 6.18). Rate constants decrease with time, because the sorption compartment is heterogeneous, as was seen already in Fig. 6.15. Desorption from the outer surface of MX-membrane was also non-linear, and linearity lasted only for 2 h (Fig. 6.20). The initial rate constant was 155 times larger than that measured with *Citrus* CM. After 2 h, rate constants decreased because diffusion in the sorption compartment became rate-limiting. Comparing the initial slope obtained with MX with that of UDIS demonstrates that PCP mobility in the outer layer of extracted CM is lower by a factor of 6 than mobility in the sorption compartment (UDIS). This difference is too small to guarantee linear desorption plots until membranes are empty. Apparently, diffusion from the anticlinal pegs to the outer layer of the MX was too

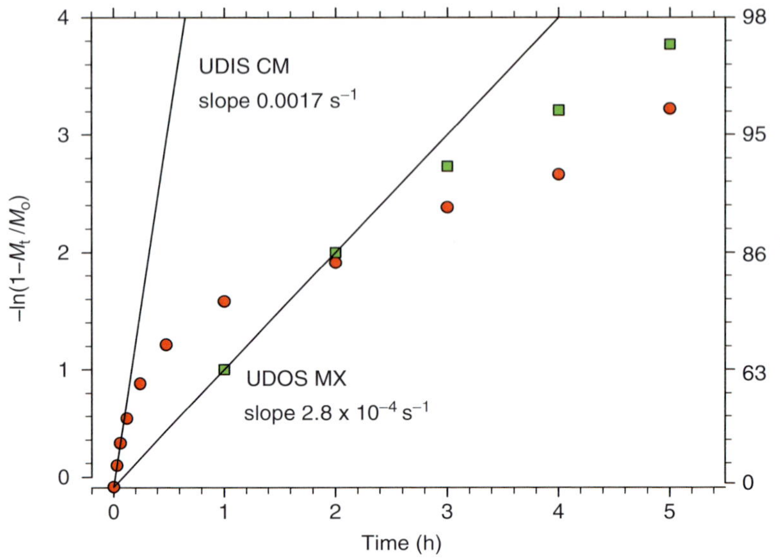

Fig. 6.20 Unilateral desorption of pentachlorophenol from *Citrus aurantium* CM or MX. Desorption from the inner surface of the CM (*UDIS*) or desorption from the outer surface (*UDOS*) of MX-membranes. Initial slopes are indicated

slow and the distance too large. However, from the initial slopes obtained by UDOS, valid mobilities can be estimated for the limiting skins of MX-membranes.

The UDOS experiment with MX demonstrates that solute mobility in the outer layer of extracted CM is lower than in the sorption compartment. There is an outer limiting layer in the MX, but mobility is much larger than mobility in the limiting skin of the CM. Waxes in the limiting skin or on top of the CM reduced PCP mobility 155-fold.

6.3.2.2 Variability of Solute Mobility among Different Plant Species

Buchholz (2006) published rate constants of bifenox measured with astomatous CM isolated from leaves of 22 plant species. Rate constants varied by more than three orders of magnitude, and ranged from 1.9×10^{-5} (*Prunus persica*) to $1.3 \times 10^{-8}\,\text{s}^{-1}$ (*Ilex paraguariensis*) (Fig. 6.21).

Extraction of cuticular waxes increased rate constants by factors of 5 (*Vanilla*) to 4,895 (*Ilex*), and weight loss on extraction varied between 10% (*Stephanotis; Melicoccus, Hedera, Vanilla*) and 30% (*Populus, Schefflera*). However, bifenox mobilities did not correlate with cuticle thickness or amounts of cuticular wax (Buchholz 2006).

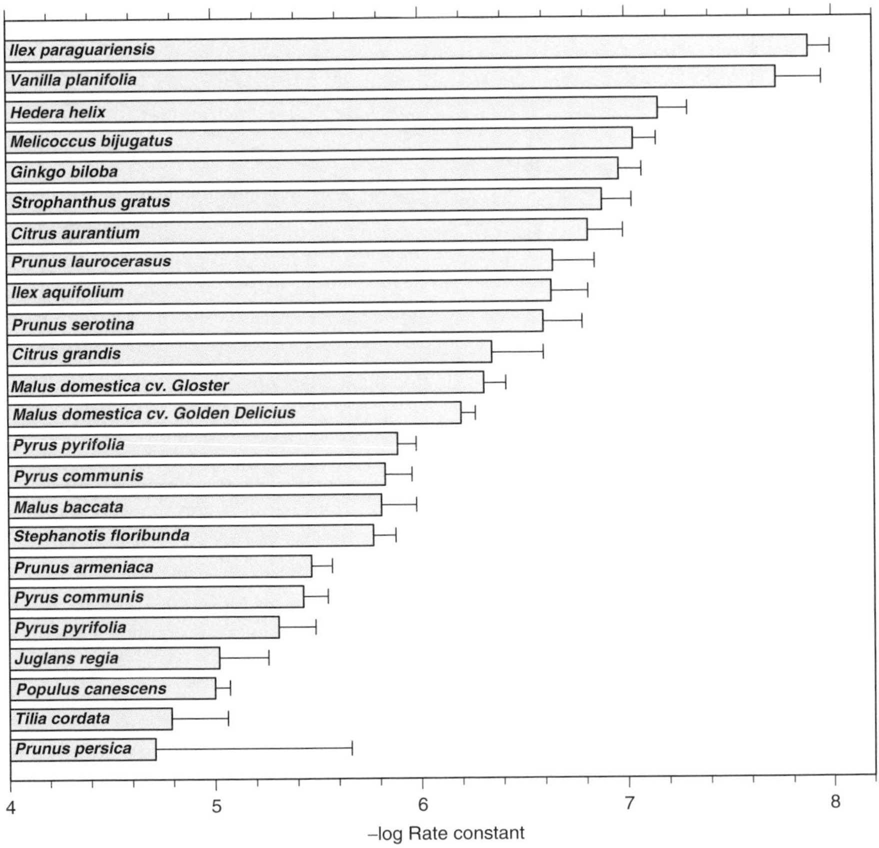

Fig. 6.21 Mobility ($-\log k^*$) of bifenox at 25°C in astomatous cuticular membranes isolated from 22 different plant species. *Error bars* represent 95% confidence intervals. (Redrawn from Buchholz 2006)

6.3.2.3 Variability of Solute Mobility with Size of Solutes

UDOS rate constants do not depend on partition coefficients, since driving force is the solute concentration in the sorption compartment of cuticles rather than concentration of an aqueous donor solution. As measure of solute size, molecular weight may be used, but we prefer the equivalent volumes (V_X cm^3 mol^{-1}) calculated according to McCowan and Mellors (1986) or Abraham and McGowan (1987) as already explained (Chap. 5). Effect of solute size on solute mobility in CM and MX membranes has been studied with cuticles from different plant species (Baur et al. 1996b; Buchholz et al. 1998). All data could be fitted to an equation of the type shown below:

$$\log k^* = \log k_0^* - \beta' \times V_X \qquad (6.21)$$

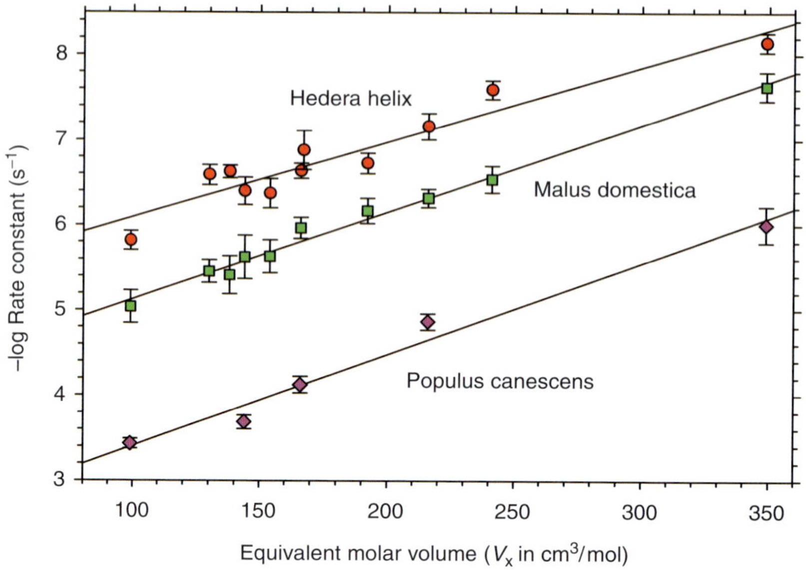

Fig. 6.22 The effect of molar volume (V_X) of solutes on solute mobility (k^*) in CM of three different plant species. Averages of 16–20 CM and 95% confidence intervals are shown. (Redrawn from Buchholz et al. 1998)

and this is demonstrated in Fig. 6.22. Slopes of the lines are selectivity coefficients (β') and the *y*-intercepts (k_0^*) are the solute mobilities of a hypothetical compound having zero molar volume (V_X). Parameters of (6.21) are given in Table 6.8. The species differed only in the *y*-intercepts, which varied from -2.33 (equivalent to $k_0^* = 4.68 \times 10^{-3}\,\mathrm{s}^{-1}$) to -5.27 ($5.63 \times 10^{-6}\,\mathrm{s}^{-1}$). Thus, rate constants for a solute with zero molar volume differed by a factor of 832. This may be compared to difference in bifenox mobility for the same range of species (Fig. 6.21), which is a factor of 770. This is an excellent agreement in view of natural variability between CM in rate constants (Fig. 6.21) and variability in *y*-intercepts (Table 6.8). Hence, differences in solute mobility among species is fully accounted for by differences in *y*-intercepts, which is a characteristic property of the species and is independent of solute properties such as K and V_X.

Size selectivity (β') varied among species between 0.007 and 0.012. Confidence intervals are relatively large, such that differences in β' are not significant. Size selectivity was the same with all species, even though their k_0^* differed by about three orders of magnitude. The mean value of β' over all species is $0.0095\,\mathrm{mol\,cm^{-3}}$, and provided k_0^* is known, the rate constant for any lipophilic solute with equivalent volume between 100 and $350\,\mathrm{cm^3\,mol^{-1}}$ can be calculated using (6.21). For instance, for log k_0^* equal to -4.25 (pear leaf), mobility of a solute having zero molar volume k^* is $5.62 \times 10^{-5}\,\mathrm{s}^{-1}$. If a solute has a molar volume of $100\,\mathrm{cm^3\,mol^{-1}}$, k^* is $6.31 \times 10^{-6}\,\mathrm{s}^{-1}$ which is smaller by a factor of 8.9. With V_X equal to 200 or $300\,\mathrm{cm^3\,mol^{-1}}$, k^* is $7.08 \times 10^{-7}\,\mathrm{s}^{-1}$ and $7.94 \times 10^{-8}\,\mathrm{s}^{-1}$

Table 6.8 Y-intercepts ($-\log k_0^*$) and size selectivity (β' in mol cm^{-3}) measured at 25 °C using various solutes differing in molar volumes (cm^3 mol^{-1})

Species	$-\log k_0^* \pm$ CI	$\beta' \pm$ CI	r^2
Populus canescens	2.33 ± 0.62	0.011 ± 0.002	0.98
Pyrus communis cv. Conference	3.66 ± 0.39	0.009 ± 0.002	0.93
Capsicum annuum	3.95 ± 0.38	0.009 ± 0.002	0.89
Stephanotis floribunda	4.11 ± 0.44	0.007 ± 0.001	0.81
Malus domesta cv. Golden Delicious	4.11 ± 0.18	0.010 ± 0.002	0.99
Pyrus communis cv. Bartlett[a]	4.25 ± 0.36	0.009 ± 0.003	0.96
Pyrus communis MX[a]	1.99 ± 0.57	0.009 ± 0.003	0.91
Citrus aurantium[a]	4.28 ± 0.53	0.012 ± 0.002	0.86
Strophantus gratus	4.94 ± 0.36	0.009 ± 0.002	0.93
Hedera helix	5.22 ± 0.41	0.009 ± 0.004	0.91
Ilex paraguariensis	5.27 ± 0.80	0.010 ± 0.003	0.95
Average		0.0095	

[a]Data from Baur et al. (1996b); all others were taken from Buchholz et al. (1998). CI is the 95% confidence interval

respectively. Doubling molar volume from 100 to 200 cm^3 mol^{-1} reduces solute mobility by a factor of 8.9, and increasing V_X threefold reduces mobility by a factor of 79.3. These factors are the same no matter what the numerical value of k_0^* is, because β' is the same for all species.

Using pear leaf cuticles, Baur et al. (1996b) studied the effect of extraction of waxes on rate constants and size selectivity. This is the only UDOS study of size selectivity in MX membranes. Extraction increased the y-intercept of (6.21), but size selectivity was unaffected, and was the same as with CM. It is astounding that extraction of waxes had no effect on size selectivity, while rate constants and the y-intercept k_0^* increased by a factor of 182 (Table 6.8).

Solute mobility in plant cuticles at 25 °C is completely determined by k_0^* and β', and this poses the question concerning their physical meaning. Size selectivity is related to the free volume available to diffusion (Potts and Guy 1992), which is the reciprocal value of $2.3 \times \beta'$. In CM and MX the free volume of diffusion is 45.77 cm^3 mol^{-1}. Free volume of diffusion is proportional to viscosity, and since it is the same in CM and MX it appears that viscosity of amorphous waxes and cutin in the limiting skin of MX-membranes are the same. If waxes embedded in the limiting skin do not reduce viscosity, which mechanism is then responsible for the effect of waxes on solute mobilities (k_0^* and k^*) in plant cuticles?

The only examples available are the pear leaf CM and MX. Extracting waxes increased k_0^* by a factor of 182, and had no effect on size selectivity (Table 6.8). Baur et al. (1996b) suggested that waxes increase the diffusion path in the CM. The upper pear-leaf cuticle contains about 30 weight percent of waxes, and a substantial portion of this occurs as epicuticular wax plates. Baur (1998) compared UDOS rate constants for bifenox in the CM of *Ilex paraguariensis* and *Pyrus pyrifolia* prior to and after stripping surface wax with cellulose acetate. Stripping did not increase rate constants, while extracting total waxes with chloroform greatly increased them.

This suggests that surface wax did not contribute to barrier properties. However, it is not known if stripping removed surface waxes completely or if a thin layer remained. We discussed this problem in Chaps. 1 and 4. Baur (1998) proposed that only waxes deposited in the limiting skin contribute to barrier properties by obstructing the diffusion path. Lipophilic solutes freely dissolve and diffuse in amorphous waxes (Sects. 6.2 and 6.5), and the only candidates for obstructing the diffusion path in amorphous waxes are crystalline wax plates (Riederer and Schreiber 1995). We return to this problem when we discuss temperature effects on solute mobility (Chap. 8).

6.4 Simulation of Foliar Penetration

When solute mobility in the limiting skin of CM is determined using UDOS, the solutes are dissolved in the sorption compartment before desorption from the outer surface is initiated. For this reason, UDOS can be used only with solutes which are sufficiently soluble in cutin, which implies that $K_{mxw} > 10$. Penetration of inorganic ions, zwitterionic organic compounds (i.e., amino acids, glyphosate) or highly water soluble neutral compounds (sugars) cannot be studied using UDOS, because they crystallise or solidify on the inner surface of the CM, when water evaporates. This is the reason why UDOS data have been collected solely for lipophilic compounds.

This limitation can be overcome in part by applying small droplets to the outer surface and desorbing them from the inner side of the CM after they have penetrated. The same apparatus as in UDOS is used (Fig. 6.17), except that the CM is inserted such that the morphological outer surface is exposed in the opening of the orifice. Small droplets (usually 2–5 μl) are pipetted on the outer surface of the CM. After droplet drying, the outer side of the chamber is sealed with adhesive tape to keep humidity at 100% and the chambers are inverted. Desorption medium is added to the receiver, and the chambers are placed into the wells of the thermostated aluminium block, which is rocked to mix the receiver solutions as described in Chap. 5 (Fig. 5.4a). The receiver solution is quantitatively withdrawn periodically and replaced by a fresh one.

This method resembles the situation in the field after spray application to the foliage, and it was initially termed (Schönherr and Baur 1994) simulation of foliar uptake (SOFU). As explained in Chap. 2 the term is unfortunate, as it implies active participation of cuticles, similar to nutrient uptake by roots. This is clearly not the case. Solutes penetrate the cuticle and the cell wall before they enter the symplast. Thus, cuticular penetration is a purely physical process, and we now use the term "simulation of foliar penetration" (SOFP), which is more appropriate.

Data shown in Fig. 6.23 can be used to explain the principles of SOFP. ^{14}C-labelled solutes were dissolved either in water (urea, tebuconazole) or aqueous lactic acid buffer of *p*H 3 (2,4-D). A small droplet (2 μl) was placed on the centre of the outer surface of the CM, and the orifice of the lid was closed with sticky transparent

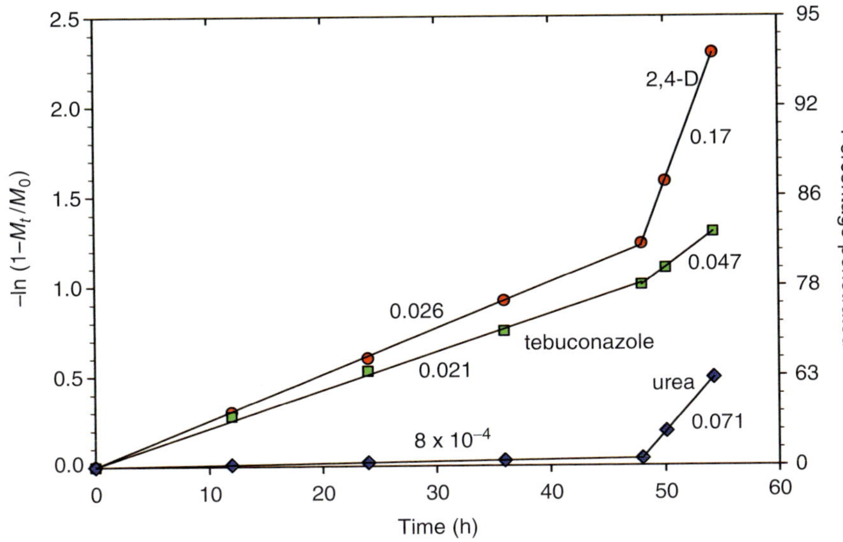

Fig. 6.23 Simulation of foliar penetration at 25 °C using astomatous isolated *Stephanotis* CM. Small aqueous droplets (2 µl) were pipetted on the centre of the outer surface of the CM, and during the first 48 h evaporation was prevented by 100% humidity over the donor droplets. At 48 h, humidity was lowered to allow evaporation of solvent water, and desorption from the inner surface was continued. Means of 15–20 CM are shown. *Numbers* on regression lines are rate constants in s^{-1}. (Redrawn from Schönherr and Baur 1994)

film (Tesafilm). A phospholipid suspension (PLS) served as receiver solution, and 0.5 ml were pipetted into the receiver compartment of the apparatus. The chambers were placed into the wells of the aluminium plate maintained at 25 °C. The receiver solutions were withdrawn and replaced by fresh ones at 12 h intervals, and radioactivity was determined by scintillation counting. After 48 h the Tesafilm was removed; the water of the droplets evaporated quickly, and desorption was resumed. At the end of the experiment the CM was cut out with a scalpel, the radioactive residues were dissolved in scintillation cocktail and counted. Data analysis was the same as in UDOS. The natural logarithms of $(1 - M_t/M_0)$ was plotted against time. The slopes of the lines are the rate constants of desorption (k), but their physical meaning differs from UDOS rate constant (k^*) because in SOFP solutes are not dissolved in cutin. In the experiment depicted in Fig. 6.23, solutes were initially dissolved in water or in lactic acid buffer. Their state after droplet drying is uncertain. 2,4-D was probably dissolved in concentrated lactic acid, and most of the very lipophilic tebuconazole was probably dissolved in epicuticular wax. Urea is hygroscopic, and all or part of it may have formed a concentrated urea solution. Due to these uncertainties, SOFP rate constants are more difficult to analyse.

Desorption plots were linear both before and after droplet drying. Slopes measured with aqueous solutions are the rate constants, and they were constant both prior to and after droplet drying. Hence, solute concentrations in droplets decreased exponentially with time. Rate constants were higher after droplet drying, but the effect of drying differed among compounds. The effect was spectacular with urea, where the two slopes differ by a factor of 93. Urea is a highly water soluble compound (1 kg urea dissolves in 1 kg water), and above 90% humidity it deliquesces. On droplet drying, it was most likely present as concentrated aqueous solution, and an increase in urea concentration by a factor of 93 implies that the volume of the donor decreased by the same factor, because according to (2.19) $C_{\text{donor}} = M/V_{\text{donor}}$. If permeance is constant and independent on urea concentration, rate constants should be proportional to $A_{\text{droplet}}/V_{\text{droplet}}$ (2.26).

Tebuconazole and 2,4-D are both lipophilic (Appendix B), and their initial rate constants were relatively high. These two compounds dissolve in the surface wax, and the situation resembles that discussed in Sect. 6.2 when we dealt with the submersion technique. Surface waxes served as intermediate donor phase. With these lipophilic solutes, equilibration between water and surface wax must have been fast, and during the first 48 h the surface wax was the donor phase, and 60% (tebuconazole) to 70% (2,4-D) of the amount in the donor droplets penetrated into the receiver in 48 h. Penetration of 2,4-D was a little faster than penetration of tebuconazole, which has a greater molar volume ($308\,\text{cm}^3\,\text{mol}^{-1}$) than 2,4-D ($241\,\text{cm}^3\,\text{mol}^{-1}$). Why rate constants of these two compounds increased on droplet drying is a matter of conjecture. With 2,4-D, *p*H and degree of ionisation probably decreased during droplet drying, while with tebuconazole the factor was only 2.2 and here an increase in droplet area during drying could have been responsible. Change in droplet area [cf. (6.14)] is of course also a factor to be considered with urea and 2,4-D.

These speculations show that simulation of foliar penetration produces very complex data, and it is difficult to explain differences among chemicals and plant species. This should not distract from the fact that SOFP can provide us with quantitative data about rates of penetration as affected by experimental conditions, which can be close to the situation in the field. The method has been used extensively to study the effects of humidity, adjuvants, plasticisers, wetting agents and temperature on rates of penetration of ionic species (Chap. 5). No other method discussed in Chap. 6 is as versatile as SOFP.

6.5 Diffusion in Reconstituted Isolated Cuticular Waxes

Simultaneous bilateral desorption (Sect. 6.3.1) of lipophilic molecules from CM (Fig. 6.15) is excellent evidence for the pronounced asymmetry of cuticles. Furthermore, it shows that the transport-limiting barrier must be located at the outer side of the cuticle. The effect of wax extraction, resulting in an increase of permeability to water (Table 4.7) and lipophilic molecules (Table 6.4) by factors of 10–10,000, convincingly shows that the transport-limiting barrier of the CM for lipophilic molecules is due to cuticular waxes. Sorption and diffusion of molecules in the waxy barrier of the CM was investigated to better understand the nature of this waxy barrier. Reconstituted cuticular waxes were used in these studies. Isolated cuticular waxes of grape berries had previously been used by Grncarevic and Radler (1967) to assess their effect cuticular transpiration. A refined method allowing the measurement of diffusion coefficients of lipophilic molecules in cuticular wax was developed with barley leaf wax (Schreiber and Schönherr 1993b).

6.5.1 Experimental Approach

Wax was extracted from barley leaves using chloroform. Aluminium disks (8 mm diameter) were immersed in the wax solution for 1–2 s. After evaporation of the solvent the solid wax adhered to the surface of the disks, and wax-covered aluminium disks were heated to 100 °C for 5 min. Thus, wax was reconstituted from the melt and not from solution. This ensured a good adhesion of the wax to the aluminium surface, which is necessary for the following desorption experiments. Scanning electron microscopy revealed that wax-covered aluminium disks had smooth surfaces, while barley-leaf surfaces are densely covered with epicuticular wax platelets. Therefore, the term "reconstituted" is preferred instead of "recrystallised" (Sect. 9.6).

Diffusion coefficients were determined by desorbing radiolabelled probes, which had been added to the wax solution before reconstitution (Schreiber and Schönherr 1993b). In some experiments, reconstituted wax samples were loaded with solutes by equilibrating them in solutions of the radio-labelled compound (Kirsch et al. 1997). This is the same method as described for the determination of wax/water partition coefficients (Sect. 6.1). Adding the radio-labelled probe directly to the wax solution offers the advantage that desorption experiments can be started directly after reconstitution of the wax.

Diffusion in reconstituted wax takes place in the amorphous wax phase, and therefore molecules must be soluble in this phase. No problems occur with lipophilic molecules (Table 6.2), which are sufficiently soluble in the amorphous wax phase when they are applied from aqueous solutions or if added directly to the wax before reconstitution. With both methods of sample preparation, the same diffusion coefficients are obtained in most cases (Schreiber and Schönherr 1993b). When highly water soluble solutes are added directly to the wax solution before reconstitution, they crystallise and are not homogeneously dissolved in the amorphous wax phase, due to their low lipid solubility. Hence, diffusion of highly water soluble solutes in cuticular waxes cannot be studied using reconstituted waxes.

Desorption of radio-labelled probes from reconstituted wax samples is initiated by immersing them in vials (5 ml) containing an inert desorption medium (e.g., phospholipid suspension) which does not affect structure of the wax layer. The desorption medium is replaced at each sampling time. The amounts desorbed and the residual amount of the probe remaining in the wax at the end of the experiment are determined by scintillation counting. The total amount of radio-labelled probe M_0 present in the wax at the beginning of the desorption experiment is calculated

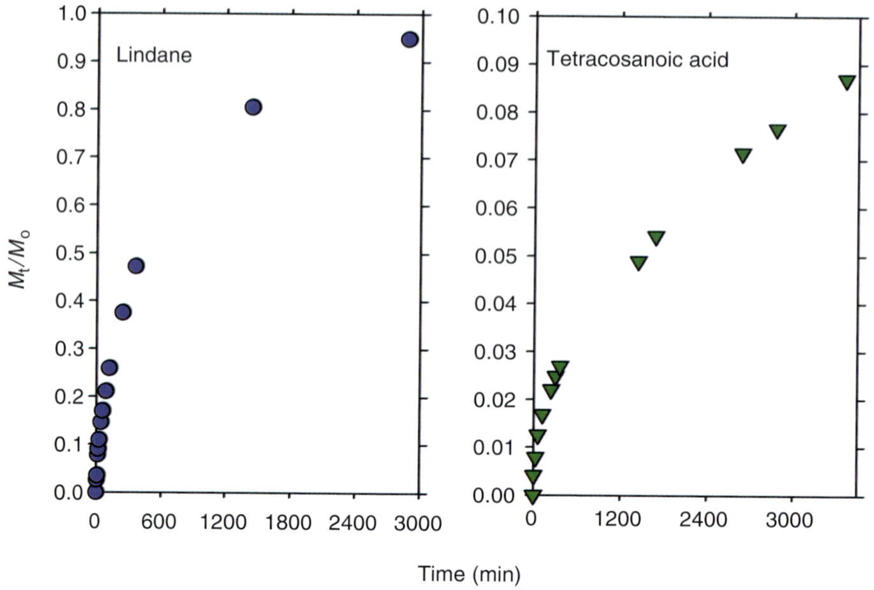

Fig. 6.24 Desorption of ^{14}C-labelled lindane and ^{14}C-labelled tetracosanoic acid from reconstituted barely wax. Relative amounts desorbed (M_t/M_0) vs time (min) are shown. Data redrawn from Schreiber and Schönherr (1993b)

by adding the amounts desorbed at each sampling time plus the residual amount in wax. M_t is the total amount desorbed at each sampling time. Non-linear desorption kinetics are obtained when relative amounts (M_t/M_0) are plotted vs time t (Fig. 6.24).

Within 48 h about 95% of the ^{14}C-labelled lindane could be desorbed, whereas only 9% of the ^{14}C-labelled tetracosanoic acid was desorbed in 72 h (Fig. 6.24). When relative amounts desorbed were plotted vs the square root of time, desorption kinetics could be linearised up to 50% desorption (Fig. 6.25).

Using (2.35) and the slopes of the linear regression lines fitted to the desorption kinetics (Fig. 6.25), diffusion coefficients D for lindane and tetracosanoic acid can be calculated if the length ℓ of diffusion in the wax layer is known. The thickness ℓ of the wax layer for each individual wax sample was determined by weighing the aluminium disks prior to and after wax reconstitution and assuming a wax density of 0.9 g cm^{-3}. On average, reconstituted wax layers had a thickness between 1.5 and 2.5 μm. With a slope of 0.0246 min$^{-1/2}$ for lindane and a wax layer of 1.67 μm thickness, the calculated D is 5.55×10^{-18} m^2 s^{-1}. For tetracosanoic acid, the wax layer had a thickness of 1.52 μm and the slope was 0.00127 min$^{-1/2}$. This results in a D of 1.22×10^{-20} m^2 s^{-1}. Thus, D of lindane is more than two orders of magnitude larger than D of tetracosanoic acid. This difference in solute mobility explains why desorption of lindane from barley wax proceeds much faster than desorption of tetracosanoic acid.

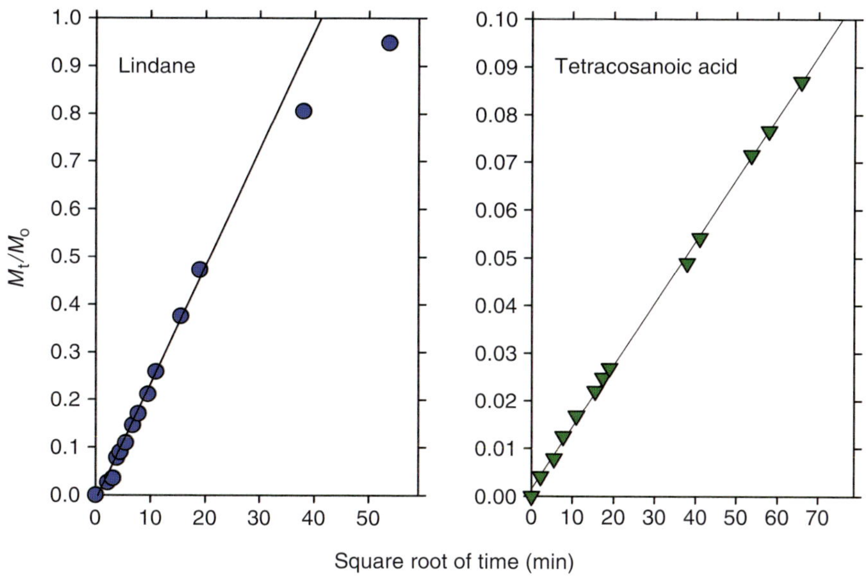

Fig. 6.25 Desorption of ^{14}C-labelled lindane and ^{14}C-labelled tetracosanoic acid from reconstituted barely wax. Relative amounts desorbed (M_t/M_0) vs the square root of time (min) are shown. Data redrawn from Schreiber and Schönherr (1993b)

6.5.2 Diffusion Coefficients in Reconstituted Cuticular Wax

Using the approach described above, diffusion coefficients of lipophilic molecules have been measured in reconstituted wax of four plant species (Table 6.9). Diffusion coefficients of lipophilic molecules were in the range of 10^{-17} to $10^{-18}\,\mathrm{m^2\,s^{-1}}$. Diffusion coefficients in solution are around $10^{-10}\,\mathrm{m^2\,s^{-1}}$ (Cussler 1984) and in phospholipid membranes they are about $10^{-12}\,\mathrm{m^2\,s^{-1}}$ (Stein 1986). Thus, D values measured in reconstituted waxes are very low, because waxes are not liquid but a partially crystalline phase at room temperature. Diffusion takes place in the solid amorphous wax phase, whereas the crystalline phase represents an excluded volume, increasing the path length of diffusion.

In the subsection "Estimating Solute Mobility from Rate Constants" within Sect 6.3.2 above, it was shown that diffusion coefficients D can also be calculated from rate constants k^* measured with UDOS experiments. D values calculated from k^* of PCP in *Citrus* and *Capsicum* CM were 1.0×10^{-18} and $2.8 \times 10^{-17}\,\mathrm{m^2\,s^{-1}}$ respectively. These D values determined with UDOS experiments (Sect. 6.3) are of the same range as the D values of similar compounds obtained by desorption experiments (Table 6.9). Comparable values are obtained by two independent methods, and this is evidence that these extraordinarily low diffusion coefficients represent genuine values for the mobility of lipophilic molecules in the cuticular wax barrier.

Table 6.9 Diffusion coefficients D and molar volumes V_x of lipophilic molecules in reconstituted wax of *Hordeum vulgare, Prunus laurocerasus, Ginkgo biloba* and *Juglans regia*

		$D \times 10^{17}$ (m^2 s^{-1})			
Substance	V_x (cm^3 mol^{-1})	*Hordeum*[a]	*Prunus*[b]	*Ginkgo*[b]	*Juglans*[b]
BA	93	–	3.54	4.37	5.59
4-NP	95	–	3.23	–	–
SA	99	2.4	2.74	3.21	3.45
2,4-D	138	1.9	1.63	2.01	2.22
AT	162	–	1.15	–	–
MET	162	1.2	0.91	–	–
TRI	219	0.36	–	–	–
TB	241	0.58	–	–	–
BIT	267	0.19	–	–	–

Benzoic acid (BA), 4-nitrophenol (4-NP), salicylic acid (SA), 2,4-dichlorophenoxyacetic acid (2,4-D), atrazine (AT), metribuzin (MET), triadimenol (TRI), tebuconazole (TB), bitertanol (BIT)
[a]Data from Burghardt et al. (1998)
[b]Data from Kirsch et al. (1997)

D of linear long-chain aliphatic molecules, which are similar or even identical to native cuticular wax molecules, was measured in reconstituted wax of *Hordeum vulgare* (Schreiber and Schönherr 1993b) *Fagus sylvatica* and *Picea abies* (Schreiber et al. 1996a). In these experiments, radio-labelled probes were reconstituted together with the wax, and thus the probes were trapped to a certain extent in the crystalline wax phase. D values ranged from 10^{-17} to $10^{-22}\,m^2\,s^{-1}$, covering a much wider range (Table 6.9). Tetracosanoic acid, which is a constituent of plant cuticular waxes, had a D of $1.22 \times 10^{-20}\,m^2\,s^{-1}$ (Sect. 6.5.1). D of dotriacontane, representing a good analogue for a C_{31} or C_{33} alkane, ranged from $1.9 \times 10^{-22}\,m^2\,s^{-1}$ in *Picea* wax (Schreiber et al. 1996a) to $4.07 \times 10^{-22}\,m^2\,s^{-1}$ in barley wax (Schreiber and Schönherr 1993b). Average chain length of cuticular wax monomers, comparable to tetracosanoic acid and dotriacontane, varies between C_{21} and C_{35} (Table 1.3). These low D values indicate that typical wax molecules forming the transport-limiting barrier of the cuticle are mobile, but their mobilities are two to four orders of magnitude lower than those of other lipophilic molecules which are not naturally contained in cuticular waxes.

In order to estimate the dependence of D on molecular size, measured D values are plotted as a function of the molar volumes V_x (cm^3 mol^{-1}) of the investigated molecules. Reasonable correlations, as shown for reconstituted *Prunus* wax and six lipophilic molecules of different molar volumes, are obtained (Fig. 6.26). An equation similar to that established describing size selectivity of rate constants k^* measured in UDOS experiments (6.21) could be established describing size selectivity of D in reconstituted wax:

$$\log D = \log D_0 - \beta' \times V_x. \tag{6.22}$$

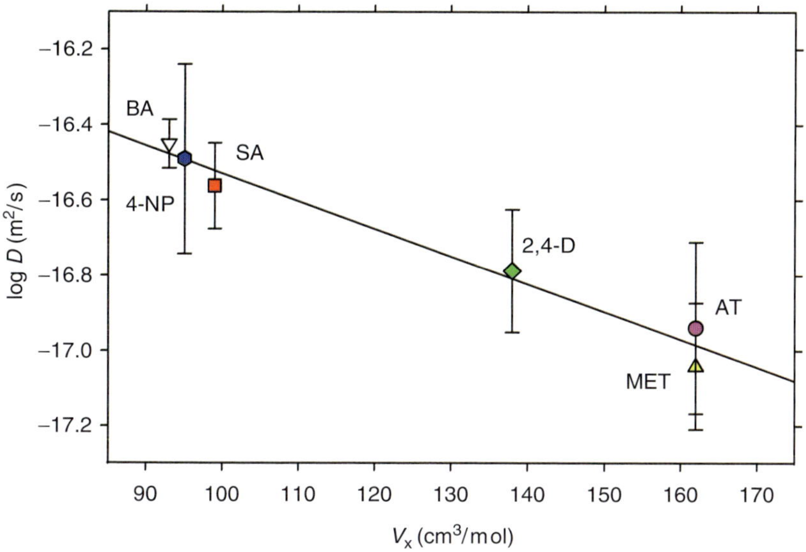

Fig. 6.26 Logarithms of the diffusion coefficients *D* of lipophilic molecules in reconstituted wax of *Prunus laurocerasus* as a function of the molar volumes V_x of the molecules. Benzoic acid (*BA*), 4-nitrophenol (4-*NP*), salicylic acid (*SA*), 2,4-dichlorophenoxyacetic acid (2,4-*D*), atrazine (*AT*), metribuzin (*MET*). Error bars represent 95% confidence intervals. Data from Kirsch et al. (1997)

Table 6.10 Slopes β', *y*-intercepts D_0 and coefficients of determination r^2 of the regression equations fitted to plots of log *D* vs V_x of the data given in Table 6.9

Species	β'(mol cm^{-3})	log D_0(m^2 s^{-1})	r^2
Hordeum vulgare[a]	−0.0065	−15.94	0.91
Prunus laurocerasus[b]	−0.0074	−15.79	0.97
Ginkgo biloba[b]	−0.0070	−15.67	0.91
Juglans regia[b]	−0.0083	−15.83	0.97

[a]Data from Burghardt et al. (1998)
[b]Data from Kirsch et al. (1997)

This model, originally developed by Potts and Guy (1992), states that the logarithm of *D* linearly decreases with increasing molar volume V_x. The slope β' (mol cm^{-3}) gives the size selectivity, characterizing the dependence of *D* on V_x, whereas D_0 (m^2 s^{-1}) represents the diffusion coefficient of a hypothetical molecule having a molar volume of zero. Regression equations fitted to plots of log *D* vs V_x gave very similar values for β' and D_0 (Table 6.10), indicating that mobility of the investigated lipophilic molecules in reconstituted wax of the four species was similar.

Calculating the means of β' and D_0 (Table 6.10), a more general equation can be established, which permits the prediction of *D* in reconstituted cuticular wax for an arbitrary lipophilic molecule of any molar volume:

$$\log D = -15.76 - 0.0076 \times V_{\mathrm{X}}. \tag{6.23}$$

This equation can be very helpful when D is not easily measured due to experimental limitations. As mentioned above, solubility of small polar molecules, such as water, urea or glucose, is too low in the amorphous phase of reconstituted cuticular wax, and D cannot be determined experimentally. However, using (6.23), D of these compounds in wax can be estimated. For example, for water having a V_{X} of $18\,\mathrm{cm}^3\,\mathrm{mol}^{-1}$, a D of $1.23 \times 10^{-16}\,\mathrm{m}^2\,\mathrm{s}^{-1}$ can be calculated.

Size selectivities measured with UDOS experiments for a similar set of lipophilic molecules and CM isolated from 11 species ranged from −0.007 to −0.012 (Table 6.8). Mean size selectivity for CM was −0.0095 (Table 6.8), which is somewhat larger than mean size selectivity of isolated wax, which had a value of −0.0076 (6.23). Nevertheless, this comparison shows that size selectivities in reconstituted cuticular wax are in a similar range to that of size selectivity for cuticular membranes. This is important evidence that D values obtained in diffusion experiments or predicted from (6.23) are reliable since they are comparable to values obtained from isolated CM. Size selectivities measured for linear long-chain aliphatic molecules in reconstituted wax of *Hordeum, Fagus* and *Picea* were somewhat higher, ranging from −0.015 to −0.020 (Schreiber and Schönherr 1993b; Schreiber et al. 1996a). However, these experiments included a very different set of compounds, and radiolabelled probes were always added to the wax prior to reconstitution. Therefore, a direct comparison with the results obtained for lipophilic molecules as they are listed in Table 6.9 should not be made.

6.5.3 Relationship Between D and P

The relationship between diffusion coefficients D ($\mathrm{m}^2\,\mathrm{s}^{-1}$) and permeances P ($\mathrm{m\,s}^{-1}$) is given by (2.18). This equation relates to a homogeneous barrier, and it is assumed that the cuticular transport barrier of wax represents a homogeneous phase. Equation (6.18) states that P can be calculated from D if the partition coefficient K and the thickness of the barrier ℓ are known. Instead of cuticle/water partition coefficients K_{cw}, wax/water partition coefficients K_{ww} should be used here, since we deal with diffusion in a waxy barrier. K_{ww} have been measured and are available (Table 6.2). The thickness of the waxy barrier in the limiting skin is used, rather than thickness of CM. This may be estimated from the total amount of wax deposited on the CM (Table 4.8) if we assume that all wax contributes to the formation of the barrier. Equation (2.18) can be rearranged to

$$D = \frac{P \times \ell_{\mathrm{wax}}}{K_{\mathrm{ww}}}. \tag{6.24}$$

Equation (6.24) states that D ($\mathrm{m}^2\,\mathrm{s}^{-1}$) and P/K_{ww}($\mathrm{m\,s}^{-1}$) should only differ by the thickness of the wax layer ℓ_{wax} (m). D (Table 6.9), K_{ww} (Table 6.2) and P have been

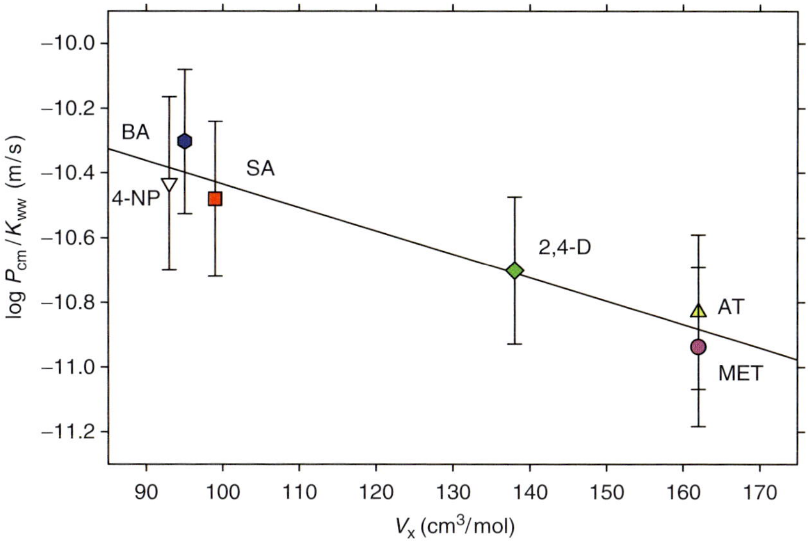

Fig. 6.27 Logarithms of the ratios P_{cm}/K_{ww} of lipophilic molecules in *Prunus laurocerasus* CM as a function of the molar volumes V_x of the molecules. Benzoic acid (*BA*), 4-nitrophenol (4-*NP*), salicylic acid (*SA*), 2,4-dichlorophenoxyacetic acid (2,4-*D*), atrazine (*AT*), metribuzin (*MET*). Data from Kirsch et al. (1997)

measured for cuticles of *Prunus, Ginko* and *Juglans* (Kirsch et al. 1997). Isolated CM (P_{cm}) or leaf disks (P_{leaf}) as described in Sect. 6.2.3 were used. As shown for *Prunus laurocerasus* CM, a plot of P_{cm}/K_{ww} vs. V_x gives a reasonable correlation (Fig. 6.27), as was also observed with diffusion in reconstituted *Prunus laurocerasus* wax (Fig. 6.26).

Using (6.22), regression equations similar to those obtained for diffusion in reconstituted wax (Table 6.10), were derived for transport across isolated CM and leaf disks of the three species (Table 6.11). From these regressions, size selectivities $'$ were obtained for the transport across CM cuticles and for non-isolated cuticles of leaf disks. Size selectivities ranged from -0.0074 to -0.013, and large differences between intact leaves and isolated CM were not observed. This is good evidence that isolation of the cuticle does not alter transport properties. The mean value for all β' in Table 6.10 is 0.0073. This is similar to the mean β' obtained with CM (Table 6.8). Cuticles appear to be very robust, since comparable results were obtained with four different experimental approaches: (1) diffusion in wax (Sect. 6.5.1), (2) UDOS (Sect. 6.3.2), (3) steady state penetration across the isolated cuticle (Sect. 6.2.1), and (4) steady state penetration into leaf disks (Sect. 6.2.3).

Path length of diffusion ℓ_{calc} can be calculated dividing D_0 in wax (Table 6.10) by D_0 obtained from transport experiments across CM and intact leaf surface (Table 6.11). Values obtained for the thickness of the transport-limiting wax barrier range from 50 nm (*Juglans* CM) to 800 nm (*Prunus* CM). These are reasonable values, since they are of the same order of magnitude as those determined from wax

Table 6.11 Slopes β', *y*-intercepts D_0 and coefficients of determination r^2 of the regression equations fitted to plots of log P_{cm}/K_{ww} and P_{leaf}/K_{ww} vs V_x. Path length of diffusion ℓ_{calc} are calculated by dividing D_0 obtained in wax (Table 6.10) by D_0 measured for CM and leaf respectively. Path length of diffusion ℓ_{meas} are calculated from wax coverage determined for the same set of cuticles used in transport experiments (Kirsch et al. 1997)

Species	β'(mol cm^{-3})	$D_0 \times 10^{10}$ (m^2 s^{-1})	r^2	ℓ_{calc}(nm)	ℓ_{meas}(nm)	ℓ_{calc}/ℓ_{meas}
Prunus laurocerasus CM	−0.0074	1.93	0.93	839	1,600	0.52
Prunus laurocerasus Leaf	−0.012	5.51	0.99	294	1,600	0.18
Ginkgo biloba CM	−0.0095	13.01	0.94	164	210	0.78
Ginkgo biloba Leaf	−0.010	20.20	0.94	106	210	0.50
Juglans regia CM	−0.013	30.13	0.97	49	570	0.09
Juglans regia Leaf	−0.011	13.14	0.93	113	570	0.20

coverage of the CM (Table 6.11). However, it is obvious that path length of diffusion ℓ_{meas}, obtained from total wax coverage is always higher than ℓ_{calc}. This could indicate that only a fraction of the total wax (0.1–0.8) contributes to the limiting barrier, while the remainder is deposited as intracuticular wax in cuticular layer(s).

Results of diffusion experiments in reconstituted wax agree fairly well with results of transport experiments using isolated cuticles and leaf disks. Size selectivities are comparable, and reasonable values for the thickness of the transport limiting barrier of the CM are obtained. This justifies the following conclusions: (1) The transport-limiting barrier of the CM for lipophilic molecules is formed by cuticular waxes deposited in/on the limiting skin, (2) water penetrates cuticles using two parallel pathways, the waxy pathway and aqueous pores (Chap. 4), and (3) penetration of ionic compounds is restricted to aqueous pores, and the waxy pathway cannot be accessed (Chap. 5).

Several micrometres away from the living epidermal cell, wax molecules spontaneously arrange themselves, which leads to the formation of an efficient transport barrier. Cuticular waxes deposited at the outer surface of the CM follow the rules of self-organisation. When wax is reconstituted on an artificial surface, barrier properties of reconstituted waxes are very similar as those in isolated CM and intact leaves. Sorption and diffusion in waxes can give valuable insights into the structure and function of the cuticular transport barrier. This experimental approach has been used to analyse the effect of plasticisers on solute diffusion in CM and waxes. This is the topic of Chap. 7.

Problems

1. The mean log K_{cw} of PCP is 4.56 and the mean log K_{ww} is 3.55 (Table 6.1). How much (mol) of PCP is sorbed in 1 mg CM and in 100 μg waxes after equilibration in a large volume of an external aqueous PCP solution of $1\,mg\,l^{-1}$? What is the ratio of the amounts of PCP in CM and wax?
2. At 25 °C, water solubility of PCP is $20\,mg\,l^{-1}$ and the mean K_{ow} is 11,749 (Table 6.1). Calculate K_{cw} from K_{ow} (6.2) and from water solubility (6.3).
3. Let the K_{cw} of two solutes be 30 or 0.03. Pieces of 1 mg CM are equilibrated aqueous solutions of $1\,mg\,l^{-1}$. How much is sorbed in the CM, and which amount is contained in a thin water film of 0.3 mg which remains on the CM after equilibration. How much of the solutes is sorbed in the CM and in the water film?
4. A piece of CM having a mass of 1 mg is equilibrated in 1 ml (=1 g) aqueous buffer which contains $20\,mg\,kg^{-1}$ non-dissociated 2,4-D. Calculate the final amounts of 2,4-D in the CM and in the solution. What is the final amount of PCP in the cuticle and in the solution, assuming the same experimental conditions?
5. In Fig. 6.2, hold-up times of up to 45 h are shown, while in Fig. 6.5 the plot amount penetrated vs time intersects the *y*-axis at a positive value. What is the reason?
6. Diffusion coefficients in leaf CM (Table 6.3) are much lower than in leaf MX membranes ($1.3 \times 10^{-14}\,m^2\,s^{-1}$). In *Citrus* this ratio is 241, while the effect of extraction on permeance was 1,767. Which other factors might have contributed to the very low P of the CM?
7. Using (6.7), data of Table 6.1 and an equivalent molar volume of 2,4-D of $138\,cm^3\,mol^{-1}$, calculate permeance for 2,4-D of *Citrus* CM. How good is this estimate?
8. With compartment sizes given in Table 6.5, calculate M_t/M_0 for compartments 1, 2 and 3 after 180 min of desorption. Use (6.10).
9. Rates of penetration of lipophilic solutes into conifer needles are proportional to sorption in surface wax and cuticles (Fig. 6.9). How can this be explained?
10. When calculated using the specific surface area of the needles, permeance increased with partition coefficient of the solutes (Table 6.6). What does this imply?
11. Half-time of penetration from a sessile droplet depends on the volume of the droplet and the contact area between droplet and cuticle (6.12) You measured permeance using droplets having contact angles of 90°. In a second experiment you added a small amount of wetting agent, such that the droplet spread on the cuticle. Half-time was reduced by factor of 0.5. Did the wetting agent increase permeance of the cuticle and how would you test this?
12. How would you interpret the autoradiographs seen in Fig. 6.14?
13. The average diffusion coefficient calculated from a desorption experiment of 2,4-D in the inner sorption compartment of the *Citrus* CM is about $2 \times 10^{-15}\,m^2\,s^{-1}$. With a steady state penetration experiment, D of $5.4 \times$

$10^{-17}\,m^2\,s^{-1}$ was determined from the hold-up time (Table 6.3). How do you explain this?

14. What is the difference between k and k^*?
15. Size selectivity β' is the same with all species investigated. What does this imply?
16. k^* values measured with CM from various plant species vary by more than three orders of magnitude (Fig. 6.21). Why?
17. Using (6.21) calculate $\log k^*$ for bifenox ($V_x = 2.67\,cm^3\,mol^{-1}$) and the species listed in Table 6.8, and compare these values with those given in Fig. 6.21. What is the ratio $\log k^*$ (measured)/$\log k^*$ (calculated) for these eight species?
18. Experimental values for $\log k_0^*$ are available only for the species listed in Table 6.8. If it is assumed that the $'$ is equal to $0.0095\,mol\,cm^{-3}$ for cuticles of all species, $\log k_0^*$ can be calculated if only one experimental value for $\log k^*$ is known for this species. How would you do that?
19. Rate constants of desorption (k^*) of PCP from the second compartment of barley leaves and conifer needles range from 1.4×10^{-4} to $3.7 \times 10^{-4}\,s^{-1}$. Rate constants of desorption from the outer surfaces of *Capsicum* and *Citrus* CM are 5.4×10^{-6} and $1.8 \times 10^{-6}\,s^{-1}$ respectively (Fig. 6.18). What could be responsible for this large difference in the two types of experiments?
20. Using (2.26), calculate urea permeance of *Stephanotis* CM prior to and after droplet drying. Assume that initial droplet volume was $2\,\mu l$ and that it decreased by a factor of 93 on droplet drying. Contact area between droplet and CM was $0.125\,cm^2$ and it did not change during droplet drying.
21. A lipophilic molecule can be desorbed from a reconstituted wax layer of $120\,\mu g\,cm^{-2}$ at a rate of $0.01\,min^{-1/2}$. What is the diffusion coefficient D in $m^2\,s^{-1}$? What is the effect on slope if the thickness of the wax layer increases by a factor of 2?
22. What is the diffusion coefficient D in $m^2\,s^{-1}$ of a molecule having a molar volume V_x ($cm^3\,mol^{-1}$) of 200?
23. What is the thickness ℓ in μm of the transport-limiting barrier of cuticular wax, if the permeance P is $5.0 \times 10^{-11}\,m\,s^{-1}$, the wax/water partition coefficient K_{ww} is 35 and the diffusion coefficient D is $8.5 \times 10^{-18}\,m^2\,s^{-1}$?

Solutions

1. The molecular weight of PCP is $266\,g\,mol^{-1}$ (Appendix B), hence the molar concentration of PCP is $3.76 \times 10^{-6}\,mol\,l^{-1}$. Using (6.1), we calculate that in 1 mg CM 1.36×10^{-7} mol PCP are sorbed, and in the wax sample of $100\,\mu g$ 1.33×10^{-9} mol PCP. The cuticle contains about 100 times more PCP than the wax sample.
2. K_{cw} calculated from (6.2) is 10,113. This value is 3.6 times lower than the experimental K_{cw} value. K_{cw} calculated from (6.3) is 3,765. This value is 9.6 times lower than the measured K_{cw} value.

3. The solute concentration in the aqueous solution is $1\,\text{mg}\,\text{l}^{-1}$ or $1\,\mu\text{g}\,\text{mg}^{-1}$. In 0.3 mg solution, 0.3 μg solute is contained. Using (6.1) we obtain a concentration in the CM of $30\,\text{g}\,\text{kg}^{-1}$, which is $30\,\mu\text{g}\,\text{mg}^{-1}$ when K_{cw} is 30. The concentration in the CM is $30\,\text{ng}\,\text{mg}^{-1}$ if K_{cw} is only 0.03. The thin film of water on the CM surface contains 0.3 μg in both cases. When K_{cw} is 30, the solute amount associated with the CM is 100 times larger than that in the water film. When K_{cw} of a compound is 0.03 this ratio is 0.1, that is, 10 times more solute is in the water film compared to the CM.
4. The mean K_{cw} of 2,4-D is 407 (Table 6.1). The total amount of non-dissociated 2,4-D in the system is 20 mg, part of which is sorbed in the CM during equilibration. We solve (6.1) for $M_{CM:}$ and obtain $M_{CM} = K_{CM} \cdot \frac{m_{CM}}{m_{water}} \cdot M_{water} = 407 \cdot \frac{1\,\text{mg}}{1000\,\text{mg}} \cdot M_{water}$. Hence, $M_{CM} + M_{water} = 20\,\mu\text{g}$ and M_{CM} is $0.407 \times M_{water}$. Combining these two equations, we obtain $20\,\mu\text{g}\text{-}M_{water} = 0.407 M_{water}$. We next divide both sides of this equation by M_{water}, and after rearranging we obtain $M_{water} = 20\,\mu\text{g}/1 + 0.407 = 14.21\,\mu\text{g}$. Thus, after equilibration the amount of 2,4-D in water is 14.21 μg and the amount in the CM is $20\,\mu\text{g} - 14.21\,\mu\text{g} = 5.79\,\mu\text{g}$. Thus, about 25% of the initial amount of 2,4-D are sorbed in the cuticle at equilibrium, and 75% remain in the external solution. With PCP having a mean K_{cw} of 36,308 (Table 6.1), it can be calculated that 19.46 μg are in the cuticle at equilibrium, whereas the external solution contains only 0.54 μg. Thus, at equilibrium 97% of the PCP will be sorbed in the cuticle. This calculation is essential for the experimental design, since in the case of PCP nearly all of the PCP will be sorbed to the cuticle and rarely will anything stay in the donor solution, making the experimental determination of K_{cw} difficult. This problem can easily be solved using an external donor solution of 50 ml instead of 1 ml.
5. In steady state experiments with CM the appearance of solutes in the receiver solution is measured, and it takes some time before the first solutes have penetrated the CM. When fluxes are measured by the submersion technique accumulation in the wax, the cuticle and the leaf tissue are estimated, and these processes are almost instantaneous.
6. In calculating D, the thickness ℓ enters as the square, and with MX membranes total thickness results in relatively good D values, as seen from the fact that $K_{calc.}$ and K_{det} for *Citrus* MX are similar. The limiting barrier of the CM is only a fraction of the total thickness, and this introduces large errors in both D and K.
7. The calculated P is $1.35 \times 10^{-10}\,\text{m}\,\text{s}^{-1}$, while the value in Table 6.3 P is $2.8 \times 10^{-10}\,\text{m}\,\text{s}^{-1}$. The agreement is quite good, considering natural variability among individual CM and CM lots of different origins.
8. For CPT_1, 0.15 is obtained, and for CPT_2, 0.086. That means that in CPT_3 is $(1 - 0.236) = 0.764$. That is, 76.4% of the PCP are desorbed in 180 min.
9. In this type of experiment, sorption in epicuticular wax and cuticles are the first steps in foliar penetration. Amounts sorbed are proportional to the amount of wax (which varied among species) and to the partition coefficient K_{ww} (which varied among solutes). Rates of penetration are proportional to permeance, and permeance is proportional to solute concentration in the waxy barrier (6.6).

10. Diffusion coefficients in the barrier must have been similar for all solutes.
11. If the ratio volume/area decreased by a factor of 0.5 and the half time by the same factor, this implies that P was constant.
12. The wetting agent improved the contact between donor solutions and cuticle. This increased contact area and uniformity of rates of penetration. As Tween 20 is not an accelerator adjuvant (Chap. 7) and Alar is an ionic solute, this is the best explanation.
13. In steady state penetration, diffusion in the waxy limiting skin is rate-controlling, while in simultaneous bilateral desorption the limiting skin is not involved because most solutes are desorbed from the inner surface of the CM.
14. They differ in the nature of the donor. With a liquid donor, (2.27) k is determined. If solutes dissolved in cutin and wax as in UDOS, the rate constant (k^*) is marked with an asterisk. k^* is independent of the partition coefficient, while k depends on K.
15. Diffusion of lipophilic solutes proceeds in the amorphous wax fraction, and fluidity of this fraction is apparently the same in waxes of all species.
16. Differences among species in solute mobility observed with the same solute are proportional to k_0^*, which reflects differences in the tortuosity factors.
17. These ratios vary between 0.87 and 1.03 and the average of these values is 0.95, which is not bad since different populations were used for these determinations.
18. According to (6.21), the product $\beta' \times V_x$ must be added to log k^*.
19. If the same solute is used for determining k^*, differences in k^* among species are proportional to the ratio A/V_{donor} (6.14) and (6.16).
20. Prior to droplet drying P was 1.22×10^{-7} m s^{-1}, and after droplet drying it was 1.28×10^{-7} m s^{-1}. Hence, permeance was independent on urea concentration.
21. A thickness of 1.33 µm can be calculated by dividing the amount of reconstituted wax of 120 µg cm^{-2} by wax density of 0.9 g cm^{-3}. Using (2.35), D is 3.47×10^{-17} m^2 min^{-1} or 5.79×10^{-19} m^2 s^{-1}. Increasing the thickness of the wax layer by a factor of 2 leads to a four-fold smaller slope, because at constant D the product of $\ell^2 \times \text{slope}^2$ must be constant.
22. Using (6.23), a D of 5.25×10^{-18} m^2 s^{-1} can be calculated.
23. Rearranging (6.24) or (2.18), a thickness ℓ of 5.95 µm can be calculated.

Chapter 7
Accelerators Increase Solute Permeability of Cuticles

Many agrochemicals are sprayed on leaves of weeds and crop plants. Foliar application minimises contamination of the soil and inactivation or degradation of active ingredients by soil microorganisms. Binding to constituents of the soil is also avoided. Systemic active ingredients must penetrate cuticles before they reach their sites of action in the leaves or in other organs of the plants following translocation (Kirkwood 1999). Pesticides are always formulated to improve wetting, deposition, rain fastness and rates of cuticular penetration. Formulations are mixtures of active ingredient, solvents, carriers, emulsifiers, wetting agents, etc. Compounds added to active ingredients are called adjuvants. Most adjuvants are biologically inert, but they improve biological performance (Kirkwood 1993). Surface active agents (surfactants) are typical adjuvants, and among other things they improve adhesion and spreading of spray droplets on the leaf surface. Surfactants may also act as emulsifiers for active ingredients having a low water solubility, and after droplet drying they can maintain the spray deposits on the cuticle in the liquid state by solvent and by hygroscopic action (Baur 1999; Baur et al. 1997b, 1999; Tadros 1987).

Some adjuvants increase diffusion coefficients of solutes in wax and cutin. This is accomplished by increasing fluidity of waxes and cutin polymer chains. As a consequence, permeability of cuticles to solutes is increased (Schönherr and Bauer 1994). In the technical polymer literature, compounds which render brittle polymers more flexible by increasing fluidity of polymer chains are called plasticisers. They intercalate between polymer chains, and they are added to the polymer melt during production (Gächter and Müller 1990). Plasticisers useful for increasing diffusion coefficients of solutes in cuticles have been termed accelerators (Schönherr 1993a, b). Some surfactants are very effective plasticisers, but surface activity is not a prerequisite for the plasticising activity. Accelerators must be added to the formulation, and they are applied to the cuticle at the same time as the active ingredient. Hence, it is a prerequisite for their usefulness that they penetrate faster into the waxy limiting barrier of the cuticle than the active ingredient, and they should remain sorbed in the wax until most of the active ingredient has penetrated the cuticle.

In this chapter we deal with the mechanistic aspects of sorption and diffusion of plasticisers in wax and in cuticles, and their effect on permeability. All other effects of adjuvants which occur on the outer surface of the cuticle (spreading, adhesion, inhibition of desiccation of the spray droplets, partitioning between the deposit and the cuticle) will not be considered, although they can significantly affect foliar penetration (Baur 1998, 1999; Baur et al. 1997b, 1999; Tadros 1987).

7.1 Sorption of Plasticisers in Wax and Cutin

Any molecule acting as a plasticiser on the transport-limiting barrier of the cuticle must be soluble in wax and cutin. Differential solubility is best described as partitioning between the lipophilic wax or cutin and the external aqueous phase (6.1). Cuticular waxes form the transport-limiting barrier of the cuticle (Chaps. 4 and 6), and the partition coefficient wax/water (K_{ww}) for plasticisers must be known. Most of the work on the effect of plasticisers on permeability of cuticles has been conducted using non-ionic alcohol ethoxylates, which are also effective surfactants. They consists of a lipophilic long-chain fatty alcohol having a variable number of carbon atoms (C_x) and a polar head composed of ethylene oxide ($–O–CH_2–CH_2–$) units (E_y). The lipophilic C_x is linked to the hydrophilic E_y by an ether bond. In addition to alcohol ethoxylates, n-alkyl esters, lacking surface activity have also been investigated. They are composed of short-chain α, ω-diacids of 8 or 10 carbon atoms. Phosphate esterified to short-chain alcohols having 1–4 carbon atoms also have good accelerating activity.

7.1.1 Sorption of Alcohol Ethoxylates in Wax

Surface active compounds exhibit a characteristic behaviour when dissolved in water. At low concentrations, surfactant molecules tend to accumulate at the air/water interface (Adamson 1976). This leads to a decrease of surface tension of water, and this is the reason why surfactants improve wetting of hydrophobic surfaces. With increasing concentrations in water, surface tension decreases, and at a certain concentration called the critical micelle concentration (cmc), surfactant molecules aggregate and form micelles within the aqueous solution (Schick 1987). At the cmc the minimum surface tension is obtained, and if more surfactants are added they form additional micelles. Micelles are spherical aggregates of surfactant monomers with the hydrophobic carbon tails buried in the centre of the micelle while the polar head groups are in contact with water. The presence of micelles affects partition coefficients.

Technical surfactants are polydisperse, that is, they are mixtures of molecules of different linear alcohols to which varying numbers of ethylene oxide groups are linked. In most of our work we have used monodisperse surfactants consisting of a specific alcohol (for instance dodecanol) and a specific number of ethylene oxide groups ($–CH_2CH_2O–$). Below the cmc, increasing amounts of $C_{12}E_8$ are sorbed in

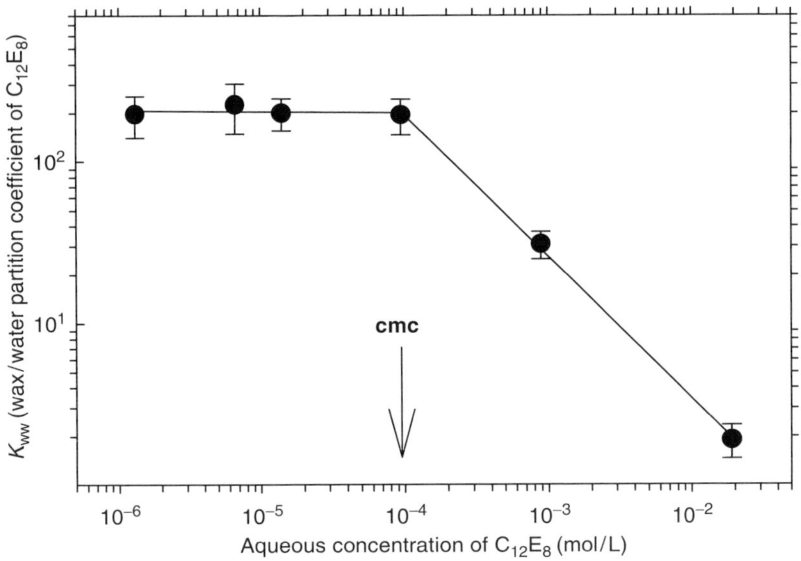

Fig. 7.1 Wax/water partition coefficient (K_{ww}) of octaethylene glycol monododecyl ether ($C_{12}E_8$) for the system reconstituted barley wax/water. Error bars represent 95% confidence intervals. Redrawn from Schreiber (1995)

the wax when external concentration of ^{14}C-labelled $C_{12}E_8$ was increased (Fig. 7.1). K_{ww} is calculated in the usual way as the ratio of the internal and external concentrations of $C_{12}E_8$ (6.1). Above the cmc, the partition coefficient K_{ww} is constant and independent of the external concentration (Fig. 7.1). This is due to the fact that above the cmc all additional surfactant molecules form micelles, and the concentration of free $C_{12}E_8$ molecules dissolved in water remains constant. Adding more surfactant increases the number of micelles, but there is no increase of $C_{12}E_8$ sorbed in the wax. Correct K_{ww} values are only obtained when cmc is used as external concentration, not total surfactant concentration.

The formation of micelles also affects the sorption of lipophilic solutes in wax. The K_{ww} measured with ^{14}C-labelled pentachlorophenol (PCP) is about 3,500 (Table 6.2 and Fig. 7.2). It is not affected by $C_{12}E_8$ as long as the cmc is not reached (Fig. 7.2). If the $C_{12}E_8$ concentrations exceeds the cmc, micelles are formed and apparent K_{ww} of PCP decreases, because PCP dissolves not only in the wax but also in lipophilic micelles (PCP is solubilised). Thus, above the cmc PCP is in equilibrium between lipophilic wax, lipophilic micelles and the aqueous phase. Increasing the number of micelles results in increasing amounts of PCP solubilised in micelles, and less is sorbed in wax. This leads to a decrease of the apparent K_{ww}. This should be taken into account when calculating partition coefficient of surface active compounds or of lipophilic compounds in the presence of micellar solutions of surface active compounds.

K_{ww} have been determined for a large number of monodisperse alcohol ethoxylates and reconstituted waxes from three different species (Table 7.1). Partition

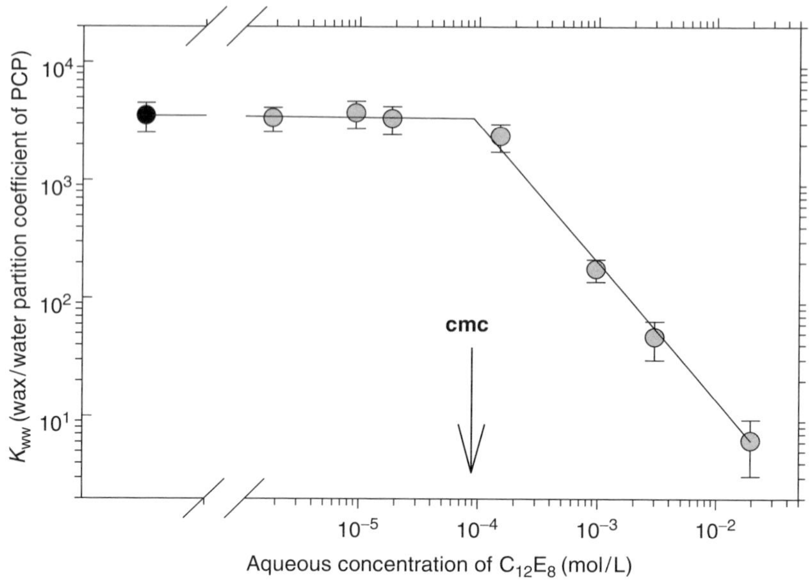

Fig. 7.2 Wax/water partition coefficient (K_{ww}) of pentachlorophenol (*PCP*) for barley wax as affected by increasing concentrations of octaethylene glycol monododecyl ether ($C_{12}E_8$) in the external solution. Error bars represent 95% confidence intervals. Redrawn from Schreiber (1995)

coefficients of the alcohol ethoxylates varied over a range of about six orders of magnitude, while the variability between reconstituted waxes from different plant species was only about one order of magnitude (Table 7.1). K_{ww} of *Stephanotis* wax were always significantly higher by a factor of nearly 10 than K_{ww} of *Hordeum* and *Chenopodium* waxes, the latter two being identical within experimental error. This pronounced difference is attributed to differences in the degree of crystallinity of the reconstituted waxes. Wax composition of *Hordeum* and *Stephanotis* has been analysed by gas chromatography and mass spectrometry (Simanova et al. 2005). Chemical composition of barley wax is very homogeneous, with nearly 90% of the wax composed of primary alcohols. *Stephanotis* wax is heterogeneous because seven different substance classes are present in about equal amounts. Therefore, barley wax has a higher degree of crystallinity than *Stephanotis* wax and thus barley wax contains viewer sorption sites. *Chenopodium* wax apparently has a similar degree of crystallinity and a similar molecular structure to that of barley wax. K_{ww} of barley wax and lipophilic solutes are also low (Table 6.2).

The large variability in K_{ww} values of monodisperse alcohol ethoxylates decreases when the number of the polar ethylene oxide units (E_y) is increased, while it increases with the number of the carbon atoms C_x in the alcohol:

$$\log K_{ww} = -2.73 + 0.54C_x - 0.23E_y \text{ (barley wax)} \quad (7.1)$$

$$\log K_{ww} = -2.43 + 0.53C_x - 0.24E_y \text{ (\textit{Chenopodium} wax)} \quad (7.2)$$

$$\log K_{ww} = 4.78 - 0.23E_y \text{ (\textit{Stephanotis} wax)} \quad (7.3)$$

Table 7.1 Wax/water partition coefficients K_{ww} of monodisperse alcohol ethoxylates in reconstituted wax of *Hordeum vulgare, Stephanotis floribunda* and *Chenopodium album*. C_x gives the number of carbon atoms of the fatty alcohol, and E_y refers to the number of ethylene oxide units in the polar part of the alcohol ethoxylates

Compound	cmc(mol kg^{-1})[a]	K_{ww} *H. vulgare*	*S. floribunda*[b]	*C. album*[c]
C_4E_2	0.589	0.083[d]	–	–
C_6E_3	0.06	0.83[d]	–	–
C_8E_4	0.0062	4.9[d]	–	–
$C_{10}E_5$	0.00063	35.7[d]	–	–
$C_{10}E_8$	0.0012	5.6[e]	–	–
$C_{12}E_2$	0.000028	2,000[e]	21,125	–
$C_{12}E_3$	0.000035	1,300[e]	–	1,610
$C_{12}E_4$	0.000043	670[e]	7,909	–
$C_{12}E_5$	0.000053	400[e]	4,439	687
$C_{12}E_6$	0.000066	201[d]	2,626	268
$C_{12}E_7$	0.000079	160[e]	–	203
$C_{12}E_8$	0.000098	104[b]	962	109
$C_{14}E_3$	0.0000029	–	–	19,700
$C_{14}E_5$	0.0000044	–	–	8,040
$C_{14}E_6$	0.0000054	–	–	3,700
$C_{14}E_7$	0.0000066	1,561[d]	–	–
$C_{14}E_8$	0.0000081	1,200[e]	–	1,410
$C_{16}E_3$	0.00000024	–	–	231,000
$C_{16}E_8$	0.00000068	12,350[d]		13,700

[a]Calculated according to Riederer and Schreiber (1995)
[b]Data from Simanova et al. (2005)
[c]Data from Burghardt et al. (2006)
[d]Data from Schreiber et al. (1996b)
[e]Data from Burghardt et al. (1998)

With all three wax samples, log K_{ww} decreased by the factor of −0.23 to −0.24 for each additional ethylene oxide (E) unit. For barley (7.1) and *Chenopodium* wax (7.2), log K_{ww} increased by the factor of 0.53–0.54 for each additional C atom. Equations for barley and *Chenopodium* wax are identical within experimental error (Burghardt et al. 1998, 2006). With *Stephanotis* the number of C atoms was constant (C_{12}), and only E_y was varied (Simanova et al. 2005). Hence, (7.3) is only available in a reduced form describing the influence of E_y units on K_{ww}. These results are consistent with data published for linear alcohols and fatty acids by Dunn et al. (1986), from which it can be calculated that for each -CH_2 group added log K_{ow} increased by a factor 0.55–0.52. Thus, the influence of the number of E and C units on log K_{ww} is similar for all three wax samples; however, the absolute values can vary by a factor of 10 depending on plant species. Using (7.1)–(7.3), K_{ww} can be estimated accurately for other monodisperse alcohol ethoxylates.

Cmc of surfactants and their lipophility (i.e., K_{ww}) are related (Table 7.1). With increasing lipophility of the surfactants, micelles form at lower concentrations. The maximum concentration (g kg^{-1}) of alcohol ethoxylates sorbed in reconstituted barley wax depends only on E_y (Burghardt et al. 1998). The contribution of C_x to maximum amounts of alcohol ethoxylates sorbed in the wax disappears from regression equations, because the increase in K_{ww} by the addition of one C atom is compensated by the decrease of the cmc (Burghardt et al. 1998, 2006):

$$\log C_{\text{wax}}^{\text{max}} = 1.29 - 0.08E_y \text{ (barley wax)} \tag{7.4}$$

A similar equation was obtained for *Chenopodium* wax (Burghardt et al. 2006).

$$\log C_{\text{wax}}^{\text{max}} = 1.61 - 0.104E_y \text{ (\textit{Chenopodium} wax)} \tag{7.5}$$

Maximum amounts of surfactant molecules which dissolve in wax are given by the product of the cmc and the K_{ww}. In most experiments, concentrations of the alcohol ethoxylates far above the cmc are used. The variability in maximum amounts of alcohol ethoxylates sorbed in reconstituted waxes is much lower than is suggested by the tremendous variability of K_{ww} (Table 7.1).

These considerations are valid only for aqueous solutions of alcohol ethoxylates. The situation changes when treatment solutions dry, which is the situation after spray applications in the field. Structures of these neat surfactant residues on the cuticle and penetration into the cuticle have not been investigated.

7.1.2 Sorption of Alcohol Ethoxylates in Polymer Matrix Membranes

Partition coefficients of alcohol ethoxylates for polymer matrix membranes (MX) K_{mxw} of *Citrus aurantium* (Riederer and Schreiber 1995) and *Prunus laurocerasus* (Burghardt et al. 2006) have also been determined. An equation which relates C_x and E_y to K_{mxw} for *Citrus* MX has been established.

$$\log K_{mxw} = -1.78 + 0.52C_x - 0.17E_y \tag{7.6}$$

This equation resembles (7.1) and (7.2). For each C atom log K_{mxw} increases again by a factor of 0.52, and for each E unit log K_{mxw} decreases by a factor of 0.17 (7.6). This is excellent evidence that the amorphous environments in cutin and wax in which alcohol ethoxylates are sorbed have similar physicochemical properties. Cutin and waxes are composed of methyl and methylene groups with small amounts of oxygen (Chap. 1). K_{mxw} values for alcohol ethoxylates are highly correlated with K_{ww} values (Fig. 7.3), but the former are higher by about one order of magnitude.

Linear regression equations quantitatively account for this correlation. Equation (7.7) applies to barley wax, and *Citrus* MX (Burghardt et al. 1998) and (7.8) to *Chenopodium* wax and *Prunus* MX (Burghardt et al. 2006):

$$\log K_{ww} = -1.06 + 1.00 \log K_{mxw} \tag{7.7}$$

$$\log K_{ww} = -1.04 + 1.05 \log K_{mxw} \tag{7.8}$$

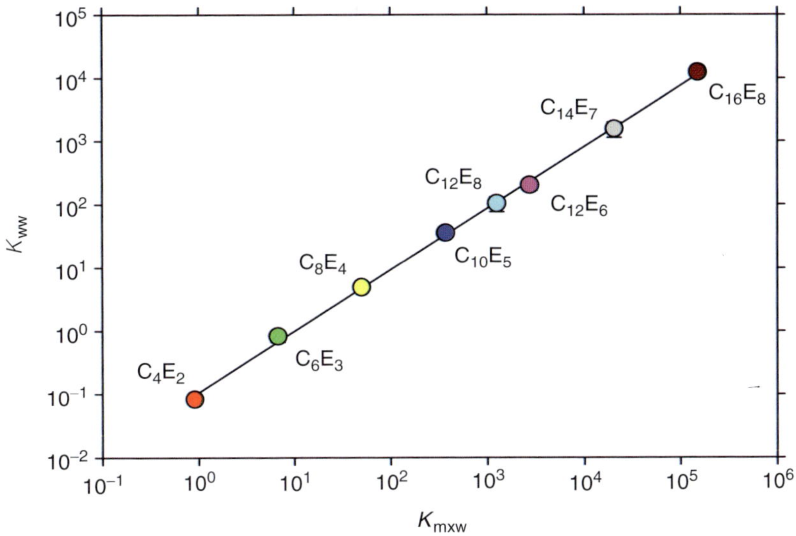

Fig. 7.3 Correlation between wax/water partition coefficients (K_{ww}) and cuticle/water partition coefficients (K_{mxw}). Redrawn from Schreiber et al. (1996b)

Reconstituted waxes from barley and *Chenopodium* leaves are solid and partially crystalline phases, and they offer fewer sorption sites for alcohol ethoxylates than the amorphous MX of *Citrus* and *Prunus*; for this reason, K_{ww} are lower than K_{mxw}. A similar observation had been made with lipophilic solutes lacking surface actvity (Sect. 6.1) where K_{ww} and K_{cw} are compared.

These results are specific for barley and *Chenopodium* wax, and should not be extrapolated to waxes of other species. With *Stephanotis* wax, considerably higher K_{ww} were observed (Table 7.1). Maximum concentrations ($\mathrm{g\,kg^{-1}}$) of alcohol ethoxylates sorbed in *Citrus* MX (7.9) are a function of E_y (Riederer and Schreiber 1995), as was previously shown for wax [(7.4) and (7.5)].

$$\log C_{\mathrm{wax}}^{\mathrm{max}} = 2.07 - 0.044 E_y \tag{7.9}$$

7.1.3 Sorption of n-Alkyl Esters in Wax

N-alkyl esters, lacking surface activity are lipophilic and have very low water solubility. Therefore, $100\,\mathrm{g\,l^{-1}}$ 1,2-propanediol was added to water to obtain homogeneous aqueous solutions without phase separation (Simanova et al. 2005). Partitioning of n-alkyl esters between these external aqueous solutions containing 1,2-propanediol and reconstituted wax was measured with *Hordeum* and *Stephanotis* wax. These partition coefficients are marked K_{wrec} (rec = receiver), instead of K_{ww}. Identical solutions of n-alkyl esters containing 1,2-propanediol were also used as receiver solutions in transport experiments (Sect. 7.3.3).

As observed with alcohol ethoxylates (Table 7.1), partition coefficients for n-alkyl esters are again a factor of about 10 higher for *Stephanotis* wax than for barley wax. With increasing number of C atoms of the dicarboxylic acids and esterified alcohols, K_{wrec} increases and the following regression equations were obtained for barley (7.10) and *Stephanotis* (7.11):

$$\log K_{\mathrm{wrec}} = 0.39C_{\mathrm{x}} - 3.79 \text{ (barley wax)} \tag{7.10}$$

$$\log K_{\mathrm{wrec}} = 0.42C_{\mathrm{x}} - 3.20 \text{ (\textit{Chenopodium} wax)} \tag{7.11}$$

This again is a quantitative structure–property relationship between log K_{wrec} and the descriptor C_{x}. The addition of one C atom increases K_{wrec} by a factor of about 0.4.

7.2 Plasticisation of Cuticular Wax: Evidence from Spectroscopy

The plasticising effect of accelerators on cuticular wax was shown for alcohol ethoxylates and reconstituted barley wax using electron spin resonance (ESR) spectroscopy (Schreiber et al. 1996b) and nuclear magnetic resonance (NMR) spectroscopy (Schreiber et al. 1997). Specific ESR- and NMR-probes were mixed with the wax which allowed direct insights into the molecular architecture of the wax phase. The molecular environment of the probes in the absence or presence of a plasticiser can be characterised.

In ESR experiments, spin labels are used which have a free unpaired electron in a side chain, which can be monitored by ESR (Sackmann et al. 1973). About 0.1% of the spin label 5-doxyl stearic acid (5-SASL) was added to barley wax, the solvent was evaporated and the remaining mixture was reconstituted for the experiment. Since the side chain of the fatty acid carrying the free electron is fairly bulky and polar, the spin label will essentially be sorbed in the amorphous phase of reconstituted wax. Further preparation of the wax samples is identical to that described for the determination of K_{ww} (Sects. 6.1.3 and 7.1.1) or D (Sect. 6.5.1), with the exception that they had to be placed in small glass tubes for ESR experiments. In order to test the plasticising effect of accelerators on the molecular architecture of barley wax, alcohol ethoxylates were added to the wax samples. The exact amount of each alcohol ethoxylate added to the wax was calculated according to (7.4). This corresponds to the maximum amount of each alcohol ethoxylate which can be sorbed to barley wax from an external aqueous solution with the alcohol ethoxylates dissolved at any concentration above the cmc.

ESR spectra were measured at temperatures between −10 and 60°C (Fig. 7.4). The characteristic parameter of the spectrum used to characterise the molecular environment of the ESR-probe is the hyperfine splitting of the spectrum indicated by the two arrows (Fig. 7.4). As long as the hyperfine splitting is visible, a significant amount of the ESR-probe is trapped in a rigid environment, which prevents free rotation of the probe around its own axis. With increasing temperature, the distance of the hyperfine splitting (measured in Gauss) decreases. In parallel, the peaks of

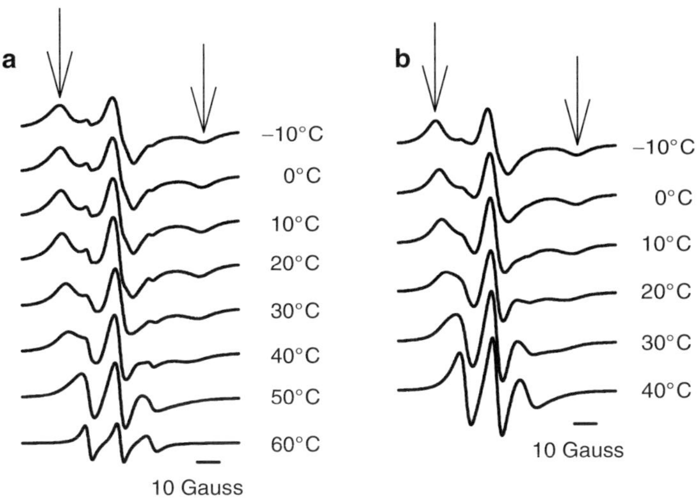

Fig. 7.4 ESR (electron spin resonance) spectra of 5-doxyl stearic acid measured between −10 and 60 °C in reconstituted barley wax in (**a**) absence or (**b**) presence of the alcohol ethoxylate triethylene glycol monohexyl ether (C_6E_3) at a concentration of $10\,g\,kg^{-1}$. *Arrows* indicate the hyperfine splitting. Redrawn from Schreiber et al. (1996b)

the hyperfine splitting continuously lose intensity until they finally disappear completely at high temperatures. A homogeneous ESR spectrum is obtained at high temperature (Fig. 7.4). This indicates that the molecular environment of the ESR-spectrum is fluid, allowing free rotation of the label around its own axis without any restriction.

The molecular environment in which the spin label is dissolved in barley wax is rigid between −10 and 40 °C (Fig. 7.4a). A transition occurs between 40 and 50 °C, and at 50 °C the spin label can freely rotate around its own axis. Increasing the temperature provides the energy for the wax molecules forming the amorphous wax phase to become fluid. This microscopic transition in the amorphous wax phase, where sorption and diffusion of lipophilic molecules takes place, is mapped by the spin label.

If C_6E_3 is present in the wax (1%), this transition from a rigid to a fluid environment takes place at much lower temperatures (Fig. 7.4b). At 10 °C the spin label is still restricted in free rotation, and the hyperfine splitting is visible. At 20 °C the transition from a rigid to a fluid environment can be seen in the spectrum, and already at 30 °C the molecular environment of the spin label is fluid and free rotation of the label is possible. The alcohol ethoxylate plasticises the amorphous phase of the wax, and it is fluid already at room temperature. In the absence of the plasticiser, the temperature needs to be 20–30 °C higher for the same degree of fluidity. Thus, adding an alcohol ethoxylate to cuticular wax mimics the effect of higher temperatures.

A series of alcohol ethoxylates was tested using ESR spectroscopy, and all were active as plasticisers compared to the control (Fig. 7.5). The plasticising effect on

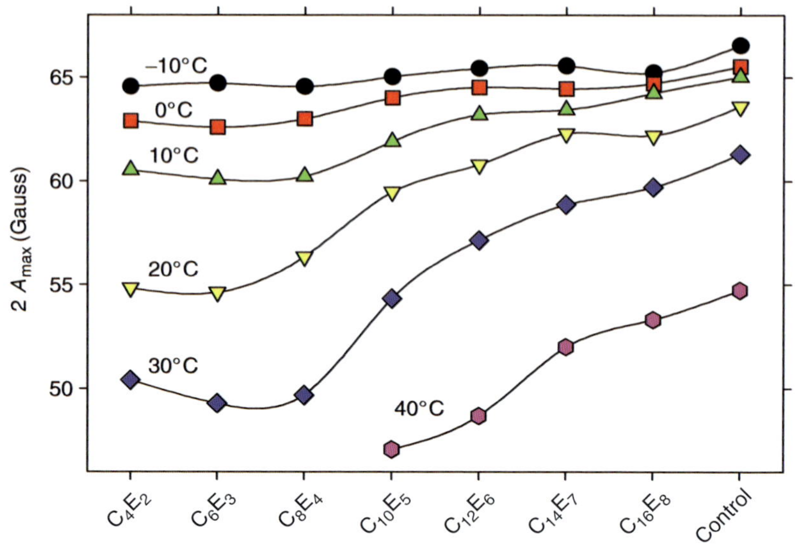

Fig. 7.5 The hyperfine splitting ($2A_{max}$) as a function of temperature for alcohol ethoxylates sorbed in barley wax at maximum concentrations varying from 10 to 5 g kg^{-1} between C_4E_2 and $C_{16}E_8$. Redrawn from Schreiber et al. (1996b)

barley wax was different among the alcohol ethoxylates. As different concentrations of alcohol ethoxylates in wax were present (7.4), intrinsic plasticising effects cannot be deduced from Fig. 7.5. The relationship between the effect of plasticisers and their concentration in the wax is presented in Sect. 7.3.

Direct observation of the plasticising effect of alcohol ethoxylates on reconstituted wax was also possible using ^{2}H-NMR spectroscopy (Schreiber et al. 1997). In this approach, two different deuterated probes, stearic acid and dotriacontane, were added to barley wax, and the molecular motion of the two probes in the absence and in the presence of C_6E_3 was observed by NMR. Most interestingly, mobility of dotriacontane, which is a very lipophilic long-chain alkane of 32 carbon atoms, was not affected by the plasticiser, whereas mobility of deuterium-labelled stearic acid significantly increased when C_6E_3 was added to barley wax. This is good evidence that dotriacontane, which is very similar in size and structure to a typical long-chain paraffin wax, is located in the crystalline phase of the reconstituted wax, which is not accessed and modified by C_6E_3. Stearic acid is located in the amorphous wax phase, which is modified by plasticiser. This is consistent with the observation that a much higher D ($1.1 \times 10^{-18}\,\mathrm{m^2\,s^{-1}}$) is measured for radio-labelled dotriacontane when loaded to the wax from an external solution, compared to the D of $1.4 \times 10^{-22}\,\mathrm{m^2\,s^{-1}}$, which is obtained when dotriacontane is recrystallised together with the wax during reconstitution (Schreiber at al. 1997). This difference was not observed with stearic acid. D was $4.0 \times 10^{-18}\,\mathrm{m^2\,s^{-1}}$ when stearic acid was loaded to the crystallised wax from solution, and it was $3.6 \times 10^{-18}\,\mathrm{m^2\,s^{-1}}$ when the radio-labelled probe was reconstituted together with the wax (Schreiber et al. 1997).

Thus, *D* did not depend on the mode of loading of the wax with the radio-labelled stearic acid. This phenomenon is also discussed in Sect. 6.5.1.

7.3 Effects of Plasticisers on Diffusion of Lipophilic Solutes in Wax

In the previous sections we have seen that accelerators sorbed in wax and in cutin (Sect. 7.1) increase fluidity of the amorphous wax phase at room temperature (Sect. 7.2). In this section, we address the question of how much diffusion and penetration of lipophilic solutes in the waxy limiting barrier are affected by plasticisers.

7.3.1 *Effect of* $C_{12}E_8$ *on Solute Diffusion in Reconstituted Wax*

Accelerating effects of plasticisers on solute diffusion in reconstituted wax can be studied using the method of paired comparison. Initially, desorption of a radio-labelled probe from reconstituted wax is measured using an inert medium. As soon as rate constants have been established, desorption is continued with a medium containing the plasticiser. Finally, desorption is again continued with the inert desorption medium.

As an example of this type of experiment, ^{14}C-labelled tetracosanoic acid ($C_{24}Ac$) was desorbed form barley wax (Fig. 7.6). The experiment was started using borax buffer at *p*H 9.0 as inert desorption medium. At this *p*H, all $C_{24}Ac$ desorbed from the wax is fully ionised. This ensures that the driving force for $C_{24}Ac$ desorption is maximum, because ionised $C_{24}Ac$ is insoluble in wax. After about an hour, $C_{12}E_8$ was added to borax buffer at a concentration well above the cmc, and within minutes rates of desorption of $C_{24}Ac$ from the wax significantly increased (Fig. 7.6). This indicates that $C_{12}E_8$ rapidly diffuses into the wax, and due to its plasticising effect it accelerates diffusion of $C_{24}Ac$. Since $C_{12}E_8$ was applied well above the cmc, it is ensured that the external concentration of free monomeric $C_{12}E_8$ molecules not associated with micelles remains constant and is not depleted by $C_{12}E_8$ sorption in the wax. Thus, the maximum plasticising effect of $C_{12}E_8$ is maintained. After elimination of $C_{12}E_8$ from the desorption medium, rates of desorption of $C_{24}Ac$ rapidly decreased to the initial rate (Fig. 7.6). This is direct evidence that plasticisation of the wax is reversible and that an irreversible change of the wax structure did not occur. Reversibility also indicates that the acceleration is not due to a solubilisation of the wax constituents, which has been suggested (Tamura et al. 2001a, b).

Reversibility of the plasticising effect has also been observed with pentachlorophenol in reconstituted wax (Schreiber 1995), with 4-nitrophenol penetration across isolated *Prunus laurocerasus* CM (Sect. 7.4.1), and in SOFP experiments studying the effect of polydisperse alcohol ethoxylates on cuticular penetration of various lipophilic solutes (Baur et al. 1999).

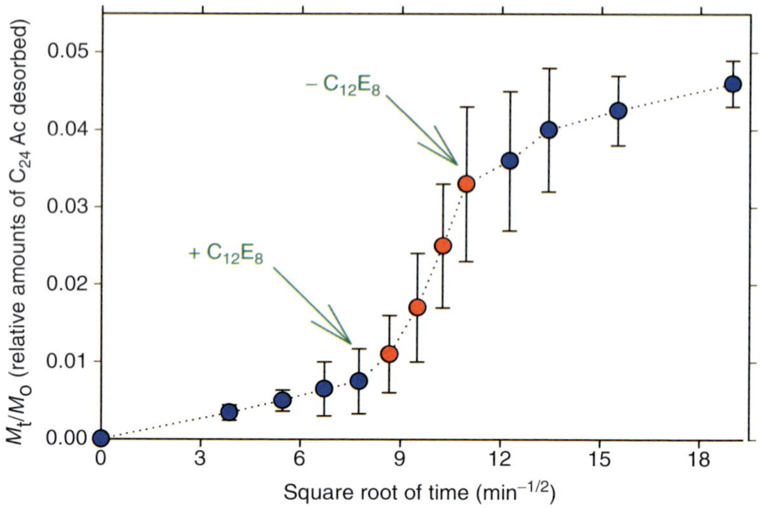

Fig. 7.6 Desorption of tetracosanoic acid ($C_{24}Ac$) from reconstituted barley wax. Within the first hour (*blue symbols*) $C_{24}Ac$ was desorbed using inert borax buffer (*p*H 9.0) as desorption medium. During the second hour (*red symbols*) $C_{24}Ac$ was desorbed by borax buffer containing $C_{12}E_8$ (10^{-2} mol l^{-1}). During the final desorption period (*blue symbols*) borax buffer was used again. Error bars represent 95% confidence intervals. Data redrawn from Schreiber (1995)

When working with aqueous micellar solutions, micelles scavenge lipophilic solutes such that their concentration in water remains practically zero. In the present case this did not occur, because $C_{24}Ac$ was ionised, but when non-ionic solutes are studied, their sorption in micelles is very useful to maintain maximum driving force.

Using (6.23) (Sect. 6.5.2) *D* in wax can be calculated from the molar volume. With the molar volume of $C_{12}E_8$ (458 cm^3 mol^{-1}), *D* is $5.74 \times 10^{-20} m^2 s^{-1}$. However, direct measurements using ^{14}C-labelled $C_{12}E_8$ result in a *D* of $8.5 \times 10^{-18} m^2 s^{-1}$ (Schreiber 1995), which is about 150 times higher than *D* estimated from (6.23). This large discrepancy between estimated and measured *D* is direct evidence that $C_{12}E_8$ not only accelerates diffusion of other solutes but simultaneously accelerates its own diffusion. This fairly high diffusion coefficient of $C_{12}E_8$ also explains the fast induction (Fig. 7.6).

These conclusions are supported by the time course of sorption in wax of $C_{12}E_8$ and desorption of $C_{24}Ac$ from reconstituted barley leaf wax (Fig. 7.7). Data were obtained by independently using either ^{14}C-labelled $C_{12}E_8$ or ^{14}C-labelled $C_{24}Ac$. It took roughly 1 hour for equilibration of the wax with $C_{12}E_8$, and the half-time ($M_t/M_0 = 0.5$) of sorption is only about 7 min (Fig. 7.7). With borax buffer, desorption of $C_{24}Ac$ is linear and very slow (Fig. 7.7). Significantly higher rates of desorption were obtained when $C_{12}E_8$ was added to the borax desorption medium. Rates increased up to 100 min. Afterwards, linear desorption kinetics were observed when $C_{12}E_8$ concentration in the wax was maximum and constant (Fig. 7.7). Comparable data were obtained with $C_{12}E_8$ and pentachlorophenol (Schreiber 1995).

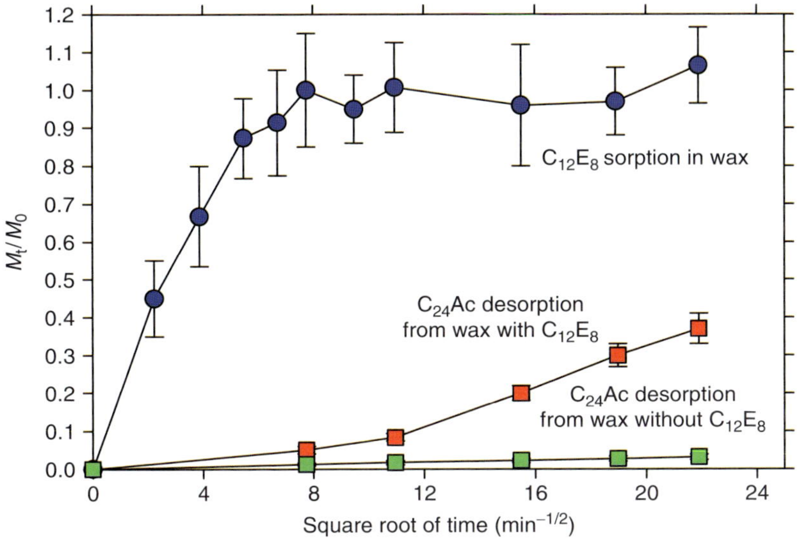

Fig. 7.7 Sorption kinetic of ^{14}C-labelled $C_{12}E_8$ (octaethylene glycol monododecyl ether) in reconstituted barley wax (*blue symbols*) and desorption kinetics of ^{14}C-labelled $C_{24}Ac$ (tetracosanoic acid) from barley wax using either borax buffer (*p*H 9.0) with $C_{12}E_8$ (10^{-2} M) (*red symbols*) or without $C_{12}E_8$ (*green symbols*). All three kinetics were determined in independent experiments. Error bars represent 95% confidence intervals. Data redrawn from Schreiber (1995)

Maximum $C_{12}E_8$ concentrations in wax are obtained when the desorption media contains $C_{12}E_8$ at concentrations well above the cmc. The relationship between different $C_{12}E_8$ concentrations and plasticising effects on diffusion of ^{14}C-labelled solutes can be investigated using desorption media with increasing $C_{12}E_8$ concentrations (Fig. 7.8).

Effects of $C_{12}E_8$ on D of $C_{24}Ac$ increased from three- to more than 40-fold when $C_{12}E_8$ concentrations in the external desorption medium were increased from 10^{-9} to 10^{-4} mol l^{-1} (Fig. 7.8). At concentrations above the cmc of 10^{-4} mol l^{-1}, effects were independent on the external $C_{12}E_8$ concentration, because sorption above the cmc does not increase (Sect. 7.1.1). A similar correlation between increasing $C_{12}E_8$ concentrations in the external desorption medium and effects on D has been shown for the desorption of ^{14}C-labelled PCP, with the main difference that effects on D only increased from 1.5- to less than five-fold, with $C_{12}E_8$ concentrations in the external desorption medium increasing from 10^{-9} to 10^{-4} mol l^{-1} (Schreiber 1995).

7.3.2 Plasticising Effects of Other Alcohol Ethoxylates

Effects of various alcohol ethoxylates of the series C_xE_y have been studied using pentachlorophenol and barley leaf wax. The alcohol ethoxylates are applied at the same external concentration, which was in all cases well above the cmc. Plasticising

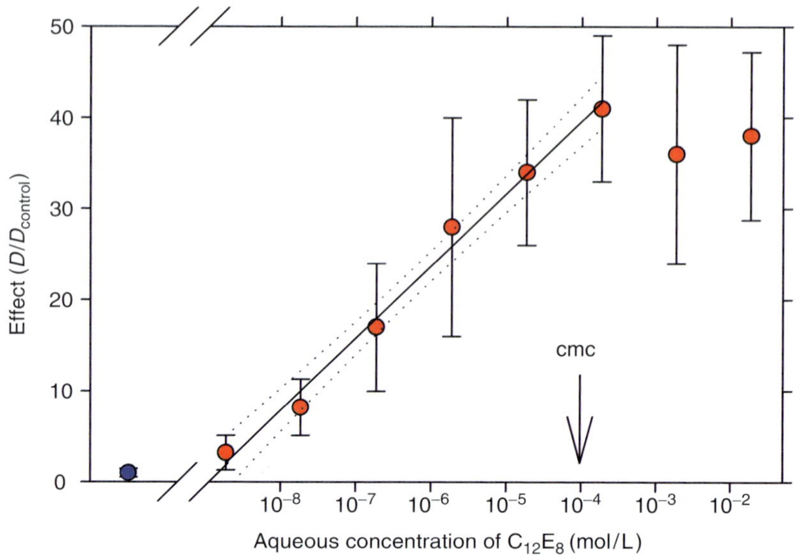

Fig. 7.8 Effect on diffusion coefficients (D/D_{control}) of C_{24}Ac (tetracosanoic acid) in reconstituted barley wax as a function of $C_{12}E_8$ concentrations in the external borate buffer. D/D_{control} is 1 in absence of $C_{12}E_8$ (*blue symbol*). Results obtained with various $C_{12}E_8$ concentrations are shown in *red*. Error bars represent 95% confidence intervals. Data redrawn from Schreiber (1995)

effects on D varied greatly depending on type of surfactant and its concentration in wax. Each alcohol ethoxylate has its specific cmc and K_{ww} (Table 7.1). As a consequence, surfactant concentration in wax and plasticising effects differed.

A plot D/D_{control} vs internal concentration of the different alcohol ethoxylates in wax resulted in a reasonable correlation (Fig. 7.9). This shows that the effects measured for the different plasticisers are not a specific property of each alcohol ethoxylate investigated. Instead, the magnitude of the plasticising effect of different alcohol ethoxylates solely depended on the total amount of plasticiser present in the wax. The plasticising effects of alcohol ethoxylates on D in wax are unspecific. At the same concentration in wax they cause the same effect, and different effects observed in the experiments are solely a consequence of differences in the cmc and the K_{ww} of the alcohol ethoxylates (Table 7.1). The same conclusions were drawn from experiments analysing the effect of different alcohol ethoxylates on D of C_{24}Ac (Schreiber et al. 1996b) and bitertanol (Burghardt et al. 1998) in barley wax, and of benzoic acid and bitertanol in *Chenopodium* wax (Burghardt et al. 2006).

7.3.3 *Plasticising Effects of* n-*Alkyl Esters*

The effect of n-alkyl esters (see Table 7.2) on D has been investigated using reconstituted waxes from barley and *Stephanotis* leaves (Simanova et al. 2005).

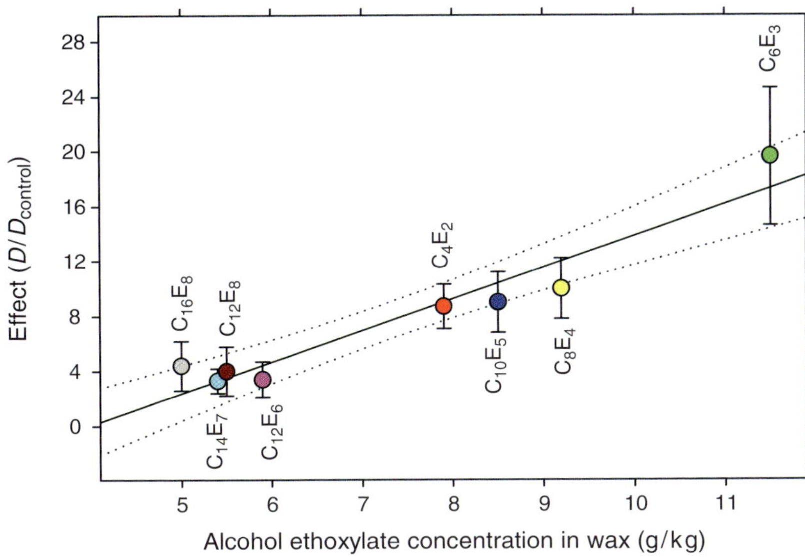

Fig. 7.9 The effect of different alcohol ethoxylates on diffusion coefficients ($D/D_{control}$) of PCP (pentachlorophenol) in reconstituted barley wax as a function of the internal equilibrium concentration of the alcohol ethoxylates in barely wax. Effects are calculated by dividing the D of PCP measured in the presence of the corresponding alcohol ethoxylate by D measured in the absence of an alcohol ethoxylate. Error bars represent 95% confidence intervals. Data redrawn from Schreiber et al. (1996b)

Table 7.2 Wax/receiver partition coefficients K_{wrec} of n-alkyl esters for reconstituted wax of *Hordeum vulgare* and *Stephanotis floribunda*

n-alkyl ester	K_{wrec} *H. vulgare*	K_{wrec} *S. floribunda*
DMSU (dimethyl suberate)	3.1	15.5
DESU (diethyl suberate)	5.9	55.7
DES (diethyl sebacate)	31.2	447
DBSU (dibutyl suberate)	503	5,319
DBS (dibutyl sebacate)	1,821	28,590

Data from Simanova et al. (2005)

Intrinsic effects of n-alkyl esters linearly increased with increasing concentrations of n-alkyl esters in the wax (Fig. 7.10). This shows again that there is no specificity in the plasticising effect between the different n-alkyl esters tested. However, effects with barley and *Stephanotis* wax differed greatly. For the same plasticising effect, about ten times higher n-alkyl ester concentrations are needed with *Stephanotis* wax than with barley wax (Fig. 7.10). Sorption capacity in *Stephanotis* wax is ten times higher than that of highly crystalline barley wax (Table 7.2).

Desorption of ^{14}C-labelled 2,4-dichlorophenoxybutyric acid (2,4-DB) from *Stephanotis* was also studied with alcohol ethoxylates (Fig. 7.10). At identical plasticiser concentrations in *Stephanotis* wax, intrinsic effects were about five-fold higher

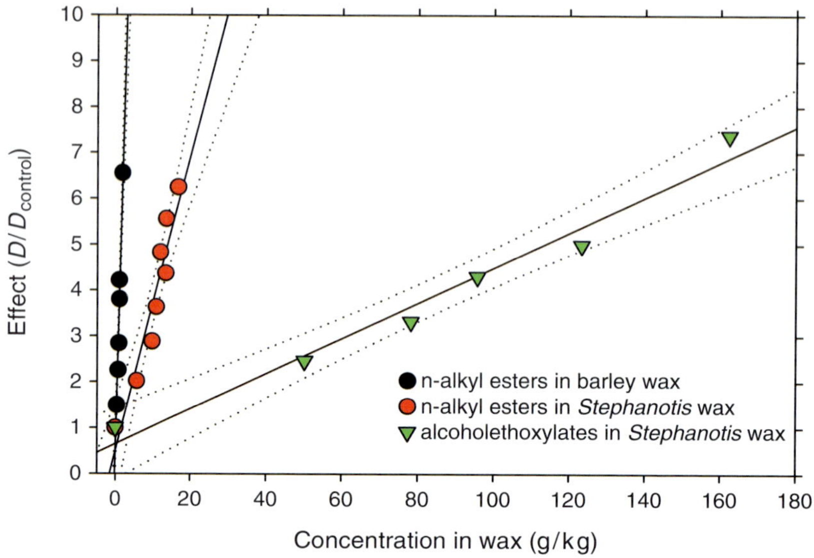

Fig. 7.10 Effects of different n-alkyl esters (DBS, DES, DBSU and DESU; full names see Table 7.2) and different alcohol ethoxylates ($C_{12}E_2$, $C_{12}E_4$, $C_{12}E_5$, $C_{12}E_6$ and $C_{12}E_8$; full names see Table 7.1) on diffusion ($D/D_{control}$) of 2–4-dichlorophenoxybutyric acid (2,4-DB) in reconstituted wax of barley and *Stephanotis* as a function of their concentrations in wax. Ninety-five percent confidence intervals for each regression line are shown as *dotted lines*. Data redrawn from Simanova et al. (2005)

with n-alkyl esters than with alcohol ethoxylates (Fig. 7.10). Thus, there are specific differences between the two different classes of plasticisers investigated.

7.3.4 Dependence of the Plasticising Effect on Molar Volume of Solutes

Diffusion coefficients depend on molar volumes of solutes (Sect. 6.5.2). With increasing molar volume (V_x), log D linearly decreases (Fig. 6.26). From this linear plot, size selectivity β' and the diffusion coefficient D_0 of a hypothetical molecule having a molar volume of zero are obtained (6.22). Size selectivity of reconstituted barley wax decreased in presence of sorbed alcohol ethoxylates (Fig. 7.11).

Slopes (β') are significantly smaller in the presence of the accelerators (Table 7.3). With $C_{12}E_8$, size selectivity decreased by a factor of 2.8, whereas with C_8E_4 size selectivity decreased by a factor of 5.9 (Table 7.3). This difference is due to the fact that about 2.1-times more C_8E_4 than $C_{12}E_8$ is sorbed to the wax (7.4).

The y-intercepts log D_0, which can be related to tortuosity (Schönherr and Bauer 1994), were not significantly affected by the plasticisers (Table 7.3). Tortuosity of the waxy pathway depends on structure of crystalline waxes. These data show that

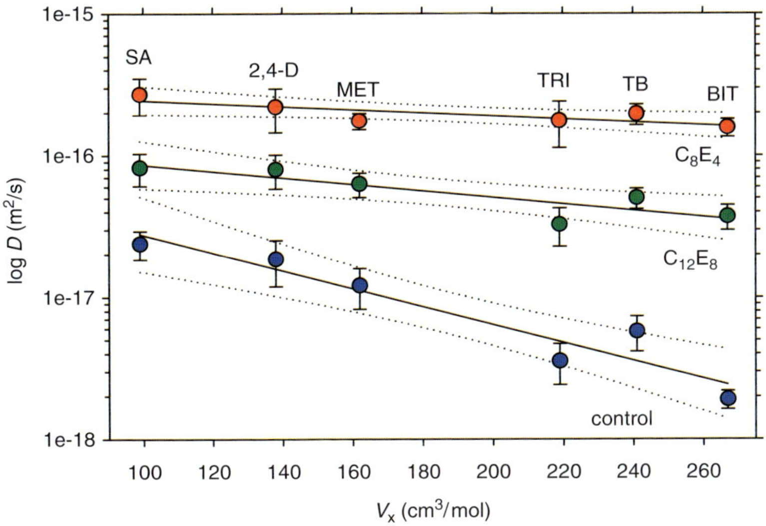

Fig. 7.11 Plots of the log D of lipophilic solutes as a function of their molar volumes V_x. D was determined either using a phospholipid suspension as inert desorption medium (control; *blue symbols*) or in the presence of the accelerators octaethylene glycol monododecyl ether ($C_{12}E_8$; *green symbols*) and tetraethylene glycol monooctyl ether (C_8E_4; *red symbols*). Salicylic acid (*SA*), 2,4-dichlorophenoxyacetic acid (2,4-*D*), metribuzin (*MET*), triadimenol (*TRI*), tebuconazole (*TB*), bitertanol (*BIT*). Error bars represent 95% confidence intervals. Data redrawn from Burghardt et al. (1998)

Table 7.3 Slopes β', y-intercepts D_0 and coefficients of determination r^2 of the regression equations fitted to data of Fig. 7.11

Desorption medium	β'(mol cm^{-3})	log D_0(m^2 s^{-1})	r^2
PLS	−0.0065	−15.94	0.91
C12E8	−0.0023	−15.85	0.76
C8E4	−0.0011	−15.51	0.67

Data from Burghardt et al. (1998)

tortuosity and wax crystallinity were not affected by $C_{12}E_8$ and C_8E_4. This result is consistent with results of ESR experiments (Fig. 7.4) and reversibility of the plasticising effect of $C_{12}E_8$ on barley wax (Fig. 7.8) described in Sects 7.2 and 7.3.

Figure 7.11 also indicates that the plasticising effect of the alcohol ethoxylates is larger with larger solutes having a lower D. This is also seen when the effect of the plasticisers (D/D_{control}) is plotted against the reciprocal of D_{control} (Fig. 7.12). The largest effects are obtained with bitertanol (which has the lowest D_{control}) while the smallest effect is obtained with salicylic acid (which has the highest D_{control}). Thus, diffusion of small solutes having a large D is less affected by plasticisers.

All the above experiments were conducted using reconstituted cuticular waxes. These data provide insights into the molecular architecture of waxes, and show that plasticisers increase fluidity of amorphous waxes. Can these findings be extrapolated to cuticles of intact leaves?

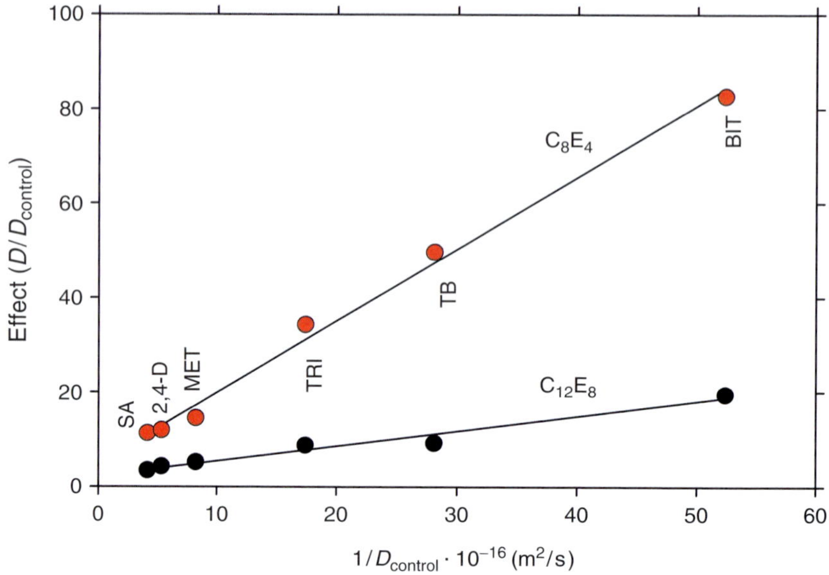

Fig. 7.12 Effects of the two accelerators octaethylene glycol monododecyl ether ($C_{12}E_8$; *black symbols*) and tetraethylene glycol monooctyl ether (C_8E_4; *red symbols*) on diffusion ($D/D_{control}$) of lipophilic solutes in barley wax. Salicylic acid (*SA*), 2,4-dichlorophenoxyacetic acid (2,4-*D*), metribuzin (*MET*), triadimenol (*TRI*), tebuconazole (*TB*), bitertanol (*BIT*). Data redrawn from Burghardt et al. (1998)

7.4 Effects of Plasticisers on Transport in Cuticles

In the field, spray droplets containing both active ingredients and adjuvants are sprayed on the foliage, and there they interact with the cuticles. Adjuvants affect both driving force (partition coefficient) and solute mobility (Chap. 6). For example, cuticle/water or wax/water partition coefficients of solutes can be greatly reduced when solutes are dissolved in a neat adjuvant phase, and this reduces driving force and rates of penetration (Schönherr and Baur 1994; Baur et al. 1997b, 1999; Marzouk et al. 1998; Baur 1998; Buchholz 2006). This effect can partially or totally mask a plasticising effect. In mechanistic studies with isolated cuticles, this problem can be avoided by applying the solutes and the plasticisers to different sides of the CM. The plasticiser to be studied can be directly used as desorption medium in the receiver compartment facing the morphological outer surface of the CM. The radio-labelled solute is applied to the inner surface of the CM (Sect. 6.4). With surfactants it is convenient to work at concentrations above the cmc as receiver solutions. The plasticiser can penetrate into the CM, while at the same time solutes which appear in the receiver are trapped in micelles (Schönherr et al. 2001).

7.4.1 Reversibility of Plasticiser Effects

Plasticiser effects observed with reconstituted waxes are reversible (Sect. 7.1). This is also true with CM. This has been studied in the steady state (Schreiber et al. 1995) and in SOFP experiments (Baur et al. 1999). In steady state experiments, cuticles were mounted between two stainless steel half-cells (Fig. 7.13), and the amount of 4-NP that penetrated from the donor compartment across the cuticle into the receiver compartment was monitored photometrically. 4-NP was buffered at *p*H 3.0, which ensures that all 4-NP is non-ionised, and it was collected in the receiver at *p*H 9.0 where 4-NP is fully ionised. This ensures that the driving force in this steady state experiment is constant and maximum.

Brij 30 is a polydisperse mixture of fatty alcohols (C_{10-12}) substituted with E_{0-16}. The effect of this alcohol ethoxylate was tested in the steady state penetration of 4-NP across isolated *Prunus laurocerasus* CM. Brij 30 was added to the morphological outer surface of the CM, which was the receiver.

When Brij 30 was added to the receiver, rates of 4-NP penetration increased after several hours by an average factor of 37 (Fig. 7.14). After Brij 30 was removed it took about 2 days before rates of 4-NP penetration decreased and this continued for several days until rates became steady and reached constant penetration rates comparable to those measured before adding Brij 30. This experiment nicely

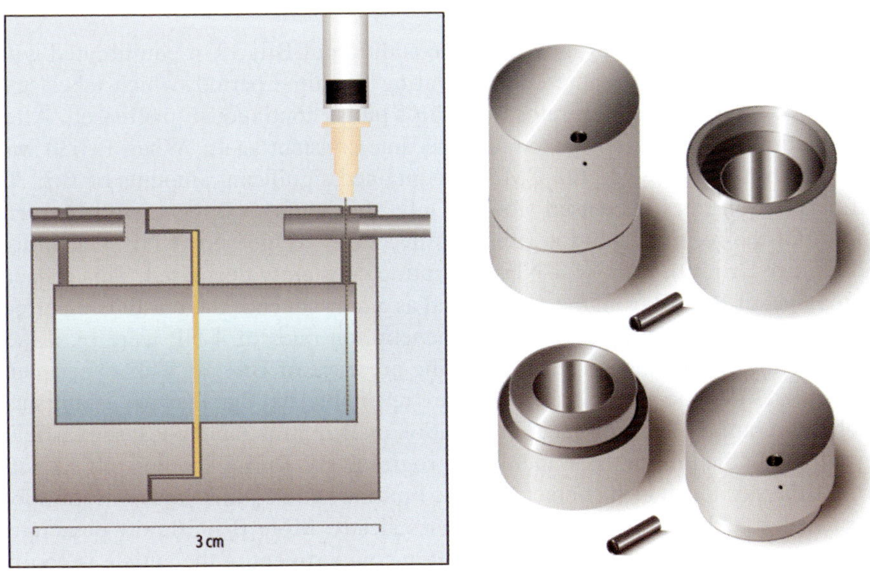

Fig. 7.13 Stainless steel transport chamber used in steady state experiments studying the effect of Brij 30 on 4-NP penetration across CM from *Prunus laurocerasus* leaves. The CM (*yellow*) separates the donor and receiver half-cells. Sampling ports permit exchange and removal of solutions. During the experiment, the sampling ports were closed by greased metal cylinders. Redrawn from Schreiber et al. (1995)

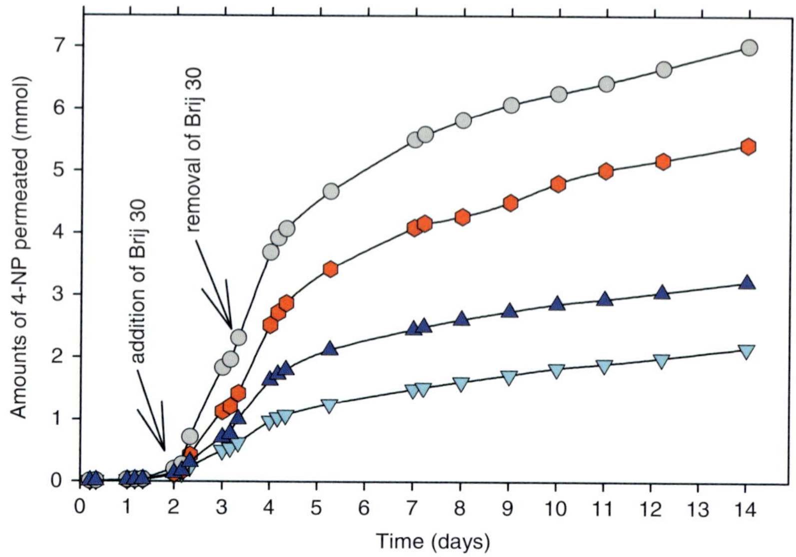

Fig. 7.14 Effects of the polydisperse alcohol ethoxylate Brij 30 on steady state diffusion of 4-nitrophenol across 4 individual *Prunus laurocerasus* CM. *Arrows* indicate the addition to and the subsequent removal of Brij 30 from the receiver solution. Data redrawn from Schreiber et al. (1995)

shows the time course of events. After the addition of Brij 30, it equilibrated with the cuticle and the wax in the cuticle. During this time period, which takes several hours, 4-NP penetration rates (slopes of plots) continuously increased. After equilibration, rates of 4-NP penetration became constant again. When Brij30 was removed from the receiver, the CM still contained significant amounts of Brij 30, and penetration rates decreased slowly. Only after about 1–2 days, with frequent renewal of the receiver solution (this was necessary to measure 4-NP penetration and remove Brij 30), rates of 4-NP penetration constantly decreased. This reflects the removal of Brij 30 from the cuticle, and as a consequence the plasticising effect decreased again. After 7 days, average penetration rates of 4-NP were by a factor of six lower than those measured in the presence of Brij 30. This experiment demonstrates that the plasticising effects are reversible, but that they are established much more rapidly than reversed. Penetration of Brij 30 is rapid because plasticising activity in the limiting skin increases progressively. During loading of CM with Brij 30, the surfactant is in contact with the limiting skin; equilibrium concentration is reached quickly, and penetration becomes steady. More Brij 30 now penetrates the plasticised limiting skin, and the sorption compartment of the cuticle is equilibrated. This has no effect on rates of 4-NP penetration. When Brij 30 is removed from the receiver, the accelerator departs from the limiting skin, *D* for 4-NP and Brij 30 decrease and it takes some time before all Brij 30 has been removed from the sorption compartment. Apart from these kinetic effects, it is clear that accelerator effects in CM are reversible, as was previously shown for reconstituted wax.

7.4.2 Effects of Plasticisers and Temperature on Solute Mobility in CM

The UDOS method is a valuable experimental tool for studying solute mobility k^* in CM (Sect. 6.3). These k^* values are directly proportional to diffusion coefficients D measured with reconstituted wax (see Sect. 6.5). In steady state experiments permeance P is obtained, which is a composite quantity (2.18) and often difficult to interpret (see Sect. 6.2). UDOS can be used to study the effect of plasticisers on mobilities of lipophilic solutes in isolated cuticles (Schönherr 1993a, b; Schönherr et al. 2001). The ^{14}C-labelled solute is sorbed in the inner sorption compartment of the CM. Subsequent desorption of the solute is carried out from the outer side of the cuticle across the transport-limiting barrier, and this arrangement assures that plasticisers and ^{14}C-labelled solutes are spatially separated (Fig. 9.5).

Using isolated *Citrus* CM, ^{14}C-labelled 2,4-D and monodisperse alcohol ethoxylates of the series C_xE_y (Schönherr 1993a, b), it was shown that the alcohol ethoxylates increased mobility in *Citrus* CM. Effects on mobilities k^* between 6 and 20 were observed (Fig. 7.15). Amounts of alcohol ethoxylates sorbed in *Citrus* MX in equilibrium with the external solutions at concentrations above the cmc can

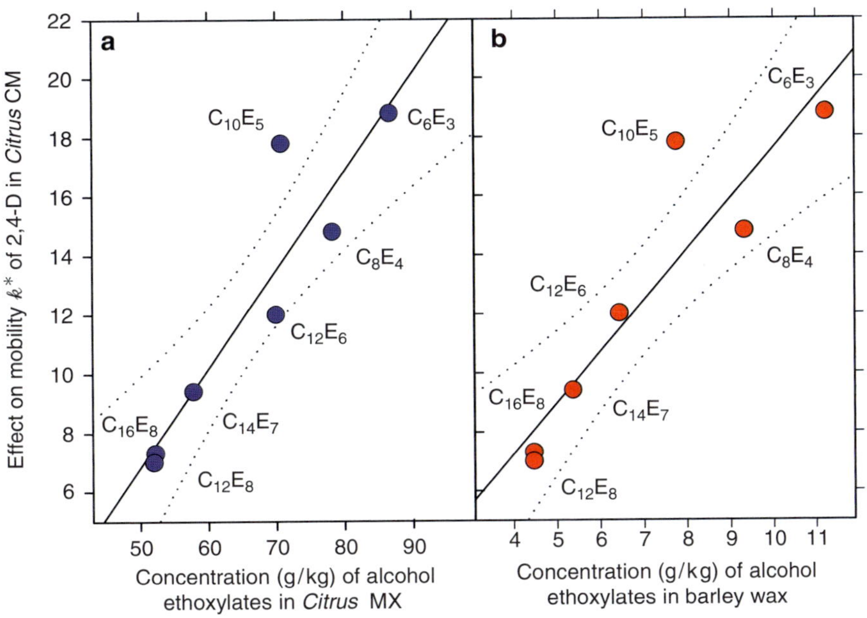

Fig. 7.15 Correlation between the effects of monodisperse alcohol ethoxylates on mobility k^* of 2,4-D in *Citrus* CM with concentrations of alcohol ethoxylates in (**a**) *Citrus* MX (*blue symbols*) and in (**b**) reconstituted barley wax (*red symbols*). Effects on mobility are taken from Schönherr (1993a); data for alcohol ethoxylate sorption in *Citrus* MX were calculated using (7.9), and for sorption in barley using (7.4)

be calculated from (7.9). A reasonable positive correlation between effects and concentrations of the respective alcohol ethoxylates sorbed in *Citrus* MX is obtained (Fig. 7.15a). Unfortunately, partition coefficients K_{ww} have not been determined for *Citrus* wax. *Citrus* wax is largely amorphous, and we expect K_{ww} to be smaller than K_{mxw}. Using (7.4), concentrations of alcohol ethoxylates sorbed in barley wax instead of *Citrus* wax can be calculated. A plot of the effects on mobilities in *Citrus* as a function of the plasticiser concentration in barley wax gives a reasonable correlation as well (Fig. 7.15b). This confirms that intrinsic effects of these plasticisers are similar, and that effects on solute penetration across cuticles are only a function of the concentration of accelerators sorbed in the transport-limiting barrier of wax.

When comparing the effects of alcohol ethoxylates with those of n-alkyl esters on diffusion in wax (Sect. 7.3.3), it was shown that n-alkyl esters were five times more efficient in plasticising waxes at the same internal concentrations in the wax (Fig. 7.10). A similar tendency can be observed with CM (Schönherr et al. 2001; Shi et al. 2005b). This is not too surprising, since above we have presented evidence showing that transport properties of cuticles and effects of plasticisers can be attributed to sorption and diffusion properties of wax.

Comparing intrinsic effects of alcohol ethoxylates with those of the n-alkyl esters tributyl phosphate (TBP) and diethyl suberate (DESU) on 2,4-DB (2,4-dichlorophenoxybutyric acid) mobility in *Stephanotis* cuticles shows that highest effects were obtained with DESU, intermediate effects with TBP, and lowest but still significant effects with alcohol ethoxylates (Table 7.4). At 15°C the n-alkyl ester DESU is about 4.4 times more effective than alcohol ethoxylates in plasticising the waxy transport barrier. This effect gradually decreases with increasing temperature from 15 to 25°C. At 30°C the difference disappears completely (Table 7.4).

Activation energies for mobility of 2,4-DB in *Stephanotis* CM are 105 kJ mol^{-1} in the absence of a plasticiser. They significantly decrease to 26 and 36 kJ mol^{-1} in the presence of DESU and of the alcohol ethoxylates respectively (Schönherr et al. 2001). TBP shows the same tendency of decreasing effects with increasing temperature (Table 7.4), and the activation energy decreases to 64 kJ mol^{-1} (Schönherr et al. 2001). Effects of plasticisers are highest at low temperatures. Increasing temperature or adding plasticisers have similar effects on fluidity of the waxy transport-limiting barrier of cuticles. ESR experiments (Sect. 7.2) support this conclusion.

Table 7.4 Effect of alcohol ethoxylates (C_xE_y), tributyl phosphate (TBP) and diethyl suberate (DESU) on 2,4-DB (2,4-dichlorophenoxybutyric acid) mobility k^* in *Stephanotis* CM at 4 different temperatures

Plasticiser	Concentration in CM (g kg^{-1})	Effect on k^* 15°C	20°C	25°C	30°C
C_xE_y	85	11	9	9	9
TBP	90	35	17	17	14
DESU	90	48	25	15	9

Data from Schönherr et al. (2001)

7.4.3 Effects of Plasticisers on the Mobility of Polar Solutes in CM

UDOS is a very useful tool for analysing the plasticising effect of chemicals on the transport-limiting barrier of cuticles, and this has been confirmed by many investigations (Schönherr 1993a, b; Schönherr and Baur 1994; Schönherr et al. 2001; Shi et al. 2005b; Buchholz 2006). Unfortunately, UDOS is limited to the investigation of lipophilic solutes, since UDOS requires that the ^{14}C-labelled solute is sufficiently soluble in the inner sorption compartment of the cuticle (Sect. 6.3). For studying cuticular penetration of polar solutes and for testing the effect of plasticisers on rates of cuticular penetration, SOFP is the method of choice (Sect. 5.2 and Fig. 9.5).

SOFP was used to investigate the plasticising effect of n-alkyl esters and alcohol ethoxylates on cuticular penetration of the highly water soluble sugar methyl glucose, in comparison to the lipophilic molecules metribuzin and iprovalicarb, which have log K_{ow} values of 1.6 and 3.18 respectively (Shi et al 2005a). Methyl glucose has a log K_{ow} of -3.0, indicating a much higher solubility in water than in the cuticle. Droplets of $5\,\mu l$ of the ^{14}C-labelled solutes dissolved in water were applied to the physiological outer side of *Stephanotis* CM and were allowed to dry. Labelled solutes were desorbed from the physiological inner side of the CM using the plasticisers DESU, TBP and alcohol ethoxylates of the $C_{12}E_y$ series with y increasing from 2 to 8 (Shi et al. 2005a).

All three types of plasticisers significantly increased rate constants k of cuticular penetration of the two lipophilic solutes metribuzin and iprovalicarb. The logarithms of the effects (k measured in the presence of the plasticizer/$k_{control}$ measured without a plasticizer) linearly increased within increasing concentration of the plasticizer in the wax, as shown for iprovalicarb and various ethoxylated alcohols (Fig. 7.16). Thus, using SOFP as a method for measuring the effect of plasticisers on rates of cuticular penetration confirms the results obtained with reconstituted waxes (Sect. 7.3) and with UDOS experiments (Sect. 7.4.2). At identical internal plasticizer concentrations in the wax, intrinsic effects of DESU were again four- to five-fold higher than for the alcohol ethoxylates, whereas the effect of TBP was intermediate (Shi et al. 2005a).

Surprisingly different results were obtained with methyl glucose. Neither DESU nor TBP had an effect on k of methyl glucose (Shi et al. 2005a). With the alcohol ethoxylates of the series $C_{12}E_n$, exactly the opposite sequence as with iprovalicarb was observed. (Fig. 7.16). The largest and most polar alcohol ethoxylate $C_{12}E_8$ had the largest effect on k of methyl glucose rates of penetration, and effects decreased with decreasing degree of ethoxylation (Fig. 7.16), that is with increasing concentration in wax. There was no statistically significant effect with $C_{12}E_4$ and $C_{12}E_2$.

We have shown in previous chapters that the plant cuticle is a heterogeneous membrane containing aqueous pores (Fig. 4.11),which are a prerequisite for penetration of ionic compounds (Chap. 5). In parallel there is a lipophilic waxy path in cuticles (Fig. 4.13), open to lipophilic solutes which are soluble in amorphous waxes (Chap. 6). All efficient plasticisers (DESU, TBP and the alcohol ethoxylates tested) are lipophilic, and they increase fluidity of the waxy pathway. They dissolve and

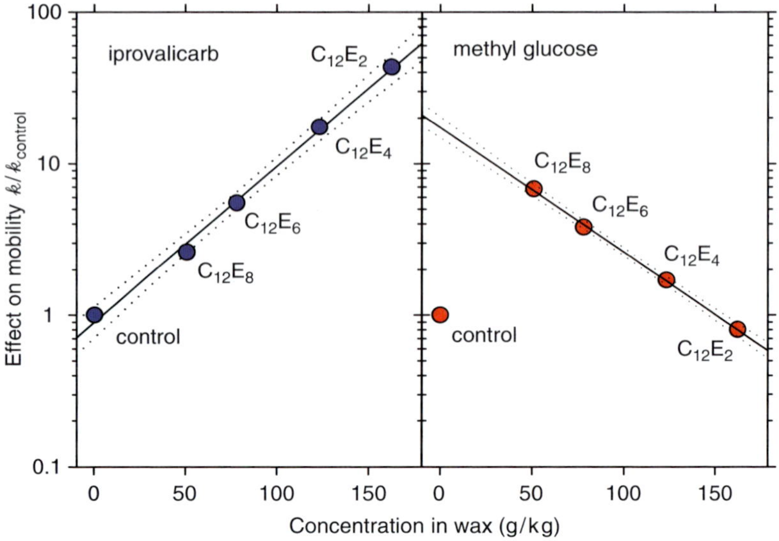

Fig. 7.16 Logarithms of the effects ($k/k_{control}$) of monodisperse alcohol ethoxylates on rate constants of penetration k across *Stephanotis* CM of iprovalicarb (*blue symbols*) and methyl glucose (*red symbols*) as a function of the concentration of the alcohol ethoxylates in *Stephanotis* wax. Dotted lines represent 95% prediction intervals for the regression lines. Data from Shi et al. (2005a)

diffuse in the amorphous wax phase, where diffusion of lipophilic solutes such as iprovalicarb takes place. They act as plasticisers, and accelerate their own diffusion and the diffusion of other solutes which can access the waxy pathway.

The fact that DESU and TBP, which are very lipophilic compounds lacking surface activity, do not enhance penetration of methyl glucose indicates that these plasticisers are spatially separated from methyl glucose during penetration of the cuticle. The plasticising effect of DESU and TBP on wax cannot affect diffusion of methyl glucose in aqueous pores. K_{ww} values measured with *Stephanotis* wax (Table 7.1) are 692 and 2,626 for $C_{12}E_8$ and $C_{12}6_6$ respectively. Hence, these alcohol ethoxylates are primarily sorbed in wax and cutin and only traces will be in the aqueous pores. If methyl glucose does not dissolve in wax and cutin, $C_{12}E_8$ and $C_{12}6_6$ should not enhance rates of penetration of methyl glucose in CM. Yet they do, and $C_{12}E_8$ is more effective than $C_{12}6_6$ even though its concentration in wax is lower. An effect on aqueous pores can be ruled out, because the overwhelming majority of these surfactant molecules are in lipid phases and not in water. How can this be explained?

Alcohol ethoxylates are surfactants, with a lipophilic and a polar domain in the same molecule, and this distinguishes them from DESU and TBP. With increasing degree of ethoxylation their polarity increases, and statistically significant accelerating effects on methyl glucose penetration have been observed only with the more polar $C_{12}E_6$ and $C_{12}E_8$ representatives. This indicates that $C_{12}E_6$ and $C_{12}E_8$ must at least partially diffuse in the same phase as methyl glucose, such that diffusion of methyl glucose can be increased by $C_{12}E_6$ and $C_{12}E_8$ to some extent.

In aqueous solutions, alcohol ethoxylates form spherical micelles as soon as the cmc is exceeded. Depending on surfactant properties these micelles may be composed of 50–200 monomers, with their hydrophobic tails forming a lipophilic phase in the centre of the micelles and the ethoxy groups in contact with water. Lipophilic solutes can dissolve in the centre of the micelle, and this is called solubilisation. If surfactants are dissolved in lipophilic solvents such as hexane, inverse micelles are formed in which the polar head groups accumulate in the centre of the micelles and the lipophilic tails are in contact with the lipophilic solvent. The centre of inverse micelles is hydrated and can solubilise polar solutes which are insoluble in the non-polar solvent. With increasing surfactant concentration, hexagonal liquid crystalline and finally liquid crystals can be formed.

There is no plausible reason why inverse micelles or liquid crystals should not form in amorphous cuticular waxes if the hydrophilic ethoxy groups impart sufficient polarity to the surfactant molecule. The surfactant concentration in total wax is fairly high, as can be calculated from the cmc and the K_{ww} (Table 7.1). For instance, C_{wax} is 0.0942 mol kg^{-1} when *Stephanotis* CM are in equilibrium with micellar solutions of $C_{12}E_8$. With a molecular weight of 539 g mol^{-1}, C_{ww} is 50.8 g kg- or 5.1% per weight. A similar calculation for $C_{12}E_6$ results in C_w of 7.8%. If the wax is only 50% crystalline, the concentrations in amorphous waxes double. These concentrations in wax are high compared to the cmc of these surfactants in water, which is 0.003 and 0.0053% by weight for $C_{12}E_6$ and $C_{12}6_8$ respectively. It appears likely that these relatively polar surfactants are not homogeneously dispersed but form separate phases in amorphous waxes.

Highly ordered surfactant phases formed in amorphous waxes could act as an additional transport pathway for water, polar and ionic solutes, and could have great importance for optimising foliar penetration of foliar nutrients and plant protection agents. This aspect has been neglected in the past, and should be investigated.

7.5 Effects of Plasticisers on Water and Ion Transport

Water is a small polar uncharged molecule. It can penetrate cuticles along both the lipophilic path and the aqueous path (Chap. 4). Compounds plasticising the lipophilic pathway should also affect cuticular transpiration. This has in fact been shown with cuticles isolated from *Citrus aurantium* and *Pyrus communis* leaves and *Capsicum annuum* and *Solanum melongena* fruits. Cuticles were mounted in transpiration chambers, and after measuring rates of cuticular transpiration, aqueous solutions of polydisperse surfactants (Triton, Brij, Tween, Renex, Ethomeen and SDS) were applied to the outer surface of the cuticles. After evaporation of the water weighing was resumed, and cuticular transpiration was measured again (Riederer and Schönherr 1990; Schönherr and Bauer 1992). All surfactants tested increased cuticular transpiration. Only sodium dodecyl sulphate (SDS) was ineffective. SDS carries a negative charge and is insoluble in the cuticle. This result obtained with SDS indicates that compounds must be soluble in waxes to develop a plasticising activity. Effects of plasticisers without surface activity (e.g., DESU or TBP) on cuticular transpiration have not been published. Rates of penetration of salts which exclusively diffuse in aqueous pores (Chap. 5) were not affected by plasticisers (Sect. 5.3.1).

Problems

1. What is the cmc and how does it affect the sorption of surfactants and lipophilic solutes to wax?
2. Alcohol ethoxylates have the general formula C_xE_y. C_x indicates the number of C atoms or lipophilic CH_2 groups of the alcohol. E_y refers to the number of the polar ethylene oxide units. How is the wax/water partition coefficient K_{ww} affected by C_x and E_y?
3. What is the K_{ww} of $C_{12}E_8$ in barley wax?
4. How much $C_{12}E_8$ is sorbed in barley wax when equilibrated with an aqueous solution above the cmc?
5. What is the K_{ww} for a solute in barley wax when the K_{mxw} of this substance is 250 in *Citrus* MX?
6. Which property of waxes is studied by ESR and NMR? How is wax structure affected by alcohol ethoxylates?
7. Mobility of lipophilic solutes in wax and cuticles is increased by the addition of alcohol ethoxylates. The effect can be reversed when alcohol ethoxylates are desorbed from wax. Is this observation consistent with solubilisation of waxes?
8. Diffusion coefficients in reconstituted waxes can be estimated from molar volumes of solutes using (6.23). For $C_{12}E_8$, a D of $5.74 \times 10^{-20}\,m^2\,s^{-1}$ is calculated. This value is 150 times higher than D measured by desorption which is $8.5 \times 10^{-18}\,m^2\,s^{-1}$. Can you explain this difference?
9. The n-alkyl esters DESU and TBP have pronounced accelerating effects on cuticular penetration of the lipophilic solute iprovalicarb, but they were ineffective with the polar compound methyl glucose. Can you explain this?

Solutions

1. The cmc is the maximum aqueous concentration of a surfactant. If surfactant concentration is increased beyond the cmc, micelles are formed and the concentration of free surfactant molecules does not increase. The maximum amount of surfactant molecules sorbed in wax is reached at the cmc. Lipophilic solutes which are sorbed in wax are also sorbed in micelles.Hence, apparent partition coefficients (K_{ww}) decrease when the surfactant concentration is increased above the cmc, and micelles are formed which compete as sorbers with the wax.

2. According to (7.1), (7.2) and (7.3), log K_{ww} of an alcohol ethoxylate increases by an increment of 0.53–0.54 with each additional C atom, and it decreases by a decrement of 0.23–0.24 with each additional ethylene oxide group.
3. Using (7.1), a K_{ww} of 81 is obtained for barley wax.
4. According to (7.4), the maximum amount of $C_{12}E_8$ sorbed in barley wax is 4.5 g surfactant per kg wax. This corresponds to 0.45 weight percent.
5. According to (7.7), log K_{ww} for barley wax is 1.34; thus, K_{ww} is 21.8.
6. ESR and NMR experiments provide information regarding the molecular architecture of wax. ESR and NMR probes directly map their environment in the wax. The main finding from ESR and NMR experiments was the observation that alcohol ethoxylates plasticise the amorphous wax phase. In the presence of sorbed alcohol ethoxylates ESR and NMR, probes in the amorphous wax phase are in a liquid environment at room temperature. In the absence of the plasticisers, the same fluidity of wax is obtained only at higher temperatures. Alcohol ethoxylates sorbed in wax mimic the effect of higher temperatures.
7. A rapid increase in solute mobility after addition of alcohol ethoxylates, and reversion of the effect after removing the plasticiser, indicate reversible changes in fluidity of the waxy pathway. Waxes are not solubilised and lost.
8. Equation (6.23) for estimating D is only valid for solutes lacking plasticising activity. $C_{12}E_8$ is a plasticiser, and increases solute mobility in wax. $C_{12}E_8$ also increases its own mobility in wax, and thus the measured D is much higher than the D value estimated for a passive solute.
9. In cuticles, an aqueous pathway and a lipophilic waxy pathway are arranged in parallel. The lipophilic iprovalicarb preferentially diffuses in the lipophilic pathway which is plasticised by DESU and TBP. Methyl glucose is not soluble in amorphous waxes, and its penetration is limited to aqueous pores. Diffusion in aqueous pores is not affected by lipophilic plasticisers.

Chapter 8
Effects of Temperature on Sorption and Diffusion of Solutes and Penetration of Water

Several processes are involved in penetration of a polymer membrane separating two aqueous solutions. Water and dissolved solutes leave the aqueous phase (donor) and enter the membrane phase in which they are sorbed. These sorbed molecules then diffuse in the membrane via a random hopping mechanism. Eventually they arrive at the opposite membrane interface facing the receiver. In cuticles, penetration is complicated, because they consist of more than one phase (cutin, waxes, polysaccharides, polypeptides and phenolic compounds) and they contain up to 8% water, part of which is clustered in aqueous pores. The polymer matrix is an ion exchanger, and concentration of counter ions depends on *p*H. Sorption, diffusion, swelling and ion exchange vary with temperature. In addition, swelling and phase transitions of membrane components can affect membrane structure and function.

8.1 Sorption from Aqueous Solutions

Riederer and Schönherr (1986a) studied sorption of 4-nitrophenol at various concentrations in isolated cuticles (CM) and MX membranes at temperatures ranging from 5 to 50 °C. The choice of the sorbate and types of cuticles was deliberate.

4-nitrophenol (4-NP) is a weak acid having a molecular weight of 139 g mol^{-1} and a pK_a of 7.25. It is a planar molecule with two permanent dipoles. The dipole moment of the $C\text{-}^{+}OH$ group is 1.65 Debye units, while the $C^{+}\text{-}NO_2$ function has 3.1–3.8 Debye units. These polar groups can form hydrogen bonds with water and with dipoles of the membranes. Van der Waals forces related to the aromatic ring affect interactions between 4-NP, water and polymer chains. The permanent dipoles are attached to the phenol ring in para position, and since sorption was studied at *p*H 3, ionisation of the acidic hydroxyl group of 4-NP was suppressed. Thus, 4-NP was neutral, and interactions with positive fixed charges of the MX did not occur (Sects. 4.2 and 4.3). The non-ionised 4-NP is sufficiently water soluble to permit sorption to be studied at concentrations ranging from 0.1 to 10^{-6} mol l^{-1}.

Tomato fruit cuticles (*Lycopersicon esculentum*) contain 7% waxes and 62% cutin. Polar polymers amount to 24% (Table 1.1). The ester cutin is of the C16 type, and dihydroxyfatty acids are the major constituents. Adaxial rubber leaf cuticles (*Ficus decora*) are composed of 56% cutin, 19% polar polymers and 25% waxes (Table 1.1). The cutin contains C16 and C18 hydroxyfatty acids, and 18-hydroxy-9,10-epoxystearic acid is a major constituent, which results in the formation of non-ester cutin (cutan) similarly to what happens in *Clivia* cutin (Fig. 1.2).

8.1.1 Sorption Isotherms and Partition Coefficients

Small pieces of cuticular material were equilibrated at constant temperature as described in Sect. 6.1. At equilibrium, the molal 4-NP concentrations (mol kg^{-1}) in the aqueous phase (C_{aqueous}) and in the CM or MX (internal phase C_{internal}) were determined. The partition coefficients K were calculated as the ratio of the molal concentrations $C_{\mathrm{internal}}/C_{\mathrm{aqueous}}$. When 4-nitrophenol concentration in the cuticle (C_{internal}) was plotted against concentration in water (C_{aqueous}), linear sorption isotherms were obtained at all temperatures and concentrations of 10^{-3}–10^{-6} mol kg^{-1}. Alternatively, the natural logarithms of these concentrations can be plotted. The slopes of these plots (k) are related to the intensity of sorption and the number of sorption sites. The parameter n characterises the slope of the isotherm. The linear portions of the plots fit the Freundlich isotherm

$$C_{\mathrm{internal}} = kC_{\mathrm{aqueous}}^{1/n} \tag{8.1}$$

or in logarithmic form

$$\log C_{\mathrm{internal}} = \log k + \frac{1}{n}\log C_{\mathrm{aqueous}} \tag{8.1a}$$

Freundlich isotherms are frequently encountered when solutes interact with heterogeneous substrates. At low concentrations the parameter n is not far from 1.0, but at higher sorbate concentrations the plots become increasingly convex to the C_{aqueous} axis. At low sorbate concentrations this type of isotherm can be classified as constant partitioning, while at higher concentrations ($>10^{-3}$ mol kg^{-1}) it resembles a Langmuir type isotherm. Linear isotherms are obtained under the following conditions: (1) the substrate consists of flexible molecules and has regions of differing accessibility to the sorbate, (2) the solute has a higher affinity to the substrate than to the solvent water, and (3) the solute is able to penetrate into initially less accessible (highly ordered) regions of the solid.

At all concentrations and temperatures, partition coefficients (K) and the Freundlich parameter k were >1. K ranged from about 15 to 250 (Fig. 8.1), and k varied between 14 and 190 (Riederer and Schönherr 1986a); that is, 4-NP is better soluble in CM and MX than in water. At low internal concentration, K is greater with the MX than with the CM. The only lipid compartment of the MX is cutin,

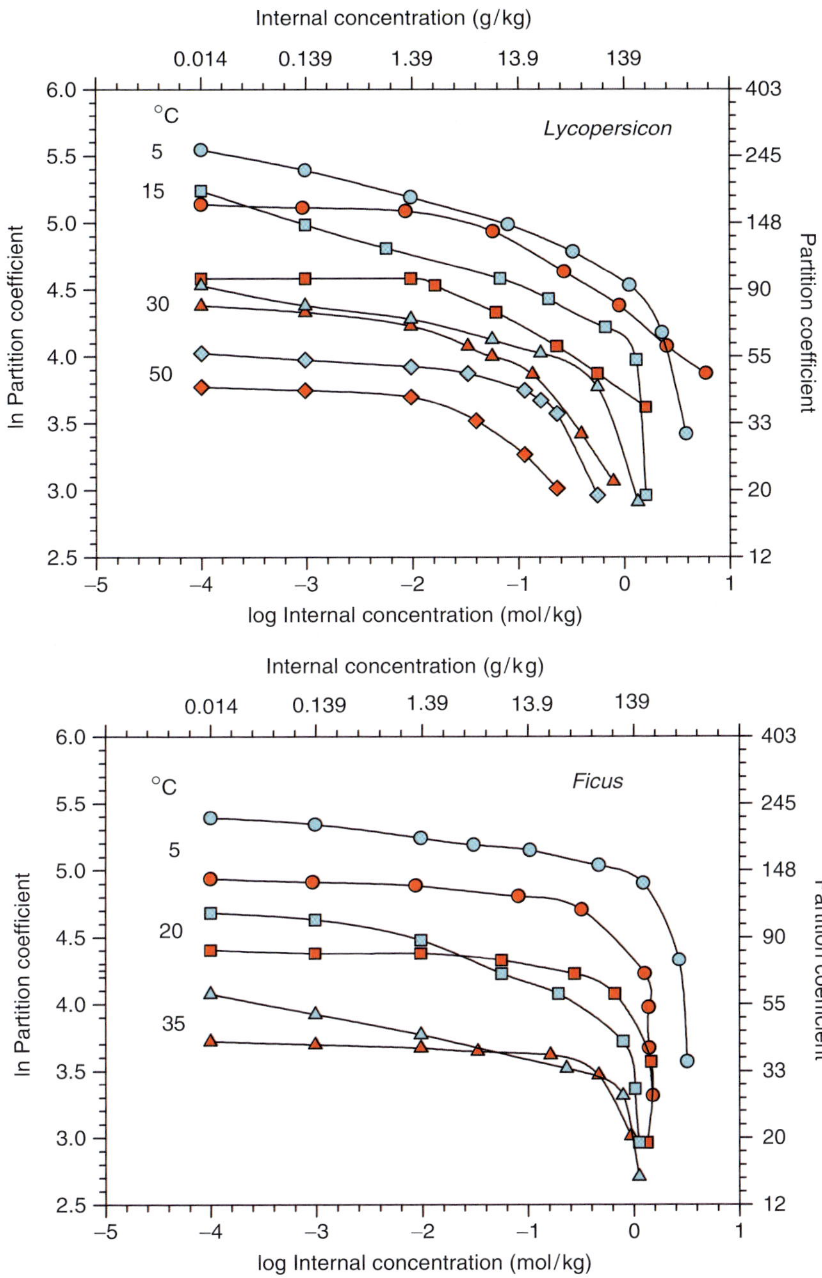

Fig. 8.1 Partition coefficients of 4-nitrophenol for CM (*red*) and MX (*blue*) from *Lycopersicon esculentum* fruits and *Ficus decora* leaves as functions of temperature and concentration of sorbed solute. (Redrawn from data of Riederer and Schönherr 1986a)

while the CM also contains waxes as additional lipid component. As shown in Sect. 6.1, wax/water partition coefficients are much smaller than cutin/water partition coefficients. Thus, cutin is the sole sorbent in the MX and the major sorbent in CM.

Partition coefficients decrease with increasing temperature and increasing internal solute concentration. Between 10^{-4} and 10^{-2} mol kg^{-1} or 1.39 and 0.014 g kg^{-1}, internal concentrations of partition coefficient were nearly constant, but at higher internal concentrations *K* decreased. This decrease is more pronounced in rubber than in tomato CM and MX. With rubber CM and MX, all sorption plots converge at an internal concentration of about 1 mol kg^{-1}. This indicates that all sorption sites are occupied and additional sorption is no longer possible. With tomato fruit cuticles, partition coefficients also decrease at higher 4-NP concentrations but plots do not converge. The difference in internal concentrations between CM and MX is highest at low sorbate concentrations, but it decreases as concentration increases. If waxes sorb little or no 4-NP, partition coefficients for MX would be expected to be higher by 8% and 25% with tomato and rubber cuticles respectively. However, differences are larger, indicating that in MX some sorption sites are occupied by waxes.

At low external concentration of 4-NP, partition coefficients were nearly constant, even though internal 4-NP concentration increased 100-fold. This indicates that additional new sorption sites become available when others are being occupied by 4-NP molecules. The flexible polymer chains open up, and sorbate molecules squeeze in between. There is a limit to this, however, and when internal concentration exceeds 10^{-2} mol kg^{-1} additional sorption sites are no longer formed and partition coefficients decrease. Partition coefficients decrease much more with rubber leaf cuticles, indicating that flexibility of polymer chains is smaller compared to tomato fruit cuticles. It appears that cutan in rubber leaf cuticle is more rigid and provides fewer sorption sites.

The temperature effect on partition coefficients depends on the temperature effect on activities of the two phases. With substituted phenols, partition coefficients between inert organic phases which are incapable of hydrogen bonding increase with temperature, while octanol/water partition coefficients decrease with increasing temperature (Korenman et al. 1977). Cuticle/water partition coefficients also decreased with increasing temperature, and this is evidence that hydrogen bonding between cutin and 4-NP was involved in sorption.

The temperature effect on sorption can be further analysed using a modified Freundlich isotherm. This makes it possible to distinguish between temperature effects on the number of sorption sites (solubility in cuticles) and on water solubility.

$$C_{\text{internal}} = k' \cdot (C_{\text{aqueous}} / C_{\text{aqueous}}^{\text{saturated}})^{1/n} \tag{8.2}$$

The term in parenthesis is referred to as the reduced concentration. Plotting log C_{internal} vs log of the reduced concentration resulted in straight lines with slopes k'. With *Ficus* CM and MX, *k* decreased with increasing temperature (8.1), while k' was independent of temperature (8.2). Hence, the temperature effect on *Ficus* partition coefficients was exclusively due to a temperature effect on solute activity in the aqueous phase, that is, on water solubility of 4-NP. With increasing temperature water solubility increased, and as a consequence partition coefficients decreased. Sorption in tomato fruit MX differed. Here, k' increased with temperature from about 5 to 14, showing that the number of available sorption sites increased with temperature. The plot k' vs T has plateaus at 7–22 °C and 22–32 °C. A gradual increase occurred between 22 and 47 °C (Riederer and Schönherr 1986a). This indicates structural changes in the cutin matrix, and the step character can be interpreted as phase transitions which lead to a loosening of the polymer chains, with a concomitant increase in number of sorption sites.

8.1.2 Thermodynamics of Sorption

Sorption in cuticles can be further characterised by Gibbs free energy (ΔG) involved in transfer of 1 mol of 4-NP from the aqueous phase to the cuticle phase.

$$\Delta G_{\text{sorption}} = -RT \ln K \tag{8.3}$$

$\Delta G_{\text{sorption}}$ is proportional to ln K, and since ln K is always >1, Gibbs free energy is always negative. This is expected for a spontaneous (exergonic) process. $\Delta G_{\text{sorption}}$ was largest at low internal concentrations (about −12 to −10 kJ mol^{-1}), and decreased with increasing internal 4-NP concentration to about −9 to −8 kJ mol^{-1}. At low internal concentration, $\Delta G_{\text{sorption}}$ was larger with MX than with CM (Riederer and Schönherr 1986a). $\Delta G_{\text{sorption}}$ depends on heat of sorption (H_{s}) and entropy (S_{s}):

$$\Delta G_{\text{sorption}} = \Delta H_{\text{s}} - T\Delta S_{\text{s}} \tag{8.4}$$

Substituting (8.3) into (8.4) leads to

$$\ln K = \frac{\Delta H_{\text{s}}}{RT} + \frac{\Delta S_{\text{s}}}{R} \tag{8.5}$$

Plotting ln K vs T^{-1} (in Kelvin) results in a straight line having the slope $\Delta H_{\text{s}}/R$, and the y-intercept is $\Delta S_{\text{s}}/R$. A large entropy is characteristic for sorption on solid substrates, which leads to a higher degree of order. For $\Delta G_{\text{sorption}}$ to be negative, the large $T\Delta S_{\text{s}}$ term must be overcompensated by the heat of sorption (ΔH_{s}). This was the case at all temperatures and concentrations (Riederer and Schönherr 1986a).

With both species, a striking difference between CM and MX exists for the effect of sorbate concentration on ΔH_{s}. At low concentration the heat of sorption of CM was constant, and decreased only if the internal concentration exceeded 10^{-2} mol kg^{-1} (Fig. 8.2a). Sorption sites in the CM seemed to be energetically homogeneous as long as the internal concentration was <1.39 g kg^{-1}. With MX membranes the enthalpy varied from the beginning, indicating that sorption sites available in the MX are more heterogeneous. With *Ficus* MX, ΔH_{s} increased from the beginning, because sorption sites evolving higher amounts of heat became

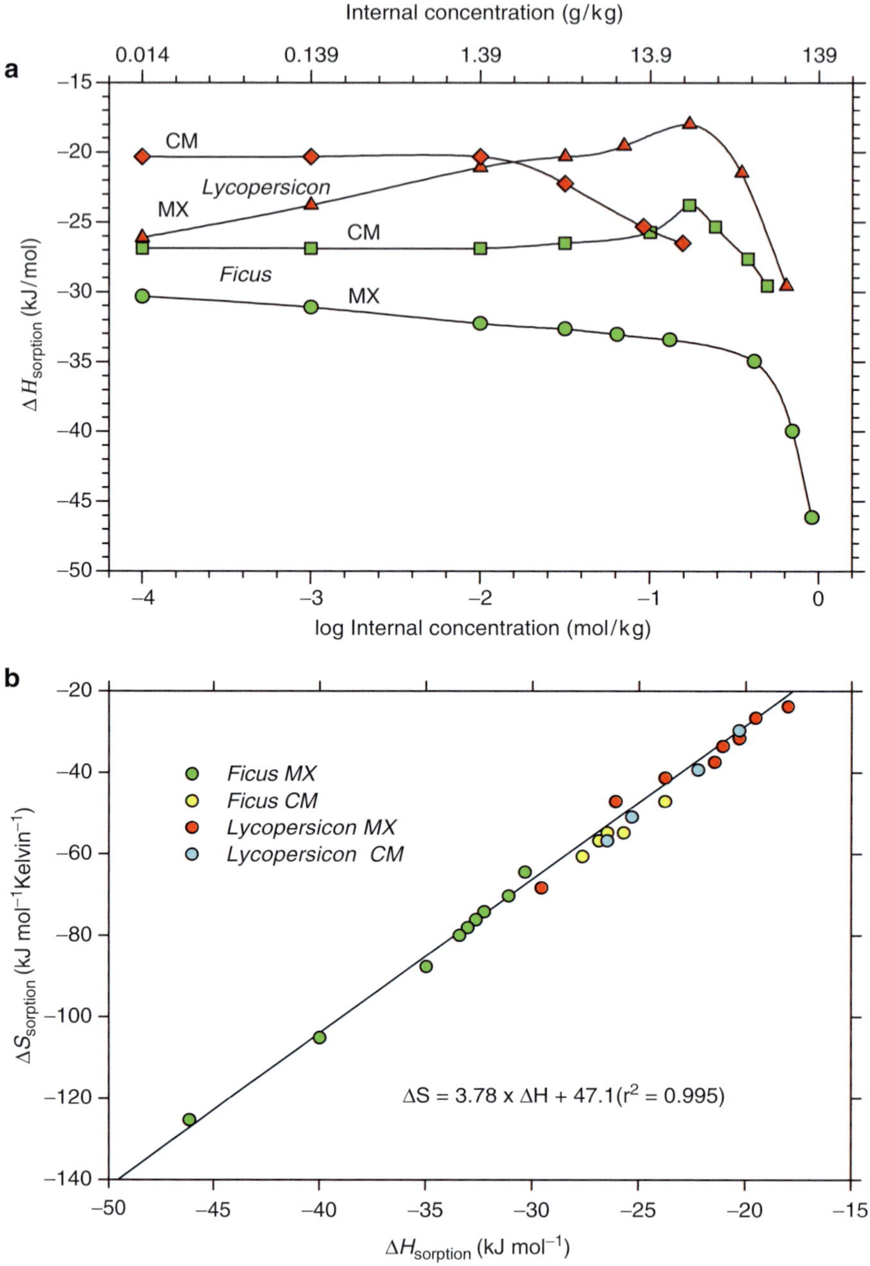

Fig. 8.2 **(a)** Change in molar enthalpy due to transfer of 4-NP from the aqueous to the cuticle phase. ΔH_S for various concentrations was calculated at 25 °C. **(b)** Entropy–enthalpy relationship at 25 °C and internal concentrations shown in **(a)**

available by an opening up or swelling of the polymer cutin once the low energy sorption sites of about 30 kJ mol^{-1} had been saturated. With tomato fruit MX, the opposite effect was observed. Heats of sorption decreased initially until the internal concentration reached about −18 kJ kg^{-1}. All sorption sites seemed to be equally accessible, but those providing higher heats of sorption were being saturated first. At higher internal 4-NP concentrations heats of sorption increased strongly, and this phenomenon can be explained by a cooperative effect between sorbate molecules inside the cuticle. After all sorption sites in cutin have been saturated, additional sorption occurs on the surface of 4-NP molecules already sorbed on the polymer. In other words, initially solute–matrix interactions dominate, while at high internal concentration sorbate–sorbate interactions become important and solid 4-NP precipitates between the polymer chains, which are forced apart. The polymer swells.

The partial molar entropy measured for transfer from aqueous solution to the cuticles has a negative sign (Fig. 8.2b). This loss of entropy can be attributed to reduced mobility of 4-NP molecules and increased degree or order on sorption on a solid substrate. The good correlation between enthalpy and entropy at all concentrations, with both species and for CM and MX, indicates that the same entropy–enthalpy relationship prevails for both species and for MX and CM. Sorption sites having a low enthalpy result in lower order of sorbed molecules and vice versa. Larger amounts of heat are evolved when solutes are sorbed at higher order and reduced mobility. Whenever solute–solute interactions dominate, multilayers of sorbate molecules may arise between the polymer chains of cutin.

It is tempting to extrapolate these data to the formation of embedded wax, that is, to the generation of wax plates in cutin. Enthalpies of transfer to cutin are very high when compared to the octanol water system, for which ΔH_S between −5 and −8 kJ mol^{-1} were measured for substituted resorcinol (1,3-benzenediol) monoethers (Sangster 1997). Sorption of 4-NP is driven by very high enthalpies, and dipole–dipole interactions are important. It is unfortunate that we have no comparable data for fatty acids, alcohols esters or alkanes. With these lipophilic solutes, van der Waals forces are probably more important than dipole-dipole interactions. For these reasons, extrapolation of the above data to formation of intracuticular waxes should be looked at with extreme caution. These highly lipophilic compounds were not included in this study because of their extremely low water solubility. Mixtures of ethanol and water would have to be used to overcome this problem.

8.2 Solute Mobility in Cuticles

We have seen that partition coefficients generally decrease with increasing temperature (Fig. 8.1). Permeance is the product of partition and diffusion coefficients (2.18), and the question concerning the effect of temperature on diffusion coefficients (D) arises. Diffusion is the consequence of Brownian motion of molecules, and rates of diffusion are proportional to the frequency of jumps and the distance travelled by the molecule. Diffusion in liquids depends on viscosity of the medium and on solute size (4.16). With increasing temperature, viscosity decreases and D increases. Diffusion in solids, for instance polymer membranes, is more complicated. Temperature dependence of D in membranes is analysed as an activated process using the Arrhenius equation:

$$D = D_o \exp^{-\frac{E_D}{RT}} \tag{8.6}$$

where E_D is the activation energy of diffusion, R the gas constant, T the Kelvin temperature and D_o is the pre-exponential factor. In diffusion across polymer membranes, the activation energy is a measure of the energy expanded against the cohesive forces of the polymer in forming gaps through which solutes can diffuse. This concept assumes that vacancies, holes or void volumes must be formed between polymer chains by thermal motion, which can accommodate solutes. These voids or holes must be large enough to accept the penetrant, and they must form very close to the original position of the solute, such that they can be reached in a single jump. Frequency of hole formation and the size of these holes determine rates of diffusion, not the jumping frequency of the molecules themselves, which is much larger. Both increase with temperature, and this is the reason why D increases with temperature and why the increase is much larger in membranes than in solutions. Based on the argument that the activation barrier in membrane diffusion is the energy necessary to form a hole of proper dimensions, Glasstone et al. (1941) proposed the transition state theory, and derived an expression for diffusion:

$$D = \lambda^2 \frac{\kappa T}{h} \cdot \exp^{-\frac{\Delta G}{RT}} \tag{8.7}$$

where λ is the mean free path of the solute in the solid, κ is Boltzmann's constant, h is the Blanck constant and ΔG is the Gibbs free energy of activation, which is

$$\Delta G = \Delta H - T\Delta S \tag{8.8}$$

where ΔH and ΔS are the enthalpy and entropy change per mole during formation of the transition state. The enthalpy of activation ΔH is related to the Arrhenius activation energy (E_D)

$$\Delta H = E_D - RT \tag{8.9}$$

and substituting (8.9) and (8.8) into (8.7) we obtain

$$D = \lambda^2 \frac{\kappa T}{h} \times \exp^{-\frac{\Delta S}{R}} \times \exp^{-\frac{E_D}{RT}} \tag{8.10}$$

which is similar to (8.6), and the pre-exponential factor D_o is

$$D_o = \lambda^2 \frac{\kappa T}{h} \times \exp^{-\frac{\Delta S}{R}}. \tag{8.11}$$

This shows that D_o is proportional to the square of the jump distances, temperature and of some physical constants. The entropy enters as an exponent.

Temperature dependence of diffusion is usually described using (8.6). D is measured at various temperatures, and ln D is plotted versus T^{-1}. If the slope is a straight line E_D can be calculated from the slope, and the y-intercept is equivalent to ln D_o. With D_o and activation energy known, D can be calculated for any other temperature (8.6).

These equations can be used to investigate dependence on temperature of solute mobility in CM and MX. Using the UDOS method (Sect. 6.3), rate constants k^* are obtained which are proportional to diffusion coefficients. In CM, their magnitude depends on properties of the waxy barrier. Temperatures studied, ranged from 15 to 70°C, and internal solute concentrations were in the order of mg kg^{-1}. At these concentrations, solute–membrane interactions predominate and solute–solute interactions in cuticles are improbable.

8.2.1 Effect of Temperature on Rate Constants k^*

Rate constants were measured at various temperatures, and k^* was calculated using (6.3). Temperature dependence was analysed based on the Arrhenius equation:

$$k^* = k^*_{\text{infinite}} \exp(-E_D/RT) \tag{8.12}$$

where k^*_{infinite} is the pre-exponential factor and E_D is the activation energy of diffusion. This equation is similar to (8.6), except that D was replaced by the UDOS rate constant k^*. In the original publications by Baur et al. (1997a), Buchholz and Schönherr (2000) and Buchholz 2006, the symbol for pre-exponential factor was k^*_o similarly as for D_o in (8.6). However, k^*_o was previously used for rate constants for solutes having zero molar volume (Buchholz et al. 1998, Sects. 6.3.2.2 and 6.3.2.3); we now define the symbol k^*_{infinite} for the y-intercept of Arrhenius plots. This y-intercept k^*_{infinite} is the solute mobility at infinite temperature. By plotting log k^* vs the reciprocal of the Kelvin temperature, a straight line is obtained and E_D can be calculated from the slope:

$$E_D = -2.3 \times \text{slope} \times R \tag{8.13}$$

The factor 2.3 is included in the above equation because published Arrhenius plots are on log scale, rather than ln scale as in (8.6) and (8.12).

Figure 8.3a shows Arrhenius plots for seven individual *Citrus* CM/2,4-D combinations. Slopes are linear, and log k^* increases with decreasing values of T^{-1}. With increasing temperature, solute mobility k^* increases by almost two orders of magnitude. Regression equations for the CM having highest and lowest rate constants are shown in the graph, and it can be seen that plots tend to converge at higher temperatures. CM having high k^* values have lower y-intercepts and smaller slopes than CM having low k^* values. Differences in k^* among membranes decrease as temperature

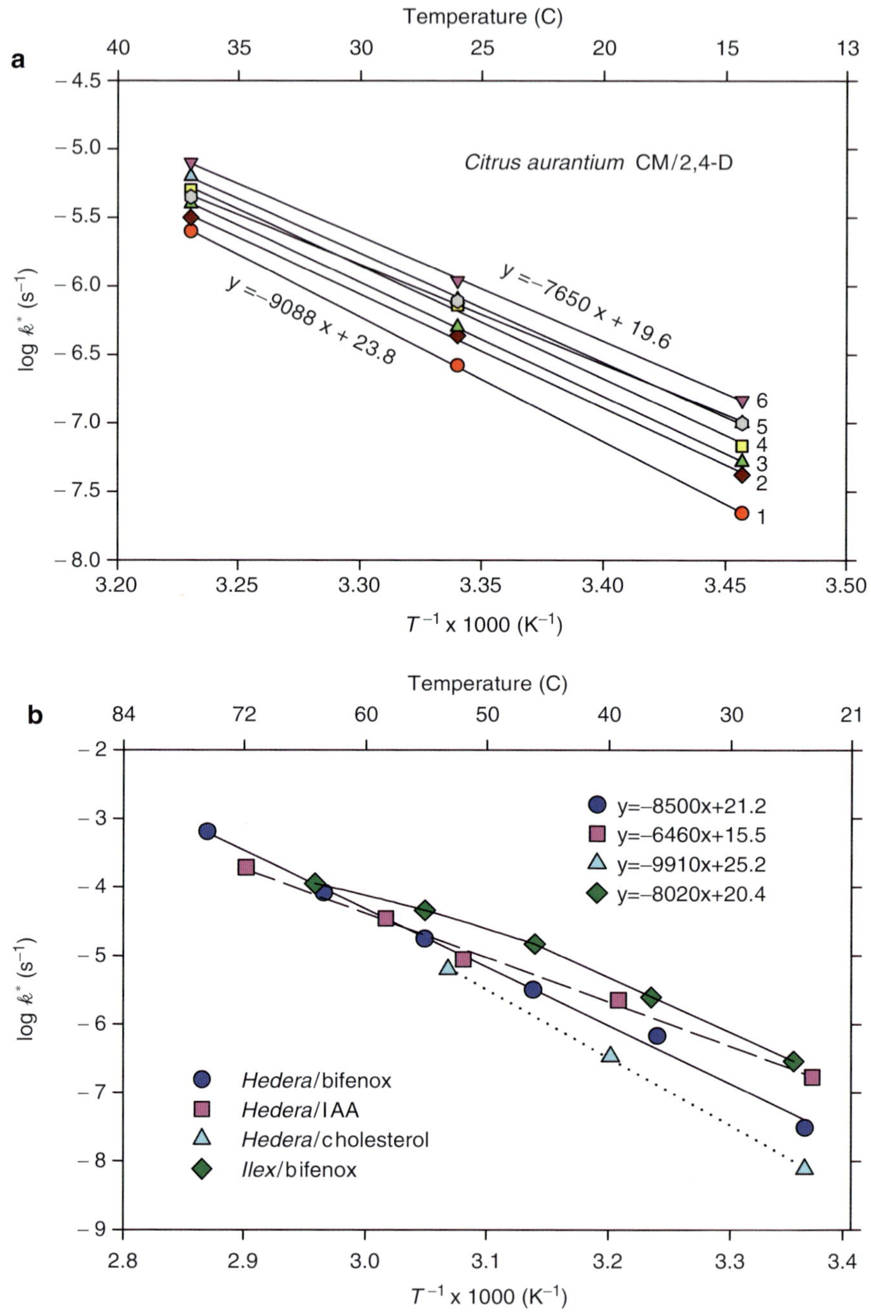

Fig. 8.3 Arrhenius plots of log k^* vs T^{-1} for CM of selected species/solute combinations. **(a)** Arrhenius plots for seven individual *Citrus* CM. **(b)** Arrhenius plots obtained with *Hedera helix* and *Ilex aquifolium* CM and the solutes bifenox, cholesterol and indoleacetic acid (IAA). Arithmetic means of k^* for 7–10 CM were plotted. Regression equations for the linear portions of the plots are given in the *upper right corner*. (Replotted from data by Baur et al. 1997a)

increases. This general feature has been observed with CM of all species and solutes (Baur et al. 1997a).

When the logarithms of average k^* values are plotted over a wider temperature range, these plots are often linear up to 70 °C, as with the combinations *Hedera*/IAA and *Hedera*/bifenox (Fig. 8.3b). With some CM/solute combinations, plots departed from linearity at temperature above 50 °C, when the melting points of the waxes were approached and the diffusion was no longer totally determined by properties of the limiting skin (Sect. 6.3.2). When rate constants for the limiting skin and the sorption compartment are similar, diffusion in the sorption compartment can become rate-limiting, as was shown for the combination *Citrus* MX/PCP (Fig. 6.20). Further, flattening of slopes is often observed with homogeneous rubbery polymers where it indicates the onset of viscous flow of polymer chains (Schlotter and Furlan 1992). For these reason, most studies have been limited to temperatures below 50 °C, which is the range experienced by most plant species under natural conditions. Generally, only linear portions of Arrhenius plots were utilised for data analysis, as shown for the combinations *Ilex*/bifenox and *Hedera*/cholesterol. In the upper right hand corner of Fig. 8.3b, regression equations for the linear portions of the Arrhenius plots are given. Slopes vary between −6,460 and −9,910, and using (8.7), E_D in the range of 123.5–185.5 kJ mol^{-1} can be calculated.

As seen in Fig. 8.3b, differences in rate constant among species also tend to decrease with increasing temperature, and above 60 °C differences have practically disappeared. Thus, differences among individual CM and differences between species/solute combinations decrease with increasing temperature, and above 50–60 °C they are small or nil. These observations are the consequences of the fact that size selectivity β', which amounts to 0.095 mol cm^{-3} at 25 °C (Table 6.8), decreases with temperature. Buchholz et al. (1998) measured k^* for ivy leaf CM and the solutes IAA and tebuconazole, which had molar volumes of 130 and 241 cm^3 mol^{-1} respectively. The ratios of the two rate constants decreased with temperature, and amounted to 24.5 at 25 °C, 10.5 at 35 °C and 2.2 at 55 °C.

8.2.2 Thermodynamics of Solute Diffusion in CM

Slopes and *y*-intercepts of Arrhenius plots vary with types of cuticle and solute, and k^*_{infinite} is related to entropy. Before discussing the physical meanings of slopes and intercepts, we take a look at their dependence on solutes and species. This was studied using leaf CM from 14 different species and varieties (Buchholz and Schönherr 2000). Barrier properties of CM can be classified based on bifenox mobility in CM at 25 °C (Fig. 6.21). *Populus canescens* and pear are species with relatively high solute mobilities. *Ilex paraguariensis* and *Hedera helix* are species having very low bifenox mobility. The other species and cultivars tested (*Malus domestica, Prunus laurocerasus, Pyrus communis, Schefflera actinophylla, Strophantus gratus*) have intermediate mobilities. All solutes were lipophilic, and their partition coefficients varied by seven orders of magnitude. They were (log K_{cw} in parentheses) IAA (1.16), NAA (2.43), tebuconazole (3.54), bifenox (4.44) and cholesterol (8.0).

From Arrhenius plots as shown in Fig. 8.3, y-intercepts (pre-exponential factor k^*_{infinite}) and the slopes (E_D/R) were obtained for individual CM, and results were plotted (Fig. 8.4). Data obtained with four species and the smallest and least lipophilic solute IAA (Appendix B) are depicted in Fig. 8.4a. There is an excellent correlation between k^*_{infinite} and the Arrhenius slopes (E_D/R). CM with steep slopes, that is, high energies of activation, have large y-intercepts and vice versa. With all species, k^*_{infinite} varied among individual CM. For instance with *Citrus* CM, k^*_{infinite} varied between 10 and 20 and (E_D/R) between 11×10^{-3} and 19×10^{-3} Kelvin. Large variability in permeance and solute mobility is a typical property of CM. In Fig. 8.4b, all available species/solute combinations are plotted for individual CM. An excellent linear correlation was again obtained, and both y-intercept and slope are very similar to those in Fig. 8.4a. This linear free energy relationship is good evidence that in CM from all species all solutes diffuse in a similar environment. This is astounding, since wax amounts and composition differ greatly among species (Tables 1.1 and 4.8) and partition coefficients ranged from 14 to 10^8.

The results characterise the lipophilic or waxy pathway, which in CM consists of methylene groups donated by amorphous waxes and cutin environment. The most polar solute included in the study is IAA; it is 14 times more soluble in CM than in water, and is still sufficiently soluble in cutin and waxes. Its solubility in water is smaller than in membrane lipids, and since water content of fully swollen CM is below 8% (Fig. 4.7) the concentration of IAA in lipid compartments is much larger. IAA exclusively diffused in a lipophilic matrix, otherwise these data would not have fallen on the straight line seen in Fig. 8.4b, and slopes and y-intercepts in Fig. 8.4a and 8.4b would have differed.

It was not possible to include highly water soluble solutes in these studies, because the UDOS method does not work with them (Sect. 6.3). However, there is no reason why results should not apply to more polar non-electrolytes, as long as they are not ionised and soluble in amorphous waxes. As shown in Sect. 4.6, water can penetrate cuticles using two parallel pathways, in aqueous pores located between polar polymers and in cutin and in amorphous waxes. E_D/R and y-intercepts for the fraction of water which diffuses in cutin and waxes most probably would fit the regression equation shown in Fig. 8.4b, but data are not available (see below).

There are a few data available for MX membranes. Extraction of waxes greatly increases solute mobility (k^*) and activation energies are much lower. Plots log k^* vs E_D/R have similar slopes than shown in Fig. 8.4 for CM, but y-intercepts were higher, that is, they had smaller negative numbers (Fig. 8.5). With the combination *Hedera*/IAA, this parallel displacement amounted to a factor of 1.97 (log scale), that is, with the MX entropy was higher by a factor of 94. With *Strophanthus*/IAA the factor was 50, and with *Schefflera*/IAA entropy was 5,011 times higher with MX than with CM. With *Pyrus*/bifenox, the MX had a 25-times higher entropy than the CM (Buchholz 2006; Buchholz and Schönherr 2000). These y-intercepts represent the entropy, when E_D/R is zero (Fig. 8.5), that is when there is no free energy and no driving force for diffusion.

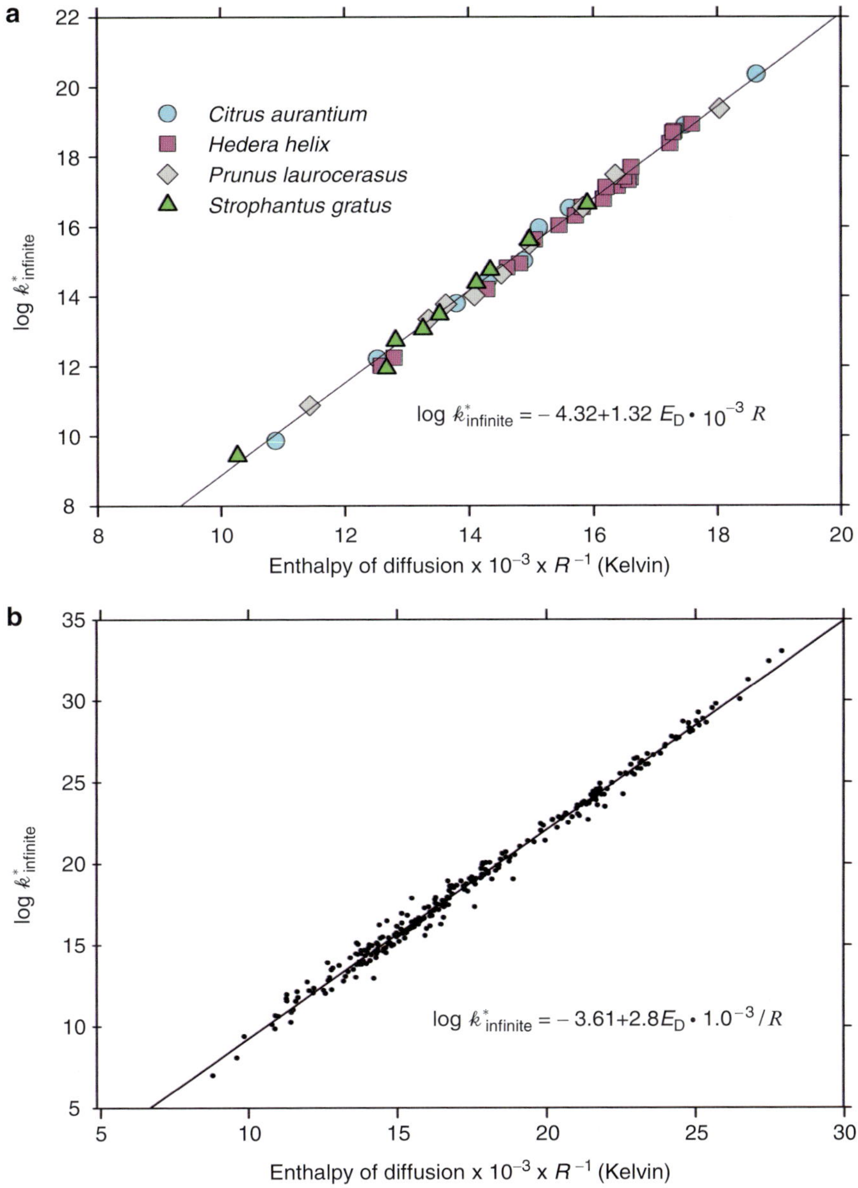

Fig. 8.4 Correlation between *y*-intercepts and slopes of the Arrhenius graphs for individual CM. **(a)** Data for IAA and leaf CM from four species are plotted. **(b)** Data for 297 leaf CM from 14 plant species (including varieties) and five test compounds (IAA, NAA, bifenox, tebuconazole, cholesterol) are plotted. (Redrawn from Buchholz et al. 1998 and Buchholz and Schönherr 2000)

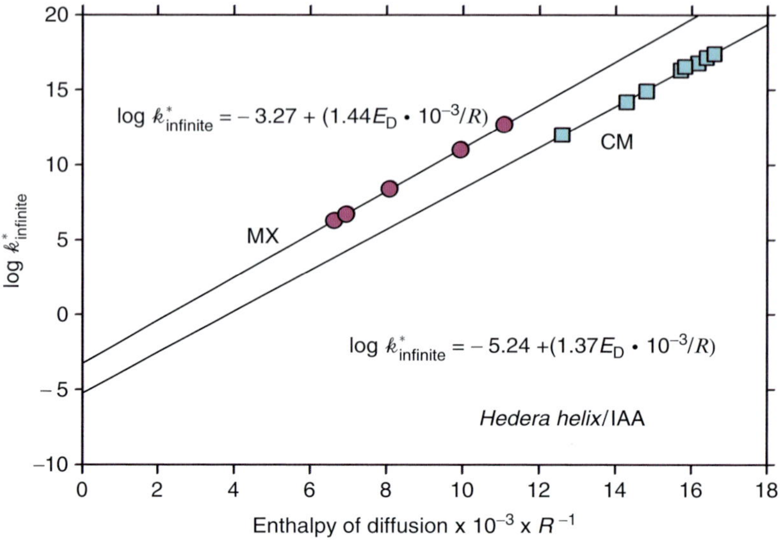

Fig. 8.5 Dependence of Arrhenius *y*-intercepts (log $k^*_{infinite}$) on slopes (E_D/R) measured with IAA and ivy CM or MX membranes

Theoretically, the mean free path of diffusion can be calculated using (8.11), but our *y*-intercept is not D_o but rather log $k^*_{infinite}$. In order to convert k^* to D we would need the thicknesses of the membrane compartments involved, which are not known and would have to be assumed (Sect. 6.3.2). This would result in arbitrary estimates of λ. Still, the *y*-intercept increased when waxes were extracted, and this increase is proportional to λ^2 and to entropy $[\exp(\Delta S/R)]$. Flexibility of the cutin polymer may well be higher than in amorphous waxes, allowing for a more frequent formation of larger voids in cutin than in amorphous waxes. This might have contributed to the differences in *y*-intercepts after extraction of waxes. However, an effect of extraction on ΔS cannot be precluded, and a separation of the two factors is not possible at this time. It is clear, however, that the micro-environment in which diffusion took place was not affected by extraction of waxes. The slopes, that is, the free energy relationship were not affected; only the free energy involved in diffusion was lower (Fig. 8.5). As in waxes, diffusion across MX membranes of lipophilic solutes takes place in the methylene group environment of cutin, but this differs structurally.

Buchholz and Schönherr (2000) suggested that the difference in *y*-intercepts between CM and MX might be due in part to a much higher tortuosity in CM. From the difference in *y*-intercepts between MX and CM, tortuosity factors ranging from 25 to 5,000 can be calculated. Factors of 25–100 might be plausible, but a factor of 5,000 is incomprehensible. Waxes are expected to decrease the length of the diffusion path, because wax crystallites embedded in cutin represent excluded volume and solutes must diffuse around them. This concept of a barrier membrane was introduced by Riederer and Schreiber (1995). It is plausible, but some of the assumptions can be questioned (Fowler 1999). In some thin cuticles (e.g. *Citrus*) there is simply not enough space in the limiting skin to accommodate many superimposed crystalline wax layers, which are separated by layers of cutin. A wax monolayer composed of a C40 hydrocarbon is 5 nm thick and has a mass of about $0.5\,\mu g\,cm^{-2}$. Total wax in *Citrus* CM is about $12\,\mu g\,cm^{-2}$, and about one third of it is crystalline, i.e. $4\,\mu g\,cm^{-2}$. This would be enough for eight superimposed wax monolayers. It is not at all obvious how they could be accommodated in cutin as large continuous platelets. Waxes also occur in cuticular layers and as surface waxes (Chap. 1). There is good evidence that waxes in the cuticular layers have very little effect on water permeance and solute mobility (Chaps. 4 and 6). When discussing effects of waxes on water permeability (Sect. 4.6), some evidence was discussed showing that one or two monolayers on top of the limiting layer could account for the effect of waxes on water permeability. A similar effect on solute mobility of superficial wax monolayers cannot be ruled out.

8.3 Water Permeability in CM and MX

In the previous sections, we have taken a look at the thermodynamics of sorption and diffusion of lipophilic solutes. For water, there are no studies of sorption or diffusion in cuticles as affected by temperature. Only permeance has been studied. Permeance depends on partition coefficients (sorption) and diffusion coefficients (2.18). This is repeated here for convenience:

$$P\ell = \mathscr{P} = DK \tag{8.14}$$

The effect of temperature on P_{wv} or $\mathscr{P}_{wv}$ can be analysed using Arrhenius plots such as

$$P_{wv} = P_o \exp(-E_P/RT) \tag{8.15}$$

where P_o is the pre-exponential factor and E_P is the Arrhenius activation energy of permeability or permeance. As long as E_P is constant and sorption isotherms are linear, we can further write

$$D = D_o \exp(-E_D/RT) \tag{8.16}$$

which is identical to (8.6), and for the partition coefficients we can write

$$K_{wv} = K_o \exp(-\Delta H_S/RT) \tag{8.17}$$

For a complete analysis of the effects of temperature on water permeability at least two of the above energies of activation are needed, and the remaining one can be calculated using the relationship

$$E_P = E_D + \Delta H_S \tag{8.18}$$

where ΔH_S is the heat of sorption as explained in Sect. 8.1. E_P has been determined for cuticles of a number of species. Later we shall take a look at the relationship between E_P, E_D and ΔH_S as determined with homogeneous synthetic polymer membranes.

Using the system buffer/CM/buffer and tritiated water, P_w was measured at various temperatures (Schönherr et al. 1979).

Water permeance of CM increased with temperature, and Arrhenius plots ln P_w vs $1/T$ had two linear portions which intersected at about 44°C. E_P was 52 kJ mol^{-1} (5–40°C) and 185 kJ mol^{-1} (50–65°C) respectively. The plot for MX was slightly convex to the abscissa, and P_w was much higher. The differences between the plots for CM and MX decrease with increasing temperature. If the straight line for the CM is extrapolated to 72°C, the plots for CM and MX converge (Fig. 8.6). The melting range of *Citrus* wax is 72–77°C (Reynhardt and Riederer 1991), and when waxes are molten and fluid the waxy transport barrier breaks down completely. Above 44°C the slope of the Arrhenius plot for the CM suddenly increases, and a kink is clearly visible. There is no discontinuity or kink in the Arrhenius plot for the MX. Similarly, the Arrhenius plot for solute mobility in CM of *Citrus* (IAA) is linear up to 50. With *Hedera* CM/bifenox, linearity was maintained up to 75°C. The kink or bend in the Arrhenius plot occurred only with P_w and CM, but not with P_w for MX or with solute mobility in CM.

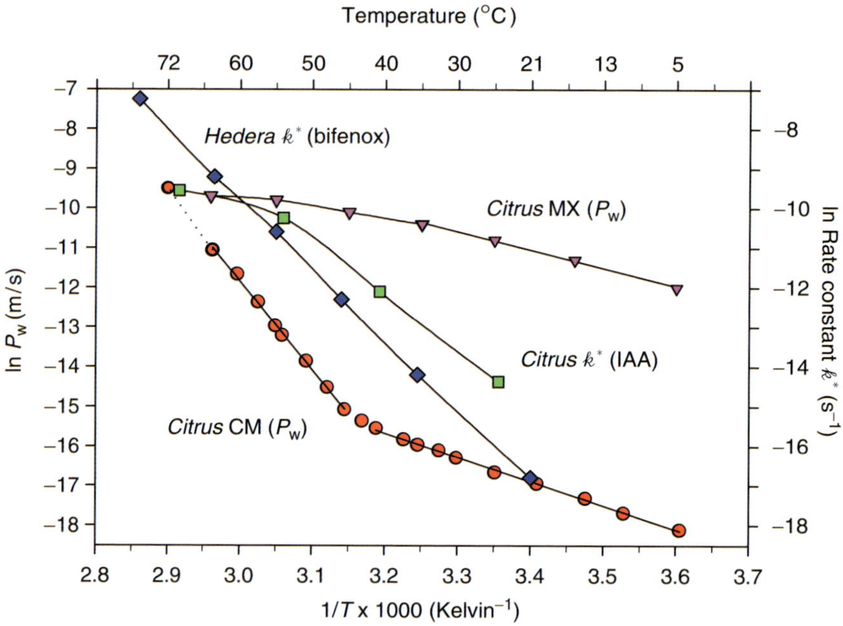

Fig. 8.6 Arrhenius plots showing the effects of temperature on water permeance (P_w) of *Citrus aurantium* CM and MX. The membranes separated aqueous buffers at *p*H 6 containing $CaCl_2$ (0.01 mol l^{-1}) and water fluxes were measured using tritiated water. For comparison rate constants (k^*) determined with *Hedera*/bifenox and *Citrus* IAA were included. Water permeance data were taken from Schönherr et al. (1979). Rate constants were taken from Fig. 8.3

The same approach was used with onion bulb scale epidermis (non-isolated CM). The kink occurred around 47 °C, and E_P below and above the kink were 58 and 177 kJ mol^{-1} respectively (Schönherr and Mérida 1981). Schreiber (2001) also used tritiated water, and compared temperature effects on P_w of isolated CM and leaf disks (*Vinca major, Prunus laurocerasus, Forsythia intermedia, Citrus aurantium, Hedera helix*). All Arrhenius plots exhibited a kink in the range of 30–39°, and no significant differences between CM and leaf disks were observed. Below the kink E_P ranged from 26 (*F. intermedia*) to 61 kJ mol^{-1} (*H. helix*). Above the kink, E_P was higher and ranged from 67 (*V. major*) to 122 kJ mol^{-1} (*F. intermedia*). E_P was higher with species having low P_w, and with increasing ln P_w the kink temperature increased. Clearly, the occurrence of the kink and the differences in E_P below and above the kink is a natural property of the CM and not an artifact related to isolation and storage of the CM. Additional data can be found by Riederer (2006), who used the cup technique to study temperature effects on P_{wv} of isolated CM of 14 species in the range of 10–55 °C. In contrast to the above methods, partial pressure at the morphological outer surface of the CM was practically zero and the CM were not fully swollen. With most species the kink occurred around 35 °C, and below the kink E_P ranged from 21 to 38 kJ mol^{-1}, while above the kink E_P was again higher and ranged from 52 and 117 kJ mol^{-1}.

High temperatures above the kink irreversibly changed permeance. With the MX in Ca^{2+} form, Arrhenius plots obtained with increasing temperatures showed decreasing slopes when heated above 45 °C (Fig. 8.7). When P_w was studied using descending temperatures beginning at 65 °C, a straight Arrhenius plot was obtained with a slope similar to the initial slope (5–35 °C) of the ascending temperature run. At all temperatures, P_w of MX membranes measured with descending temperatures were significantly lower than before heating the MX. The opposite was observed with CM. The Arrhenius plot obtained with decreasing temperatures starting at 65 °C had significantly higher P_w values, and the kink occurred around 35 °C (Fig. 8.7). After heating the CM once to 65 °C, the descending and ascending temperatures runs were superimposed and the system became reversible. What causes the kink in the Arrhenius plots for P_w with the CM and its absence with the MX?

The kink in the Arrhenius plots of the CM and its absence with the MX has been interpreted as solid/liquid phase transition of cuticular waxes (Schönherr et al. 1979; Eckl and Gruler 1980). Differential scanning calorimetry (DSC) of moist CM and MX at a rate of 0.1 °C min^{-1} revealed a number of first-order phase transitions, while with dry CM no phase transitions were observed between 0 and 200 °C. At low temperatures a phase transition with a latent heat of 4.7 J g^{-1} cuticle at 16.3 and 17.6 °C was observed with MX and CM respectively. The transition at 38 °C occurred only with CM. Its latent heat was about 0.4 J g^{-1} CM. Above 40 °C, three additional first-order phase transitions at 41, 46 and 49 °C in MX and CM were observed. Latent heat was around 0.3 J g^{-1} MX for each transition (Eckl and Gruler 1980). Between 35 and 40 °C, a sudden change in volume of the *Citrus* CM was also observed (Schönherr et al. 1979). The only phase transition which can be attributed to waxes is that observed at 38 °C, while all others were observed with MX and CM and can be assigned to the polymer matrix.

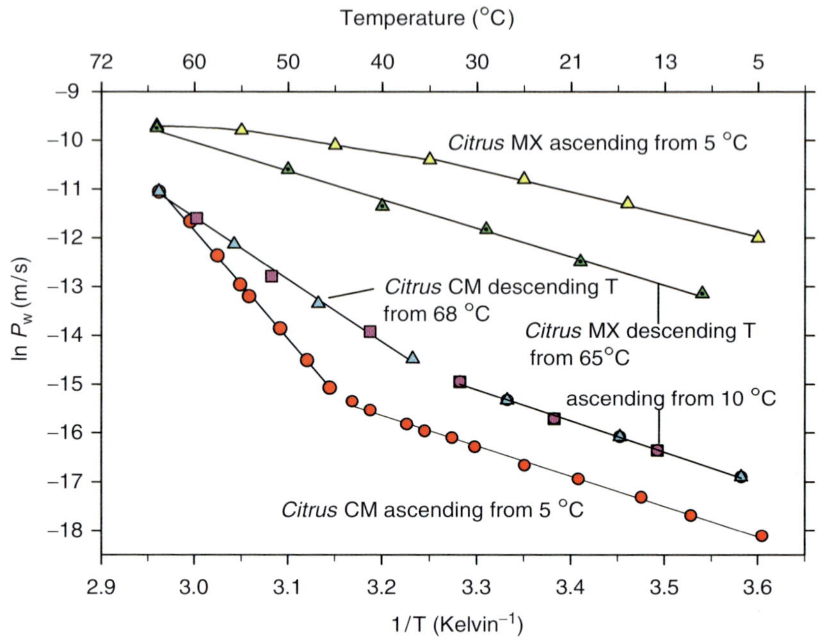

Fig. 8.7 The effect of heating *Citrus* CM and MX on water permeance and on Arrhenius graphs. The same methods and experimental conditions as in Fig. 8.6 were used. (Redrawn from data of Schönherr et al. 1979)

In CM and MX we have two parallel pathways for water, the lipophilic cutin/wax pathway and the aqueous pathway formed by pores associated with polar polymers (Sects. 4.5 and 4.6). Diffusion of lipophilic solutes in the cutin and in the waxy pathways did not reveal any phase transition (Sect. 8.2), as Arrhenius plots were linear at least up to 50 °C. At higher temperatures Arrhenius plots occasionally became convex to the abscissa, as seen in Fig. 8.6 for *Citrus*/IAA. A sudden increase in slope and E_D above 35 to 40 °C was never observed. Hence, a phase transition of waxes can be ruled out. The transition at 38 °C observed with *Citrus* CM certainly had no effect on solute diffusion in waxes. With this in mind, the kink in the plots ln P_w vs $1/T$ cannot be attributed to changes in wax structure. All evidence (DSC, thermal expansion) points to changes in structure of the polymer matrix, that is, temperature affected the aqueous rather than the waxy pathway. Above the kink, the aqueous pathway underwent a phase transition or a new aqueous pathway opened up. The aqueous pathway had been studied and characterised only at 20–25 °C (Sect. 4.5). It appears that at temperatures below the kink the bulk of the water diffuses along the waxy pathway. Co-permeation of lipophilic solutes and water are excellent evidence (Sect. 4.6.2.3). Above the kink, an increasing fraction of water uses the aqueous pathway(s).

High temperatures irreversibly decreased P_w of the polymer matrix in Ca^{2+} form (Fig. 8.7), while in Na^+ form P_w decreased only slightly (Schönherr et al. 1979).

This again points to a change in structure of the MX rather than to an effect of temperature on structure of waxes. Schreiber and Schönherr (1990) took a closer look at thermal expansion of CM, MX and cutin.

8.4 Thermal Expansion of CM, MX, Cutin and Waxes

Thermal expansion was measured by stuffing wet cuticular materials in a specific gravity bottle ($1\,cm^3$), and after exhaustive aspiration to remove all air bubbles from cuticles and water the bottle was closed with a very fine calibrated capillary. Gravity bottle and capillary, with the exception of its tip, were submerged in a water bath at constant temperature ($\pm 0.01\,°C$). The bath temperature was raised from 5 to 65 °C at a rate of $0.25\,°C\ min^{-1}$, and the position of the meniscus in the capillary was continuously read using a travelling microscope (cathetometer). From the total volume expansion, the volume of the cuticular materials was calculated by subtracting the volume of the water in the bottle.

Specific volume ($cm^3\ kg^{-1}$) was plotted vs temperature, and the volume expansion coefficient ($cm^3\ kg^{-1}\ K^{-1}$) was calculated from the slope of the plots. Total volume expansion in the range of 5–65 °C of *Ficus, Capsicum* (fruit) and *Citrus* increased in the order CM < MX < cutin. Plots for cutin from *Ficus, Citrus* and *Capsicum* (fruit) were linear over the entire temperature range. Phase transitions were observed only with CM and MX. With *Ficus* and *Citrus* CM, two phase transitions were detected in the range of 17–20 °C and 42–50 °C. All other CM and MX investigated had only one phase transition between 40 and 55 °C (Table 8.1).

Table 8.1 Thermal expansion coefficients for the temperature range 25–65 °C

Polymer	PTT	Volume expansion coefficient ($cm^3\ kg^{-1} K^{-1}$)		
	°C	below 42 °C	above 42 °C	above/below
Citrus CM	17 and 42	0.53	0.74	1.40
Citrus MX	21 and 44	0.50	0.65	1.30
Citrus cutin	none	0.95	–	1.0
Ficus CM	18 and 50	0.55	0.95	1.73
Ficus cutin	none	0.96		
Capsicum CM	42	0.56	0.60	1.07
Capsicum cutin	none	0.82	–	1.0
Lycopersicon CM	46	0.57	0.69	1.21
Pyrus CM	46	0.62	1.41	2.27
Nerium CM	39	0.39	0.62	1.59
Olea CM	55	0.45	0.70	1.56
PEMA[a]	65	0.275	0.27	0.98
Polyvinylacetat[a]	32	0.24	0.45	1.88
PVA-Cl copolymer[a]	30	0.20	0.24	1.20
PETP amorph[a]	67	0.19	0.80	4.21

PTT is the phase transition temperature
[a] Taken from Stannett and Williams (1965). All other data from Schreiber and Schönherr (1990)

Volume expansion coefficients were always larger by a factor of 1.07–2.27 above the phase transition. Expansion coefficients of cutin were much larger than with CM and MX.

Volume expansion of cutin is linear in the range of 5–65 °C, hence it cannot be responsible for the phase transition observed with CM and MX. Volume expansion was larger with MX than CM, but the difference is related to the fraction of wax contained in the CM. Correcting for the wax fraction, volume expansion of CM is the same as that of MX. Volume expansion of MX is the sum of expansion of its components cutin and polar polymers. Expansion coefficients of cutin are much higher than those for MX (Table 8.1), and it follows that volume expansion of MX is hindered by polar polymers, probably by embedded cellulose.

Volume expansion of cuticular waxes was not measured. Rosenberg and Brotz (1957) measured volume expansion of natural and synthetic waxes, and found them to be linear. Relative expansion was 0.057% per °C, which is lower than that for *Citrus* MX (0.074% per °C above 45 °C) and cutin (0.098% per °C) measured by Schreiber and Schönherr (1990). Relative expansion of crystalline cellulose is 0.057% per °C (Banderup and Immergut 1966), which is also lower than expansion of MX and cutin. These data indicate that the kinks observed in Arrhenius plots of water permeance (ln P_w vs. T^{-1}) may have been caused by strains arising from different expansion of waxes, cutin and polar polymers in MX. Volume expansion of amorphous and crystalline waxes could not keep pace with that of cutin and MX, and this induced micro-defects in the wax. These cracks are new pathways for diffusion and water, and permeance increases much more rapidly than below 40 °C. No kinks were observed with *Citrus* MX membranes, because in these membranes waxes are absent and new defects cannot arise at higher temperatures. The convex curvature of the Arrhenius graphs at high temperatures (Fig. 8.6) could have been caused by reduced sorption of water in cutin and polar polymers (Sect. 8.5).

Cracks in paraffin wax films formed by sudden cooling led to relatively high water permeability, but these cracks disappeared during storage and water permeability decreased significantly (Table 4.12). Water permeance of freshly isolated CM from *Citrus* leaves decreased by factors of two to three during storage (Geyer and Schönherr 1990). Water permeance of *Citrus* CM at 25 °C increased upon heating to 65 °C, and after subsequent cooling to 25 °C, P_w was larger by a factor of 2.8 than before heating (Schreiber and Schönherr 2001), indicating that wax structure differed after re-solidification. When these CM were stored for 100 days in the dry form, permeance was greatly reduced and nearly reached permeances measured before heating. These results very much resemble those observed with paraffin waxes, and demonstrate the capacity of waxes to change structure when stored at room temperature. Reduction in free energy probably is the driving force for these structural changes (Fox 1958).

Four synthetic polymers are included in Table 8.1. Their thermal expansion coefficients are a little smaller below the PTT, while above the PTT expansion coefficients are similar as with CM. The phase transition temperatures given for the synthetic polymers refer to a transition from the glassy to the rubbery state, and this poses the question as to whether the transition observed with CM also represent a glassy/rubbery transition. At room temperatures, cutin has the consistence of chewing gum and it did not undergo a phase transition. It is in the rubbery state from the beginning. *Citrus* waxes associated with cutin exhibited a first-order phase transition at 38 °C when CM were moist during DSC measurements (Eckl and Gruler 1980). The latent heat associated with this transition was small (0.4 kJ kg^{-1} CM), and it is not clear if this transition is due to waxes or wax/cutin interactions. Reynhardt and Riederer (1991) also used DSC, but their samples were under dry nitrogen and temperature was increased at a rate of 10 °C min^{-1}. This is 100 times faster than that of Eckl and Gruler, and it probably reduced resolution. Further, Eckl and Gruler (1980) did not find any phase transitions with CM and MX when samples were dry, and there are doubts that DSC with dry CM scanned at very high rates can produce useful data for interpreting water permeability of CM. With this meagre data base, the role of waxes in the kink of Arrhenius plots of P_w remains obscure.

The absence of kinks in thermal expansion of cutin and their presence with CM and MX leaves little doubt that they are related to polysaccharides. Double refraction of cuticles is evidence that some of the polysaccharides are crystalline (Sect. 1.4.2), most probably cellulose. Unfortunately, we have no data as to the type of structural changes which occur in cellulose of cuticular origin between 35 and 50 °C, and as to which role water plays in these changes. Crystal structure of cellulose involves intermolecular hydrogen bonds, and it is probable that water is involved in structural changes. This needs to be investigated, and before this has been done it is futile to speculate if the kinks seen in Arrhenius plots of water permeance reflect a glass/rubbery phase transition.

8.5 Water Permeability of Synthetic Polymers as Affected by Temperatures

Water permeance depends on sorption and diffusion of water in the constituents of cuticular membranes. Neither sorption of water nor its diffusion in waxes or MX has been studied so far, and this prevents any thermodynamic analysis of water permeability of cuticles. In an attempt to overcome these limitations, Riederer (2006) studied permeance of isolated CM and assumed that all water diffuses in aqueous pores. No attempt was made to demonstrate the presence of aqueous pores in the CM studied. Enthalpy of sorption in CM was assumed to be identical to that measured for wetting of wood. Wood consists of lignin and cellulose, but lignin is not a constituent of cuticles. These assumptions are arbitrary and in conflict with current knowledge (Chap. 4). There can be no doubt that a large fraction of water diffuses across the cutin and waxy pathways. Due to the heterogeneity of cuticles, it is extremely difficult to actually measure diffusion and sorption in the constituents of cuticles such as waxes, cutin and polar polymers. In an attempt to better understand penetration of water across cuticles, we searched the literature for information obtained with homogeneous synthetic membranes.

8.5.1 E_P, E_D and ΔH_S Measured with Synthetic Polymers

Doty et al. (1946) measured water vapour permeability of a number of synthetic polymers in the temperature range between zero and 60 °C. P_{Hg} was calculated from the steady state increase in vapour pressure on the receiver side. We converted these P_{Hg} values to $\mathscr{P}_{wv}$ as explained in Chap. 3. Temperature effects were analysed using the Arrhenius equation (8.15). Water permeability of most polymers increased with temperature, but slopes of Arrhenius plots differ greatly (Fig. 8.8). All slopes are linear, and there are no kinks visible. Polystyrene has the highest permeability ($\mathscr{P}_{wv}$), and between 25 and 45 °C permeability was not affected by temperature. Polyvinylidene chloride has the lowest $\mathscr{P}_{wv}$ and the steepest Arrhenius slopes. $\mathscr{P}_{wv}$ varied among polymers by a factor of 440. E_P varied from about zero (polystyrene) to 85 kJ mol^{-1} (polyvinylidene chloride), and the pre-exponential factor log $\mathscr{P}_o$ varied among polymers by orders of magnitude from 4 to -9 (m^2 s^{-1}).

There was a good ($r^2 = 0.95$) negative linear correlation between E_P and log $\mathscr{P}_{wv}$. The regression equation $E_P = -31.7 \times \log \mathscr{P}_{wv} - 291$ was calculated from the data in Fig. 8.8. A similar dependence was observed with plant CM, where *Hedera helix* had the lowest permeance and the highest E_P (62 kJ mol^{-1}), while *Forsythia* and *Vicia* had the highest permeance and the lowest E_P of 27 kJ mol^{-1} (Schreiber 2001).

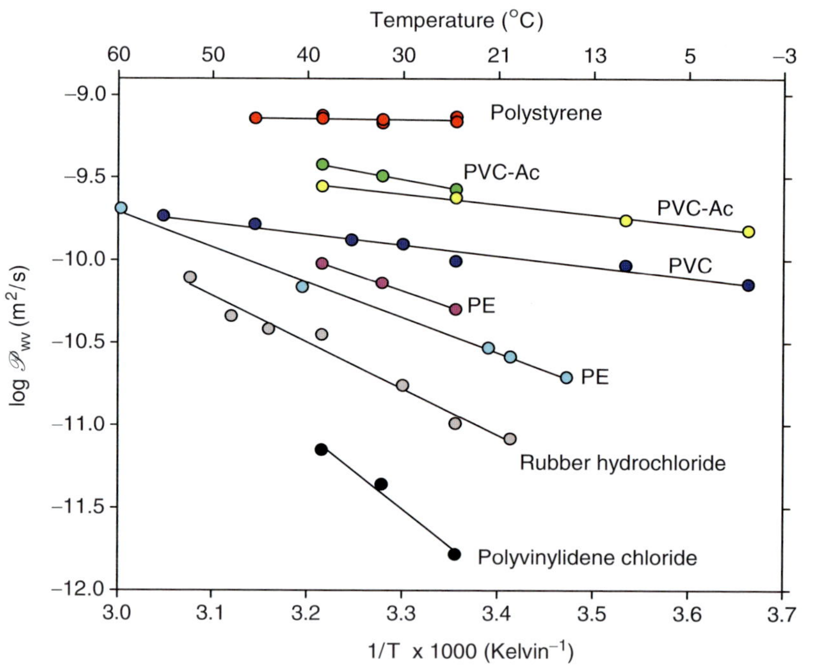

Fig. 8.8 Arrhenius plots showing the effect of temperature on water vapour permeability ($\mathscr{P}_{wv}$ in m^2 s^{-1}). Two lots of polyvinyl chloride-acetate copolymer (PVC-Ac) and polyethylene (PE) were used. PVC is polyvinyl chloride. Data from Doty et al. (1946) were re-calculated and re-plotted

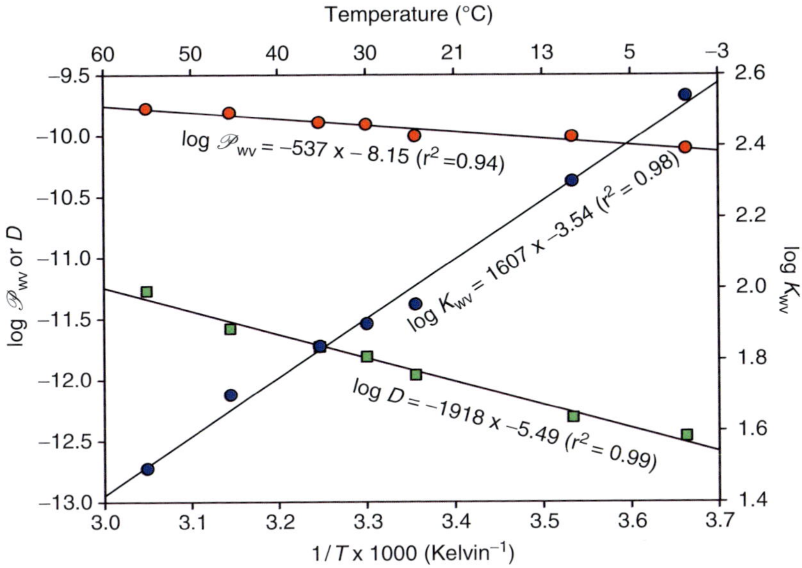

Fig. 8.9 Temperature dependence of $\mathscr{P}_{wv}$, D, und K_{wv} for polyvinyl chloride. Data taken from Doty et al. (1946)

For polyvinyl chloride, D was also determined by Doty et al. (1946) using the hold-up time method (2.5). With increasing temperature (1 to 55 °C), D increased slightly (7.8×10^{-11}–17×10^{-11} m^2 s^{-1}), while P_w increased from 2.2×10^{-13} to 54×10^{-13} m^2 s^{-1} (Fig. 8.9). The partition coefficient K_{wv} can be calculated from these data using (3.17). K_{wv} decreased with increasing temperature from 246 to 31 cm^3 vapour per cm^3 polymer.

From the slopes given in Fig. 8.9, the energies of activation can be calculated. For E_P and E_D 10.3 and 36.7 kJ mol^{-1} are obtained, and by difference (8.18) the enthalpy of sorption or the heat of solution ΔH_S of −26.4 kJ mol^{-1} is obtained. The enthalpy change for transfer of water into PVC is negative, indicating it to be a spontaneous process, but with increasing temperature decreasing amounts of water are sorbed (c.f. Sect. 8.1). For the other polymers in Fig. 8.8, D was not measured and E_D remains unknown. With polystyrene E_P is zero, and this implies that E_D and ΔH_S are numerically equal but have opposite sign.

In Table 8.2, some additional data including rather polar polymers (EC, CA, PEMA) have been collected. At 25 °C, some of the polymers are in the glassy state (PET, Nylon, PVA, EC and PEMA). In this collection, $\mathscr{P}_{wv}$ was highest with cellulose acetate (which is used for desalination membranes) and lowest with polyvinyl acetate. They differ by a factor of 251. PVA has the lowest diffusion coefficient of 5.1×10^{-15} m^2 s^{-1}. For the other polymers, D varies from 1.1×10^{-11} (EC) to 1.2×10^{-12} m^2 s^{-1} (nylon, rubber HCl) by a factor of 9. Partition coefficients (K_{wv}) in the range of partial pressures where sorption isotherms are linear (3rd column) ranged from 7,255 (PVA) to about 8 for PE.

Table 8.2 $\mathscr{P}_{WV}$, D, and K_{WV} at 25°C and E_P, E_D and ΔH_S for selected synthetic polymers cellulose acetate (CA), polyvinyl butyral (PVB), polyethylene (PE), polyethylene terephthalate (PET), rubber hydrochloride (rubber), polyvinyl acetate (PVA), ethyl cellulose (EC) and polyethyl methacrylate (PEMA)

Polmer	T_g °C	p/p_o	$\mathscr{P}_{WV}$ $m^2 s^{-1}$	D $m^2 s^{-1}$	K_{WV}	E_P kJ mol^{-1}	E_D kJ mol^{-1}	ΔH_S kJ mol^{-1}
CA[a]	–	0.1–0.6	9.3×10^{-9}	3.1×10^{-12}	3,000	0	50.2	−50.2
PVB[a]	–	0.1–0.6	1.5×10^{-9}	1.3×10^{-12}	1,153	−8.8	45.5	−54.3
PE[a]	–	0.4–0.8	5.4×10^{-11}	6.8×10^{-12}	8	33.4	80.2	−46.8
PE[b]	–	0.1–1.0	–	–	–	–	59.5	–
PET[b]	67	0.1–1.0	–	–	–	–	43.5	–
PP[b]	–	0.1–1.0	–	–	–	–	68.7	–
Nylon[a]	–	0.1–0.5	3.3×10^{-10}	1.2×10^{-13}	2,750	11.7	55.6	−43.9
Rubber[a]	–	0.1–0.6	5.4×10^{-11}	1.2×10^{-13}	450	26.4	72.0	−45.6
Rubber[b]	–	–	–	–	–	–	58.6	–
PVA[a]	30	0.1–0.5	3.7×10^{-11}	5.1×10^{-15}	7,269	962	59.8	−50.2
PVC[c]	–	0.1–0.9	9.8×10^{-11}	1.1×10^{-12}	89	10.2	36.7	−31.0
EC[d]	50	0.2–0.5	1.9×10^{-8}	1.9×10^{-11}	1,000	−6.3	26.3	−32.6
PEMA[e]	65	0.1–0.5	2.9×10^{-9}	1.1×10^{-11}	264	2.1	36.4	−34.3

T_g is the glass transition temperature, p/p_o partial water vapour pressure used for measurements
[a]Hauser and McLaren (1948)
[b]Yasuda and Stannett (1962). Original data (P_{Hg}) were converted to SI units as explained in Chap. 3. Hauser and McLaren (1948) determined sorption gravimetrically and calculated D as $\mathscr{P}_{WV}/K_{WV}$: The other diffusion coefficients were obtained from the hold-up time and ΔH_S was calculated as $E_D - E_P$
[c] Doty et al. 1946
[d] Wellons et al. (1966)
[e]Stannett and Williams (1965)

E_P ranged from −8.8 to 33.4 kJ mol^{-1}. Negative values indicate that permeance slightly decreased with increasing temperature. E_P for CA was zero, that is, E_D and ΔH_S were 52 and −52 kJ mol^{-1} respectively. With this data set, which includes relatively polar polymers, high permeability was again associated with low E_P and vice versa, but the correlation between log $\mathscr{P}_{WV}$ and E_P (log $\mathscr{P}_{WV} = 0.06 \times E_P - 8.78$) was much weaker than with the data set of Fig. 8.6. E_D and ΔH_s varied much less than E_P. (Fig. 8.10).

E_D increased with increasing E_P, but the correlation is too weak to use the linear regression equation shown in Fig. 8.10 for the purposes of prediction. $\mathscr{P}_{WV}$ was determined at relatively low partial pressure, when water content of the polar membranes increased linearly with partial pressure. Water content of EC, PEMA and EC greatly increases when partial pressures are higher than 0.5–0.6. Hence, the data should not be extrapolated to 100% humidity.

The enthalpy of sorption (ΔH_S) was very similar for all polymers, and did not depend on either E_D or E_P. The mean enthalpy was −47 kJ mol^{-1}. It is negative and in the vicinity of the heat of evaporation or condensation, which at 25°C is 44 kJ mol^{-1}. Again, it should be remembered that these values characterise sorption

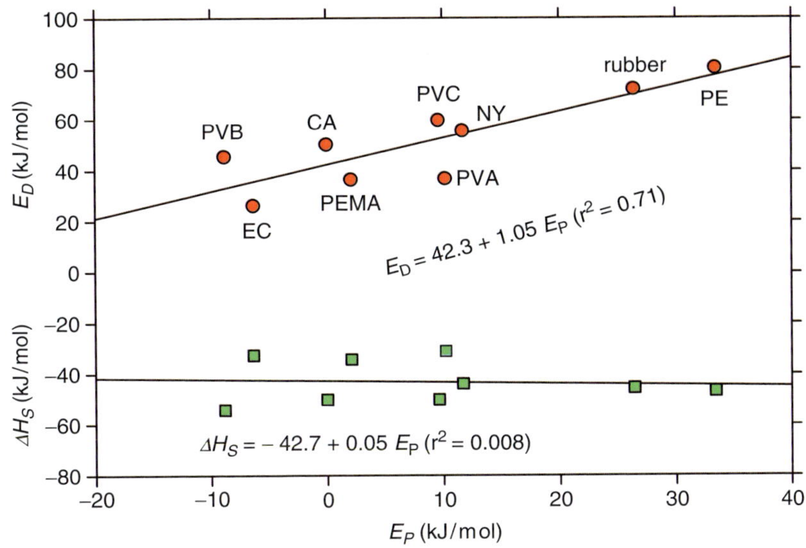

Fig. 8.10 The relationship between E_D and ΔH_S with E_P. Data from Table 8.2 were plotted

of water at low partial pressures, where clustering of water and swelling of the polar polymers are not involved.

The above data obtained with homogeneous polymers leave little doubt that temperature effects on water permeability of cuticles cannot be analysed if only activation energy for penetration (E_P) is known. The problem cannot be solved by analogy, because cuticles are heterogeneous, and data on temperature effects on water vapour sorption in waxes, cutin and MX have not yet been determined. Sorptive properties of polar polymers in the MX can be obtained by comparing sorption in cutin and in MX. Some of these aspects have been discussed when dealing with sorption of 4-nitrophenol in CM and MX (Sect. 8.1), but comparable data for water vapour sorption as affected by temperature are not available. Such measurements can be made. They may be difficult, but there are no insurmountable obstacles. As long as these data are not available, speculations concerning the magnitudes of E_D and ΔH_S in CM should be avoided.

Problems

1. What is the difference between the partition coefficient K and the parameter k of the Freundlich isotherm? For $n = 1$ calculate k when $K = 90$.
2. When all sorption sites are occupied $K = 1$. What is Gibbs free energy change in this situation?
3. How large are ΔH and $T\Delta S$ when $\Delta G = 0$?
4. Given the data of Fig. 8.2b, what is ΔS when $\Delta H = -30\,\text{kJ mol}^{-1}$?

5. Using the Arrhenius equations given in Fig. 8.3a for the CM number 1 and 6, calculate the rate constants k^* at 10°C and 50°C?
6. Calculate the energy of activation for the species/solute combinations shown in Fig. 8.3b.
7. Why does E_D with ivy leaves increase in the order IAA, bifenox, cholesterol?
8. What is the difference between D_o and P_o?
9. Water permeance of *Citrus* CM at 5 and 35°C is 1.36×10^{-8} and 1.17×10^{-7} m s^{-1} respectively. For MX, the respective values for P_w is 6.14×10^{-6} and 3.04×10^{-5} m s^{-1} (Fig. 8.7, ascending T). How much larger is P_w of the MX at 5 and 35°C? What are the Arrhenius equations for CM and MX?
10. From the slopes given in Fig. 8.9, E_D and E_P can be calculated and 10.3 and 36.7 kJ mol^{-1} respectively are obtained. $\Delta H_S = -26.4$ kJ mol^{-1} can be calculated by difference (10.3–36.7) as specified in (8.18). If ΔH_S is calculated from the slope of the plot log K_{wv} vs $1/T$ we obtain -30.7 kJ mol^{-1}. How do you explain this difference between the two ΔH_S values?

Solutions

1. K is the slope of a plot internal vs external concentration, while k is obtained when the logarithms of the two variables are plotted. When $n = 1$, then $K = k = 90$.
2. $\Delta G = 0$.
3. $\Delta H = T\Delta S$.
4. $\Delta S = 66$ kJ mol^{-1} K^{-1}.
5. For CM 6, we obtain 3.7×10^{-8} (10°C) and 8.2×0^{-5} s^{-1} (50°C) respectively. For CM 1 we have 4.9×10^{-9} (10°C) and 4.61×10^{-5} s^{-1} (50°C). With CM 6 having the initially highest k^*, the increase with T was 2,216-fold. With CM 1 the increase was even 9,408-fold. The k^* ratio between the two CM is 7.55 at 10°C and 1.78 at 50°C.
6. E_D is 123.5 kJ mol^{-1} (*Hedera*/IAA), 162.5 kJ mol^{-1} (*Hedera*/bifenox), 189.5 kJ mol^{-1} (*Hedera*/cholesterol) and 153.4 kJ mol^{-1} (*Ilex*/bifenox).
7. Because solute size increases from 130, 216 to 349 cm^3 mol^{-1} (Appendix B) and the activation energy is larger when larger holes must be formed in the membrane matrix.
8. E_o depends on jump distance within the polymer chains and entropy (8.11). P_o depends in addition on thickness ℓ of the cuticles and on sorption of water in the polymer and in waxes.
9. The ratio of P_w is 451 at 5°C and 260 at 35°C. The Arrhenius equation for CM is $P_o = 3.98–6{,}142 \times 1/\mathrm{T}$, and for the MX $P_o = 4.44 - 4{,}571 \times 1/\mathrm{T}$.
10. The individual data points for log K_{wv} were obtained from individual data of log D and log P_w at the respective temperatures. The difference is due to rounding errors.

Chapter 9
General Methods, Sources of Errors, and Limitations in Data Analysis

In this book, we have focussed on the physical meaning of transport parameters such as permeance (P), rate constants (k^*) and diffusion coefficients (D) measured with plant cuticular membranes. Experimental methods and statistics have been kept to a minimum, as we felt this might distract the reader from the arguments regarding chemistry–structure–property relationships. This does not imply that we consider methods and statistics to be matters of minor importance. They are of importance, and during the course of our investigations new methods had to be devised. We had to overcome problems related to the variability inherent to biological materials.

Anyone trying to study permeability of plant cuticles has to cope with a discouraging problem. Permeability among individual cuticles, leaf disks and leaves varies tremendously. Differences between lowest and highest rates often amount to one or even two orders of magnitude, and the problem increases with decreasing permeability. How can such data sets be characterised, and how can treatment effects be tested for statistical significance? A normal symmetric distribution of data is the prerequisite for calculating arithmetic means, standard deviations and confidence intervals. The distribution of data must be established using large populations, and it must be known to justify elimination of outliers. In most of our investigations, we have used isolated cuticular membranes to study the effects of humidity, temperature, adjuvants, and solute properties on permeability. With this approach, the permeability of a large population of 50–100 CM can first be characterised and treatment effects can be established using the method of paired observation. The same CM is used for various treatments, and it can be tested whether the treatment effect depends on initial permeability of the CM. If this approach is used, it must be established that treatments do not change properties of cuticles.

9.1 Isolation of Cuticular Membranes

Enzymatic isolation at ambient temperatures is the mildest method for isolating cuticles. One of the first to study enzymatic isolation of cuticles was Orgell (1955),

L. Schreiber and J. Schönherr, *Water and Solute Permeability of Plant Cuticles*.

and initially we isolated cuticles according to his directions (Schönherr and Bukovac 1973). Initially, we used crude preparations of pectinase and cellulase of fungal origin, which can be obtained from suppliers of biochemicals. As these enzymes are rather expensive, we later changed to pectinases used by the fruit and wine industry (Erbslöh, Geisenheim, Germany). The results were identical.

In an attempt to find optimum conditions, Schönherr and Riederer (1986) used fungal pectinase (2–6,841) and cellulase (catalogue number 2–3,056) obtained from Roth (Germany). Leaf discs (15 mm in diameter) were punched from young but fully expanded rubber-tree leaves (*Ficus elastica* var. *decora*). Effects of enzyme concentration, *p*H and temperature on time needed to detach CM were studied. Based on this investigation and some 30 years of experience, some basic rules for successful isolation of cuticles have evolved.

Best results are obtained with young but fully expanded leaves. With increasing age it takes much longer to detach cuticles from the epidermal wall, and often some cellular remnants cannot be removed. With some species (e.g. *Populus canescens, Liriodendron tulipifera, Juglans regia*), isolation of old cuticles becomes more and more difficult or they cannot be isolated at all. Integrity of young leaves is superior, because they have not been exposed to pests and diseases and damage by wind and storm. However, there are many species from which astomatous cuticles cannot be isolated enzymatically. With hypostomatous leaves, cuticles from the abaxial surface are perforated by stomatal pores, and they cannot be used for transport measurements. They can easily be sorted out if they are stained with a water-insoluble marker. We have used waterproof felt pens.

The activity of our pectinases was optimum between *p*H 3 to 4. Cuticles could not be isolated when *p*H was 5 or 6 and, we usually used a 0.01 mol l^{-1} citric acid buffer adjusted with KOH. NaN_3 at a final concentration of 10^{-3} mol l^{-1} must be added to suppress growth of micro-organisms. Proteins can spoil glass electrodes, and for this reason it is better to add the enzyme to the buffer rather than adjusting the *p*H of the enzyme solution. Time needed to detach cuticles decreases with increasing enzyme concentration, but in most cases 1–2% worked well. Crude pectinases often have some cellulase activity, and adding cellulase was not necessary nor beneficial.

Pectinase molecules cannot penetrate cuticles, and for this reason leaf discs must be vacuum-infiltrated, with the pectinase solution entering the leaf from the cut edges. Infiltrated leaves appear dark green and glassy. If this step is omitted, the enzyme can work only at the cut edges, and it may take weeks before it has reached the centre of the leaf disc. During disintegration of the leaf tissue, phenolic compounds are often liberated which inhibit enzyme activity. If the enzyme solution turns brown, it is advantageous to decant it and add fresh enzyme solution periodically. The bottles with the leaf discs and the enzyme solution should never be shaken or agitated in any other way, as this will damage many cuticles, especially when they are thin and fragile. Enzymes inside the intercellular space digest the middle lamellae and pectins in the outer epidermal wall. They do not benefit from shaking external solutions. Enzyme activity depends on temperature, and we obtained best results at 30–40°C. As all cuticles tested so far underwent a phase transition between 35 and 50°C, temperatures higher than 30°C must be avoided.

After 1–20 days most leaf discs are digested, and the upper astomatous cuticles can be selected from the slurry. It is better to use a blunt pair of tweezers, as this avoids puncturing the cuticles accidentally. Cuticles are collected in deionised water, which is changed carefully a couple of times to remove enzymes and cellular debris.

The cuticles of some species contain epoxy fatty acids which covalently bind lipophilic organic chemicals with carboxyl groups. Treating the isolated cuticles with 1 mol l^{-1} aqueous HCl converts these epoxy groups to vicinal diols and eliminates binding (Riederer and Schönherr 1986b, 1988). There are no indications that any other property of cuticles is affected, and for this reason treatment with HCl for a day, followed by washing with deionised water until free of chloride ions, is a good routine.

During disintegration of the leaf tissue, fatty acids and phenolic compounds are liberated from cells and sorbed in the cuticles. Most of the sorbed phenolic compounds can be removed by repeated washing with deionised water or with aqueous borax buffer (*p*H around 9.0), but some C16 and C18 fatty acids cannot be removed. These sorbed materials have no effect on permeability of cuticles, but if waxes are to be extracted from isolated cuticles, identified and quantified by GC-MS it should be kept in mind that some of the short-chain fatty acids detected may have been sorbed during isolation (Schönherr and Riederer 1986).

Isolated clean cuticles must be flattened and dried for storage and better handling. Cuticles floating in water are pulled on small pieces of Teflon and flattened with a stream of pressurized air. Initially the stream must be directed to the centre, and from there it is slowly moved towards the edges. This takes care and some experience. Cuticles dry quickly when exposed to ambient air. When dry, thin cuticles stay reasonably flat, and they are collected into vessels which are covered to protect them from exposure to chemicals in the air of the laboratory. Thick fruit cuticles (pepper, tomato, apple) tend to roll up when dry, and it can be difficult to unroll them in order to insert them into the transport apparatus.

It is also essential to distinguish the physiological outer side from the physiological inner side of the isolated cuticle. With some species the outer side appears more shiny, whereas the inner side is more dull. However, this is not the case with all species. In case there are difficulties in identifying the two sides of an isolated cuticle, as an unambiguous test a small drop of water (10–20 μl) can be added on the surface of the cuticle. Water will spontaneously spread when added on the physiological inner side (due to carbohydrates present), whereas water forms droplets with finite contact angles on the outer side (due to the presence of lipophilic cutin and wax).

9.2 Testing Integrity of Isolated CM

Any attempt to relate transport parameters to structure and function of cuticles requires that membranes are intact and that holes and other defects are absent. Before experimenting, all CM should be checked with the light microscope for holes or fissures. However, small holes having diameters of 1–2 μm diameter are easily overlooked. Such tiny defects often occur in *Citrus* CM over oil glands.

Initially we used porometry to test integrity of the CM. The CM was attached to the transport chamber (Fig. 5.4a) with high-vacuum silicon grease (Baysilon, Bayer, Germany) but the cover was not yet applied. A rubber suction cup attached to a 1 ml glass pipette was affixed to the sampling port of the receiver chamber, and in a horizontal position a 1 cm mercury column was injected into the top part of the pipette. With the index finger pressed on the top opening, the pipette with the chamber attached to it was brought into a vertical position, and the CM was slowly and carefully positioned on a piece of soft tissue paper. Only then was the index finger removed. If the mercury column remained stationary for more than 2 min, the membrane was accepted as intact and free of holes. In this case, the pipette was detached from the sampling port, the greased cover was placed on the CM and the transport units were set aside for subsequent experimentation.

As porometry is a rather laborious procedure, we looked for a more convenient method. As before, the CM were mounted to the receiver chamber using silicon grease. A 20 μl droplet of absolute ethanol was placed on the morphological outer surface of the CM. Ethanol perfectly wets the CM, that is the contact angle is zero, and the droplet spreads. If a defect exists, the ethanol quickly penetrates through the hole and spreads on the inner surface of the CM. This can be seen easily by eye, as in incident light the area around the imperfection turns dark. Both tests work only with dry CM, and they yielded identical results. Ethanol did not affect water permeability of the CM, and since the ethanol test is much easier and quicker, it represents the method of choice. Acetone was also tried but turned out unsuitable, as it has a higher surface tension and forms finite contact angles on the outer surface of the CM. With some species (e.g., *Stephanotis*), most CM proofed intact, but with some lots of pear and poplar CM only 20–30% of the CM turned out free of holes and defects.

9.3 Effects of Holes on Permeance, Rate Constants and Diffusion Coefficients

In most experiments, membrane area was about $0.75\,\mathrm{cm}^2$, and a hole of 1 μm radius has an area of $3.14 \times 10^{-8}\,\mathrm{cm}^2$. The area fraction of this hole is 4.2×10^{-8}. The permeance of a water film of 3 μm thickness as calculated from (2.18) is $1.67 \times 10^{-4}\,\mathrm{m\,s^{-1}}$ if D for solutes is $5 \times 10^{-10}\,\mathrm{m^2\,s^{-1}}$ and K is 1. Multiplying this by the area fraction of the hole, we obtain $7 \times 10^{-12}\,\mathrm{m\,s^{-1}}$ for the permeance of the hole. The hole constitutes a parallel pathway, and in this case permeances are additive (Chap. 2). If the permeance measured (P_{total}) is $1 \times 10^{-10}\,\mathrm{m\,s^{-1}}$, the permeance for the hole must be subtracted, and we obtain $9.3 \times 10^{-11}\,\mathrm{m\,s^{-1}}$ for the permeance of the CM. The contribution of the hole to total permeance is 7% for a single tiny hole. Permeances of CM to lipophilic solutes are rarely smaller than $1 \times 10^{10}\,\mathrm{m\,s^{-1}}$ (Tables 6.3 and 6.6), and the contribution of one hole having a radius of only 1 μm to total permeance may be acceptable. But holes larger than that, and more numerous ones, may contribute significantly to P_{total}, and this may go unnoticed if hold-up times are not determined. This is the case when foliar penetration is measured with only one time point.

In some studies, k^* was determined using UDOS (Sect. 6.3), and with many species k^* was as low as 10^{-7} and $10^{-8}\,s^{-1}$ (Fig. 6.21). These rate constants can be converted to permeances P^* by multiplying k^* by membrane thickness ℓ (6.16). If k^* is $1 \times 10^{-8}\,s^{-1}$ and ℓ is $3\,\mu m$, P^* is $3 \times 10^{-14}\,m\,s^{-1}$. P can be obtained by multiplying P^* by the partition coefficient (6.17). For bifenox log K is 4.4, and we obtain P equal to $7.5 \times 10^{-10}\,m\,s^{-1}$. This calculation shows that with lipophilic chemicals small holes in the CM result in few problems, as long as the partition coefficient is very high. With polar non-electrolytes or with ionic solutes having a $K < 1$, P values as small as $10^{-15}\,m\,s^{-1}$ or lower are obtained. Such CM are essentially impermeable to these solutes when penetration is restricted to the waxy pathway.

From the data of Table 6.8 and (6.21), k^* for *Pyrus communis* can be calculated. When the molar volume is $100\,cm^3\,mol^{-1}$ we obtain $3.5 \times 10^{-6}\,s^{-1}$, and a calculation similar to that above yields a permeance P^* of $1.05 \times 10^{-11}\,m\,s^{-1}$ if ℓ is $3\,\mu m$. For ionic solutes K is at least 0.01 or smaller, and this results in a P of $1.05 \times 10^{-13}\,m\,s^{-1}$. A hole in this CM having a permeance of $7 \times 10^{-12}\,m\,s^{-1}$ would be a disaster, as the flux across the hole would be 67 times larger than the flux across the CM. This calculation assumes that aqueous pores are absent from this CM. In this case, almost the total ionic flux observed would be the flux across the hole. With pear CM, ionic solutes penetrate across the aqueous pathway, having rate constants in the range of 1×10^{-5}–$1 \times 10^{-6}\,s^{-1}$ (Fig. 5.7). Hence, P ranges from 3×10^{-11} to $3 \times 10^{-12}\,m\,s^{-1}$. This is of the order of the permeance of the hole of $7 \times 10^{-12}\,m\,s^{-1}$, and by no means negligible.

Extremely low diffusion coefficients (Chap. 6) in the range from 10^{-15} to $10^{-20}\,m^2\,s^{-1}$ or even lower are typical for CM. As a consequence [cf. (2.5)], very long hold-up times in the order of hours or even days are observed (Fig. 6.2). If plots amount diffused vs time intersect the origin or even the positive part of the ordinate, this is an indication for membrane defects, resulting in water–water continuity between donor and receiver. D of water in water is about $2.5 \times 10^{-9}\,m^2\,s^{-1}$, and with a $3\,\mu m$-thick CM the hold-up time for a water layer of this thickness would be $6 \times 10^{-4}\,s$. If cuticular or foliar penetration is studied using only one sampling interval, the hold-up time is not known and cannot be used as an integrity criterion.

9.4 Distribution of Water and Solute Permeability

The tremendous variability of permeance and rate constants among individual CM of a given species poses the question as to the appropriate statistics for characterising populations. This was studied by Baur (1997),who investigated distribution of permeance and rate constants of CM, MX and leaf discs for water and lipophilic solutes.

Population size ranged from 50 to 750 CM. With MX membranes distribution were normal, and arithmetic means, standard deviations or confidence intervals can be used to characterise these MX membranes. With CM from all species, histograms of frequency distributions of k^* or P were always skewed and had a pronounced tail at high values. Skewness tended to be more pronounced with small populations. After log transformation, symmetrical histograms with normal distributions were obtained. If populations are skewed, arithmetic means are higher than geometric means obtained by log transformation of the data. For this reason, it is a good practise to use large sample sizes of 50–100 CM and to subject data to log transformation for calculating geometric means and confidence intervals. Log transformation of normal distributions presents no problems, as in this case arithmetic and geometric means are numerically identical.

9.5 Very High or Very Low Partition Coefficients

Determination of partition coefficients, although technically easy and straightforward, can cause serious experimental problems with highly lipophilic ($K_{cw} > 10^4$) or very polar ($K_{cw} < 1$) compounds. These problems have been dealt with in detail in Sect. 6.1.6. Possible solutions when working with highly lipophilic and polar compounds, insurmountable difficulties, and problems in data interpretation have been presented and discussed.

It is very difficult or impossible to determine K_{ow} for emulsifiers, because in their presence oil in water emulsions can be formed and phase separation may not be possible. No such problems arise with MX or CM, and reliable partition coefficients have been determined (Riederer et al. 1995). As there is an excellent correlation between K_{ow} and K_{CM} at low solute concentrations, the problem of emulsification can be circumvented by using pieces of MX rather than octanol as lipid phase.

9.6 Cutin and Wax Analysis and Preparation of Reconstituted Cuticular Wax

Gas chromatography with flame ionisation detection (GC-FID) was used for the quantification of cuticular waxes (Riederer and Schneider 1989) and cutin monomers after depolymerisation (Hauke and Schreiber 1998). Identification of constituents by gas chromatography coupled to mass spectrometry is the method of choice. Experimental problems arise when cutan occurs in the polymer matrix (incomplete depolymerisation) or when high molecular weight oligomers occur in wax. They have been discussed in Sect. 1.3.

Solute sorption and diffusion in wax was studied using reconstituted wax preparations (Sects. 6.1.3 and 6.5). The experimental setup is shown in (Fig. 9.1).

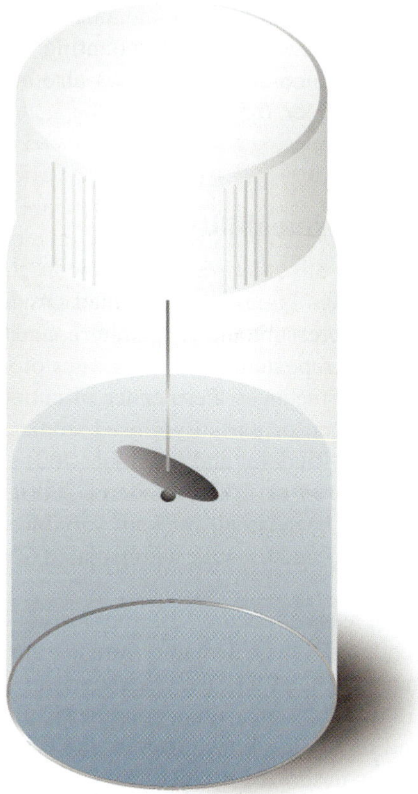

Fig. 9.1 Experimental setup used for measuring sorption or diffusion coefficients. An aluminium disk covered with reconstituted wax was affixed with a needle to the screw cap of a glass vial. The aluminium disk was immersed in an aqueous solution for either loading the wax with a radio-labelled compound (determination of K_{ww}) or for desorption of a radio-labelled compound from wax (determination of D)

Aluminium disks (diameter 8 mm) affixed to stainless steel needles were immersed in a chloroform solution of wax (20–50 mg ml^{-1}). After evaporation of chloroform, the aluminium disks were uniformly covered with microcrystalline wax (150–250 µg per 100 mm^2). However, when immersed in an aqueous desorption medium the wax crystallites were rapidly detached from the aluminium disks. This was overcome by heating them for 5 min to 100 °C in an oven. The wax melted, and at room temperature it solidified and formed a uniform smooth layer of reconstituted wax, which adhered to the aluminium disk during subsequent experimentation.

In most cases, radio-labelled solutes were added directly to the chloroform/wax solution and were reconstituted together with the wax. These samples can be used for the determination of D by desorption experiments. This type of loading of wax is called "internal loading". For the determination of K_{ww}, wax samples were reconstituted, and loaded with radio-labelled solutes by immersing them in solutions with the radio-labelled solutes. After equilibration, the amounts of radioactivity in both phases were determined by liquid scintillation counting. This type of loading is called "external loading". Diffusion coefficients may also be determined following external loading of waxes (see Sect. 6.5.1).

9.7 Measuring Water Permeability

Water loss from detached leaves is easily determined gravimetrically. Leaves are exposed to a constant vapour pressure and temperature, and rates of water loss are determined by weighing them repeatedly. Dividing rates of water loss by leaf area and driving force results in permeances. Permeance of cuticles is obtained by correcting for water loss from open or partially closed stomata (Kerstiens 1996a, b; Schreiber and Riederer 1996a; Riederer and Schreiber 2001). When using detached leaves or leaf disks (Schreiber et al. 2001; Riederer 2006), a mechanistic analysis of permeances is difficult, because all relevant variables cannot be controlled. This is the reason why we preferred to work with isolated CM and MX membranes whenever possible.

The cup method is the easiest method of measuring water permeability of CM (Schönherr and Lendzian 1981). Isolated cuticles are mounted in stainless steel transpiration chambers (Fig. 9.2). The chambers are kept at constant temperature, and the loss of water is measured gravimetrically. Linear transpiration kinetics are obtained, and regression lines intersect the origin of the axis (e.g. see Fig. 4.19). Hold up-times as described in Sect. 2.1.1 (Fig. 2.2) cannot be detected by this method. After inserting the CM, transpiration chambers are incubated until

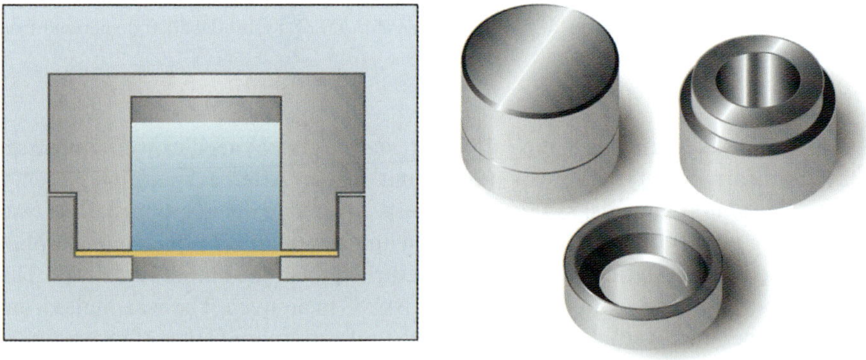

Fig. 9.2 A stainless steel transpiration chamber consisting of donor chamber and lid. The area of the orifice in the lid is 1.13 cm^2. The chamber is filled with 800 μl water (*blue colour* inside the chamber). An isolated CM (*yellow*) is mounted between lid and donor chamber using silicon grease. Chambers are placed in boxes containing silica gel beads, boxes are closed and incubated at constant temperature. Water loss from the chamber is determined using a balance (accuracy ±0.1 mg)

temperature equilibrium and steady state are established before weighing is started. Depending on water permeability, chambers are weighed at intervals of hours or even days to get a significant loss of mass. These intervals are too long for detecting extrapolated hold-up times in the range of 4–15 min (Table 4.10).

When we used the cup method, the CM were inserted such that the morphological inner side of the CM was in contact with water while the outer side was exposed to dry air. In this arrangement, driving force was maximum ($\Delta a_{\mathrm{w}} = 1.0$; $\Delta C_{\mathrm{wv}} = 23.05\,\mathrm{g\,m^3}$ at 25 °C). With most plant species, swelling of cuticles and permeance depend on partial vapour pressure (Fig. 4.6). When in equilibrium with silica gel the limiting skin is not or very little swollen, and the cup method results in minimum permeances. Humidity in the box does not have to be zero. If constant humidity saturated salt solutions (cf. Greenspan 1977; Kolthoff et al. 1969; Appendix B) or glycerol/water mixtures (Slavik 1974) are used instead of silica gel, humidity can be varied and effects of humidity on permeance can be studied. We have not tried this, however. The cup method has been used extensively because it is simple, cheap, and very accurate, and large sample sizes can be handled (Geyer and Schönherr 1990; Mérida et al. 1981; Riederer and Schönherr 1990; Schönherr and Bauer 1992; Schreiber and Riederer 1996a, b).

The cup method resembles the situation in nature, except that in the field humidity is rarely close to zero. It has some other limitations. Water fluxes at 100% humidity cannot be measured, because driving force is zero and no net flux occurs. In this situation, tritiated water (THO) was used as tracer (Schönherr and Schmidt 1979). An aqueous buffer containing THO served as donor in contact with the morphological inner surface of the CM. Air of constant humidity obtained by the dew point method was blown over the outer surface of the CM. The moisture containing the THO which had penetrated was trapped in a dry-ice cold trap (−70 °C), and radioactivity in the cold trap was assayed by scintillation counting (Figs. 4.6 and 4.8). Schreiber et al. (2001) also studied the effect of water activity of the receiver using a slightly different approach. The stainless steel transpiration chamber, with THO as donor solution facing the morphological inner side of the cuticle, was attached upside down to a scintillation vial (Fig. 9.3). Humidity of the receiver (the scintillation vial) facing the morphological outer side of the CM was varied using 100 µl glycerol (2% RH), pure water (100% RH) or mixtures of glycerol and water having humidities between 2 and 100% RH (Slavik 1974). The surface of the CM was in contact with the equilibrium vapour pressures of the glycerol/water mixtures, but not with the solutions themselves, which were on the bottom of the vials (500 µl). THO which penetrated the cuticle was absorbed in the glycerol, water or glycerol/water mixtures, and after adding scintillation cocktail, radioactivity was determined by liquid scintillation counting.

The effect of cations and *p*H on water permeability of CM or MX membranes was studied using the system buffer/CM or MX/buffer and tritiated water as tracer (Schönherr 1976a, b). If a fine capillary is attached to the receiver containing an osmotic solute, this system can be used to measure osmotic water transport in MX membranes (Fig. 4.9), but it is not sufficiently sensitive for CM (Schönherr 1976a; Schönherr and Mérida 1981).

Fig. 9.3 Experimental setup used for measuring the effect of humidity on water permeability. Stainless steel transpiration chambers containing radio-labelled water were mounted upside down on scintillation vials. Inside the vials, various humidities were established using glycerol (2% humidity), glycerol water mixtures (humidity between 2 and 100%) or pure water (100% humidity)

9.8 Measuring Solute Permeability

Solute permeability depends on partition coefficients and diffusion coefficients in the cuticle (2.18). Cuticles are heterogeneous membranes, and waxes determine solute permeability. Hence, mobility and sorption of solutes in waxes must be studied, but interactions between waxes and the polymer matrix are also important (Chap. 6). The role of aqueous pores in penetration of ionic solutes is discussed in Chap. 5, and problems related to partitioning and sorption are treated in

Sects. 6.1 and 9.5. During our studies of permeability of plant cuticles, a number of new methods have been developed, some which proved useful and are summarised here. Further details not mentioned in Chaps. 1–8 are also provided.

Mass flow can be studied either in the steady (Sect. 6.2.1) or in the transient state (Sect. 6.3). Detached leaves or conifer needles can be studied using the submersion technique (Sect. 6.2.2). The well technique is more convenient with leaf disks from broad-leaved plants (Sect. 6.2.3). Merits and drawbacks associated with the above methods are discussed in Chap. 6, and need not be repeated here.

Our analysis of barrier properties of cuticles draws heavily on data collected using isolated cuticular membranes (CM). This makes it possible to measure penetration in the transient state (Sects. 2.5, 6.3 and 6.4). Solute mobility in the limiting skin and in the sorption compartment of the CM can also be studied quantitatively. These methods are very powerful for analysing the reasons for differences in permeability among species and solutes (Sect. 6.3.2). The role of accelerators (plasticisers) in cuticular penetration also relies on these methods (Chap. 7). This warrants a closer look at these new methods. All of them rely on radio-labelled solutes.

The CM are equilibrated with an aqueous solution containing the radio-labelled solute. Solutions must be buffered if the solute is a weak acid or base. After equilibration most solutes are contained in the sorption compartment, which has also the largest mass (Sect. 1.4). The limiting skin is very thin, and only a small percentage of the total amount sorbed is located there (Fig. 9.5). Solutes sorbed in the CM can be desorbed using an apparatus shown in Fig. 9.4. For simultaneous bilateral desorption (BIDE) the CM is inserted into the apparatus, and the two compartments are filled with the desorption medium (water, buffer or phospholipid suspension, depending on properties of solute). The receiver solutions are withdrawn periodically and exchanged for fresh ones. Radioactivity of receiver solutions is determined by scintillation counting. Typical results are shown in Fig. 6.15. From the desorption kinetic, the average rate constants (6.13) and diffusion coefficients in the limiting skin and the sorption compartment can be estimated [(2.33) and (2.35)]. Desorption from the inner surface is very rapid, and samples must be taken in very short intervals.

If desorption media are added only to the half-cell facing the morphological inner surface of the CM, the process is called unilateral desorption from the inner surface (UDIS). Diffusion of solutes into the sorption compartment (soco) of the CM is very rapid,and sampling intervals of 1–3 min are necessary to estimate diffusion coefficients from the initial rates of desorption (Fig. 6.20). These methods are depicted schematically in Fig. 9.5.

Solute mobility in the limiting skin is measured using unilateral desorption from the outer surface (UDOS). A simpler method of loading the CM with solutes is used. The CM are inserted in a simple apparatus (shown in Fig. 6.17). A droplet (200 μl) of aqueous donor solution is added to the inner surface of the CM exposed in the orifice of the cover and allowed to dry. During evaporation of water at ambient temperature and humidity, the solute is driven into the sorption compartment of the CM, as solute concentration in the droplet increases during evaporation of water. An experiment with ^{14}C-labelled solute and THO showed that the ratio $^{14}C/^{3}H$

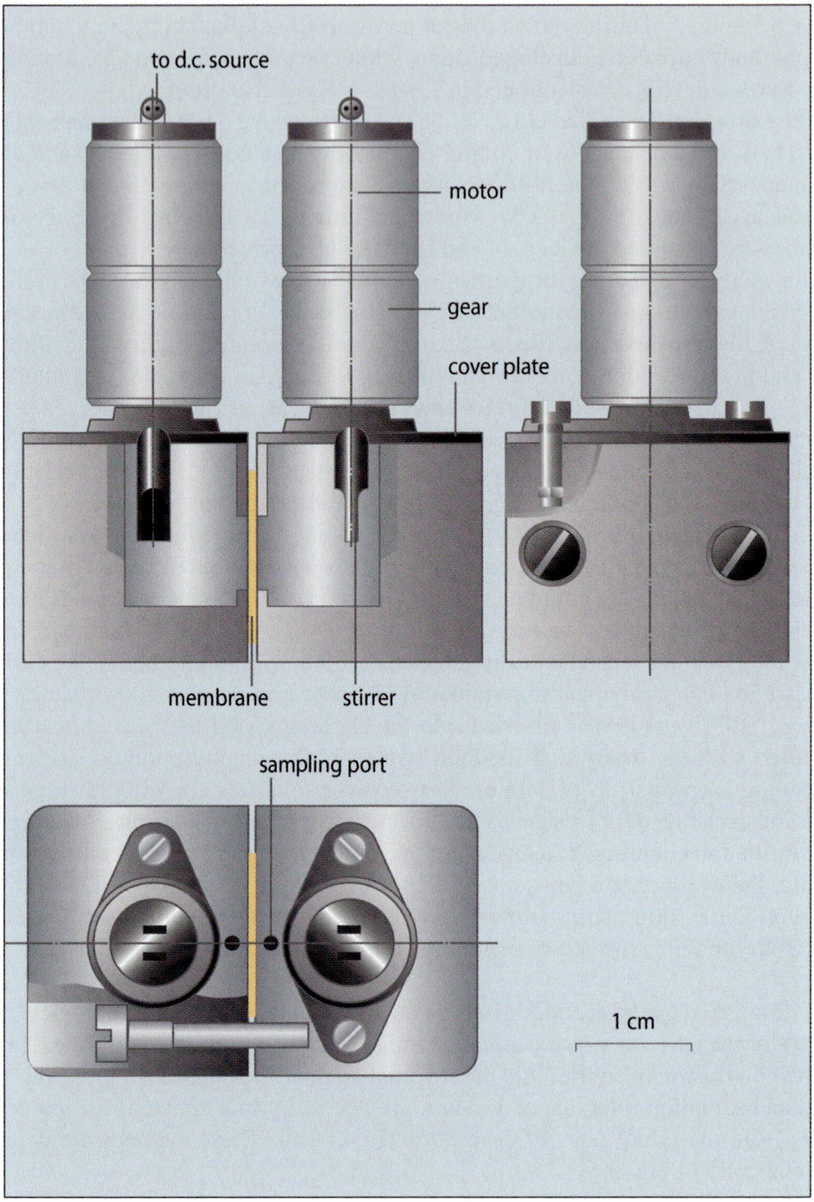

Fig. 9.4 Apparatus made of stainless steel used for unilateral or bilateral desorption (UDIS or UDOS). The CM (*yellow*) loaded with radio-labelled solute is inserted between the two halves of the apparatus, desorption media are added through the sampling ports and solutions are stirred for mixing. Desorption media are withdrawn periodically and fresh ones are added. The apparatus in maintained at constant temperature

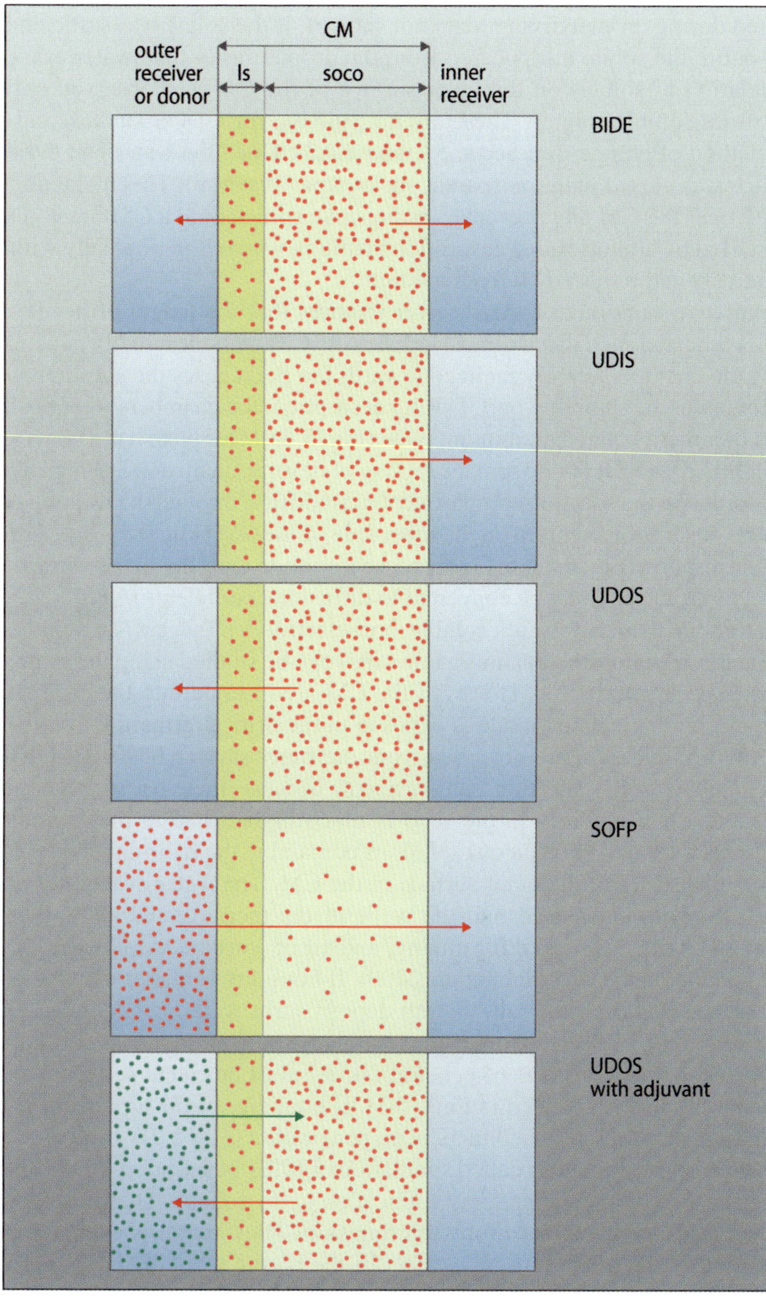

Fig. 9.5 Schematic drawing showing initial distribution of solutes (red dots) in CM consisting of the limiting skin (*ls*) and the sorption compartment (*soco*). In BIDE, UDIS and UDOS experiments, most of the solute molecules are initially in the soco and very few in the limiting skin. In SOFP, the solutes are added to the outer surface of the limiting skin and they form a concentrated solution when they deliquesce. The effect of adjuvants (green dots) is studied by adding the adjuvant to the desorption medium. The directions of solute (red arrow) and adjuvant flows (green arrow) are indicated by *arrows*

decreased during droplet drying (data not shown). If the solute was sufficiently soluble in cutin, the solute disappeared from the droplet faster than water evaporated. Accumulation of solutes on the inner surface of the CM was observed only with highly water soluble solutes. These are not suitable for UDOS studies, and SOFP is the method of choice (see Sects. 5.2 and 7.4.3). Once the water had evaporated the orifice was closed with transparent sticky tape (Tesafilm). This maintains 100% humidity over the CM and prevents a radioactive spill should a CM break. Desorption is started by adding 500 µl desorption medium, which is periodically withdrawn quantitatively and replaced by fresh medium.

Quantitative sampling of the receiver requires that the length of needle of the syringe is adjusted such that it almost touches the CM without actually puncturing it. Holding the chamber slightly inclined towards the port causes the sorption medium to collect under the sampling port. During desorption, the chambers were positioned in wells of a thermostated aluminium plate (Fig. 5.4a). Our plate had 100 wells, and it was lightly rocked at 60–80 rpm for mixing of desorption media. During UDOS all solutes sorbed in the sorption compartment must diffuse through the limiting skin, as there is no desorption medium on the inner side of the CM (Fig. 9.5). Rate constants (k^*) of desorption (Fig. 6.18; 6.13) characterise solute mobility in the waxy limiting skin (Sect. 6.3.2), and may be converted to diffusion coefficients (6.20).

Penetration of highly water soluble non-electrolytes (sugars) or ionic solutes (amino acids, glyphosate salts, inorganic salts) can be studied using the same chambers and experimental setup. However, the CM are inserted into the chamber such that the morphological outer side is exposed in the orifice of the lid. Solutes penetrating the limiting skin are desorbed from the inner surface of the CM. They do not accumulate in the sorption compartment, because their partition coefficients are <1 and solute mobility in the soco is much higher than in the limiting skin. Small droplets (2–5 µl) of aqueous solutions containing the radio-labelled solute are placed on the centre of the outer surface of the CM. The water evaporates quickly, and the chambers are placed into the wells of the thermostated aluminium plate (Fig. 5.4). The plate is rocked for mixing, and air of constant humidity is blown on the salt residue on the CM (Schönherr 2000). If humidity is higher than the point of deliquescence (POD) of the salt, the salt deposit sorbs water and deliquesces. Penetration into the CM occurs from a concentrated donor of unknown concentration. For this reason, rate constants of penetration (k) must be calculated from amounts (M) of salt rather than concentrations (5.1). If humidity of the air stream is below POD, the salt remains solid and there is no penetration (Fig. 5.7).

This type of experiment is called simulation of foliar penetration (SOFP) because it resembles the situation after spray application to the foliage. These rate constants (k) characterise permeability of the limiting skin. They are proportional to permeance (P) rather than D because the effect of the partition coefficient is not known and cannot be corrected for (6.14)–(6.17). In these experiments, temperature control is critical because humidity of the air stream depends on temperature of the aluminium plate. When POD of the salt is close to 100%, the donor side can be closed with Tesafilm and humidity over the salt residue quickly reaches 100% due to penetration of water from the receiver. In SOFP experiments, area after droplet drying is not known exactly. As the area is not needed to calculate rate constants (5.1), this is no problem. Adding a surfactant to the donor solution improves contact between the aqueous solution and the lipophilic waxy cuticle. This greatly increases rate constants (Fig. 5.6).

Surfactants and other adjuvants affect partition coefficients (Sect. 7.1) of lipophilic solutes. If surfactants and solutes are both contained in the donor, data interpretation can be difficult, because adjuvants affect both partition coefficients and solute mobility in the limiting skin. These effects can be separated by adding the adjuvants to the desorption media. When solutes are desorbed, the adjuvants penetrate the CM from the outer surface and quickly accumulate in the limiting skin and the sorption compartment. They accumulate in the CM, from which they cannot escape (Fig. 9.5). There is no effect of adjuvants on driving force, because solute concentration in the soco is the driving force (Fig. 9.5). The adjuvant concentration in the CM can be calculated from the partition coefficient and the concentration in the receiver (Schönherr et al. 2001; Shi et al. 2005a, b). If solute mobility in the control (no adjuvant) is smaller than in the treatment (adjuvant), the adjuvant is called accelerator (Chap. 7).

Appendix A

SI units used, fundamental constants, abbreviations, and symbols

SI units used

Quantity	SI unit	Symbol	Magnitude
Energy	Joule	J	$kg\,m^2\,s^{-2}$
Force	Newton	N	$kg\,m\,s^{-2} = J\,m^{-1}$
Pressure	Pascal	Pa	$N\,m^{-2}$
Electric potential	Volt	V	$J\,A^{-1}s^{-1}$

Fundamental constants

Constant	Symbol	SI unit	CGS
Avogadro's constant	N_0	$6.022 \times 10^{23}\,mol^{-1}$	$6.022 \times 10^{23}\,mol^{-1}$
Boltzmann's constant	κ	$1.381 \times 10^{-23}\,J\,K^{-1}$	1.381×10^{-16} erg degree^{-1}
Molar gas constant	R	$8.314\,J\,K^{-1}\,mol^{-1}$	$1.987\,cal\,mol^{-1}\,K^{-1}$
Faraday constant	$\mathcal{F}$	9.649×10^4 coulomb mol^{-1}	$2.306 \times 10^4\,cal\,mol^{-1}\,V^{-1}$
Planck's constant	h	$6.626 \times 10^{-34}\,J\,s$	6.626×10^{-27} erg s

Abbreviations

Abbreviation	Description
CA	Cellulose acetate
CL	Cuticular layer(s), layer composed of cutin, cutan and polar polymers located between CM and cell wall
CM	Isolated cuticular membrane

(*continued*)

Abbreviations (*continued*)

Abbreviation	Description
cmc	Critical micelle concentration
CP	Cuticle proper, outer layer of cuticles, often laminated
CU	Cutin, obtained by hydrolysing MX in 6 N HCl at 120°C
DSC	Differential scanning calorimetry
EC	Ethyl cellulose
ESR	Electron spin resonance
GC	Gas chromatography
MS	Mass spectroscopy
MX	Polymer matrix, obtained by extracting waxes from CM
NMR	Nuclear magnetic resonance
PE	Polyethylene
PEMA	Polyethyl methacrylate
PET	Polyethylene terephthalate
PMA	Polymethyl methacrylate
POD	Point of deliquescence
PP	Polypropylene
PVA	Polyvinyl acetate
SCL	Soluble cuticular lipids or cuticular waxes
SOFP	Simulation of foliar penetration
TLC	Thin layer chromatography
UDIS	Unilateral desorption from the inner surface
UDOS	Unilateral desorption from the outer surface

Symbols

Symbol	Quantity	Magnitude
a	Activity	–
a_{w}	Activity of liquid water	–
a_{wv}	Activity of water vapour	–
A	Area	m^2
C	Concentration	$\mathrm{mol\,m^{-3}}$ or $\mathrm{mol\,l^{-1}}$
C_{s}	Solute concentration	$\mathrm{mol\,m^{-3}}$
C_{w}	Concentration of water	$\mathrm{kg\,m^{-3}}$ or $\mathrm{g\,cm^{-3}}$
C_{wv}	Concentration of water vapour	$\mathrm{kg\,m^{-3}}$ or $\mathrm{g\,cm^{-3}}$
$C_{\mathrm{w}}^{\mathrm{polymer}}$	Concentration of water in polymer	$\mathrm{kg\,m^{-3}}$ or $\mathrm{g\,cm^{-3}}$
D	Diffusion coefficient	$\mathrm{m^2\,s^{-1}}$
d	Ratio of D of water in a helium and nitrogen atmosphere ($D_{\mathrm{wH}}/D_{\mathrm{wN}}$)	–
δ	Density	$\mathrm{kg\,m^{-3}}$ or $\mathrm{g\,cm^{-3}}$
δ(STP)	Density of water vapour at standard temperature (0°C) and pressure (10^5 Pa)	$8.03\times10^{-4}\ \mathrm{g\ cm^{-3}}$
E_{D}	Energy of activation of diffusion	$\mathrm{kJ\,mol^{-1}}$
E_{P}	Energy of activation of permeance	$\mathrm{kJ\,mol^{-1}}$
F	Flow	$\mathrm{mol\,m^{-2}}$ or $\mathrm{g\,cm^{-2}}$
γ	Activity coefficient	–
ΔG	Change in Gibbs free energy	$\mathrm{kJ\,mol^{-1}}$

(*continued*)

Symbols (*continued*)

Symbol	Quantity	Magnitude
ΔH	Enthalpy change	$kJ\,mol^{-1}$
η	viscosity	Pa s
J	Mass flux	$kg\,m^{-2}\,s^{-1}$
J_v	Volume flux of gas or vapour	$m^3(STP)m^{-2}\,s^{-1}$
J_s	Solute flux	$mol\,m^{-2}\,s^{-1}$
J_w	Mass flux of water	$kg\,m^{-2}\,s^{-1}$
J_{wv}	Mass flux of water vapour	$kg\,m^{-2}\,s^{-1}$
K	Partition coefficient	–
K_{Hg}	Water partition coefficient calculated as P_{Hg}/D	–
K_w	Water partition coefficient	cm^3 liquid water cm^{-3} polymer
K_{wrec}	Solute partition coefficient between CM, MX, or wax and aqueous solutions	–
K_{wv}	Water vapour partition coefficient	cm^3 vapour cm^{-3} polymer
K_{ow}	Octanol/water partition coefficient	–
K_{cw}	CM/water partition coefficient	–
K_{mxw}	MX/water partition coefficient	–
K_{cuw}	Cutin/water partition coefficient	–
K_{ww}	Wax/water partition coefficient	–
$\mathscr{k}$	Mass transfer coefficient	s^{-1}
$\mathscr{k}^*$	UDOS mass transfer coefficient	s^{-1}
$\mathscr{k}^*_o$	UDOS mass transfer coefficient of a solute having zero molar volume	s^{-1}
$\mathscr{k}^*_{infinite}$	Pre-exponential factor of the Arrhenius equation ln $\mathscr{k}^*$ vs. $1/T$	s^{-1}
k	Slope of the Freundlich isotherm	
k'	Slope of the modified Freundlich isotherm	
ℓ	Membrane thickness	m
λ	Mean free path of solute in polymers	m
m	Mass	kg
M	Amount	mol
μ	Chemical potential of solvent or solute	–
N	Mole fraction	–
$\mathscr{P}$	Permeability of a membrane of 1 m thickness	$m^2\,s^{-1}$
$\mathscr{P}_w$	Water permeability of a membrane having 1 m thickness calculated using a_w as driving force	$m^2\,s^{-1}$
$\mathscr{P}_{wv}$	Water vapour permeability of a membrane having 1 m thickness calculated using C_{wv} as driving force	$m^2\,s^{-1}$
P	Permeance of a membrane	$m\,s^{-1}$
P_{Hg}	Permeance calculated using a vapour pressure of 1 cm Hg as driving force. Usually for a membrane having $\ell = 1\,cm$	cm^3 vapour (STP) $cm^{-2}\,s^{-1}$
P_w	Water permeance calculated using a_w as driving force	$m\,s^{-1}$
P_{wv}	Water vapour permeance calculated using water vapour concentration (C_{wv}) as driving force	$m\,s^{-1}$

(*continued*)

Symbols (*continued*)

Symbol	Quantity	Magnitude
p	Pressure or vapour pressure	Pa or cm Hg
pK_a	Negative logarithm (log) of ionisation constants of acidic groups	–
pK_b	Negative logarithm (log) of ionisation constants of basic groups	–
p/p_o	Partial pressure	–
$\mathscr{R}$	Resistance	$s\,m^{-1}$
S_{water}	Water solubility	$mol\,kg^{-1}$ or $mol\,l^{-1}$
ΔS	Change in entropy	$kJ\,mol^{-1}$
$\mathscr{S}$	Vapour sorption coefficient calculated as P_{Hg}/D	–
σ'	Size selectivity for ionic compounds	$mol\,cm^{-3}$
β'	Size selectivity for neutral compounds	$mol\,cm^{-3}$
t_e	Hold-up time	S
$\bar{t}$	Transference number	$eq/\mathscr{F}$
T	Temperature	°C or Kelvin
τ	Tortuosity factor	–
V	Volume	m^3 or l
$\overline{V}$	Partial molar volume	$cm^3\,mol^{-1}$
V_x	Equivalent molar volume of solutes	$cm^3\,mol^{-1}$
y	$(D_{wH}/D_{wN})/(D_{wH}/D_{wN}-1)$	–
z	Valence or number of charges of ions	–

Appendix B

Physico-Chemical Properties of Chemicals

Physico-chemical properties of organic chemicals

Name	Abbreviation	CAS	MW[a]	V_x^b	log K_{cw}
Atrazine	AT	1912-24-9	216	162	2.16[c]
Benzoic acid	BA	65-85-0	122	93	1.71[d]
Betaine	$C_5H_{11}NO_2$	107-43-7	117	88	–
Bifenox	$C_{14}H_9Cl_2NO_5$	42576-02-3	342	216	4.44[e]
Bitertanol	BIT	55179-31-2	337	267	3.87[f]
Cholesterol	CHOL	57-88-5	387	349	~8[g]
2,4-Dichlorophenoxyacetic acid	2,4-D	94-75-7	221	138	2.61[c]
2,4-Dichlorophenoxy-butyric acid	2,4-DB	94-82-6	249	205	3.05[g]
Diethylhexylphtalate	DEHP	117-81-7	390	340	7.69[c]
2,4,5-Trichloro-phenoxyacetic acid	2,4,5-T	93-76-5	255	166	3.16[c]
Hexachlorobenzene	HCB	118-74-1	285	222	5.77[c]
Indoleacetic acid	IAA	87-51-4	175	130	1.16[g]
Iprovalicarb	IPRO	140923-17-7	320	296	3.18[h]
Lindane	LIN	58-89-9	291	158	3.80[i]
Methyl glucose	M-Gluc	13224-94-7	194	153	−0.87[e]
Metribuzin	MET	21087-64-9	214	162	1.48[d]
1-Naphthaleneacetic acid	NAA	86-87-3	186	144	2.25[f]
4-Nitrophenol	4-NP	100-02-7	139	95	1.87[c]
Paraquat	$C_{12}H_{14}N_2$	4685-14-7	186	157	–
Pentachlorophenol	PCP	87-86-5	266	139	4.56[c]
Perylene	PER	189-55-0	252	263	6.36[c]
Tebuconazole	TB	107534-96-3	308	241	3.54[g]
Salicylic acid	SA	69-72-7	138	99	2.03[d]

(*continued*)

Physico-chemical properties of organic chemicals (*continued*)

Name	Abbreviation	CAS	MW[a]	V_x^b	$\log K_{cw}$
Octadecanoic acid	$C_{18}AC$	57-11-4	284	272	8.20[j]
Tetracosanoic acid	$C_{24}AC$	557-59-5	369	356	11.33[j]
Triadimenol	TRI	55219-65-3	296	219	3.37[f]
Urea	$(NH_2)_2CO$	57-13-6	60	47	−2.11[k]

[a]Molecular weight ($g\,mol^{-1}$); [b]Equivalent molar volume ($cm^3\,mol^{-1}$); [c]Kerler and Schönherr (1988a); [d]Kirsch et al. (1997); [e]Baur et al. (1997b); [f]Sabljic et al. (1990); [g]Buchholz et al. (1998) and Buchholz and Schönherr (2000); [h]Shi et al (2005a); [i]Schreiber and Schönherr (1992b); [j]$\log K_{ow}$ calculated from data by Dunn et al. (1986); [k]$\log K_{ow}$ from Sangster (1997)

Anionic salts and ionic organic chemicals

Name	Abbreviation	CAS	MW (g)	POD (%)
Ammonium ferric citrate	$FeC_6H_8O_7NH_3$	1185-57-5	254	~100[a]
Berberin sulfate	BS	633-66-9	769	–
Betaine	$C_5H_{11}NO_2$	107-43-7	117	<50[b]
Calcium acetate	$Ca(H_3C_2O_2)_2$	62-54-4	158	>80[a]
Calciumchloride × $6H_2O$	$CaCl_2$	10043-52-4	111	29[c]; 32[a]
Calcium dicitrate	$Ca_3(C_6H_5O_7)_2$	5785-44-4	499	~100[a]
Calcium gluconate	$C_{12}H_{22}CaO_{14}$	299-28-5	430	~100[a]
Calcium glutamate	$C_{10}H_{16}CaN_2O_8$	19238-49-4	332	~100[a]
Calcium heptagluconate	$C_{14}H_{26}CaO_{16}$	10030-53-2	490	~100[a]
Calcium hydroxide	$Ca(OH)_2$	1305-62-0	74	–
Calcium lactate	$C_6H_{10}CaO_6$	41372-22-9	218	>80[a]
Calcium lactobionate	$(C_{12}H_{21}O_{12})_2$ Ca	110638-68-1	755	~100[a]
Calcium nitrate × $4H_2O$	$Ca(NO_3)_2$	35054-52-5	164	50[c]; 56[a]
Calcium pantothenate	$C_{18}H_{32}CaN_2O_{10}$	137-08-6	477	~100[a]
Calcium propionate	$(CH_3CH_2COO)_2Ca$	4075-81-4	186	90[d]
Ferric citrate	$FeC_6H_8O_7$	2338-05-8	263	~100[a]
Ferric chloride	$FeCl_3 \times 6H_2O$	7705-08-0	162	44[d]
Ferric ethylenediaminetetraacetic acid	FeEDTA	15708-41-5	367	~100[a]
Ferric imidosuccinic acid	FeIDHA	–	393	~100[a]
Ferric ligninsulfonic acid	Natrel	–	507	~100[a]
Mercury chloride	HgCl	7487-94-7	272	–
Isopropyl amine glyphosate	IPA-GLY	38641-94-0	228	<50[a]
Potassium carbonate × $2H_2O$	K_2CO_3	584-08-7	138	45[c]
Potassium chloride	KCl	7447-40-7	75	85[c]
Potassium nitrate	KNO_3	7757-79-1	101	94[c]
Potassium phosphate	KH_2PO_4	7778-77-0	136	95[d]
Proline	$C_5H_9NO_2$	609-36-9	115	70[d]
Putrescine	$C_4H_{12}N_2$	110-60-1	88	78[d]
Sequestrene 138Fe	FeEDDHA	–	368	~100[a]
Silver chloride	AgCl	7783-90-6	143	–

[a]Estimated from dependence on humidity of permeance; [b]POD at 25°C; [c]Kolthoff et al. (1969); [d] Our own measurements at 20–25°C

Monodisperse surfactants and other plasticisers

Name	Abbrev.	CAS	MW^{a}	V_x^{b}	$\log K_{ww}^{c}$
Monodisperse alcohol ethoxylates					
Diethylene glycol monobutyl ether	C_4E_2	112-34-5	162	141	−1.08
Triethylene glycol monohexyl ether	C_6E_3	25961-89-1	234	213	−0.08
Tetraethylene glycol monooctyl ether	C_8E_4	19327-39-0	306	266	0.69
Pentaethylene glycol monodecyl ether	$C_{10}E_5$	23244-49-7	379	332	1.55
Octaethylene glycol monodecyl ether	$C_{10}E_8$	24233-81-6	511	430	0.75
Diethyleneglycol monododecyl ether	$C_{12}E_2$	3055-93-4	274	254	3.30
Triethyleneglycol monododecyl ether	$C_{12}E_3$	3055-94-5	318	288	3.11
Tetraethylene glycol monododecyl ether	$C_{12}E_4$	5274-68-0	363	322	2.82
Pentaethylene glycol monododecyl ether	$C_{12}E_5$	3055-95-6	407	356	2.60
Hexaethylene glycol monododecyl ether	$C_{12}E_6$	3055-96-7	451	390	2.30
Heptaethylene glycol monododecyl ether	$C_{12}E_7$	3055-97-8	495	424	2.20
Octaethylene glycol monododecyl ether	$C_{12}E_8$	3055-98-9	539	458	2.02
Triethylene glycol monotetradecyl ether	$C_{14}E_3$	26826-30-2	347	316	–
Pentaethylene glycol monotetradecyl ether	$C_{14}E_5$	92669-01-7	435	384	–
Hexaethylene glycol monotetradecyl ether	$C_{14}E_6$	5157-04-0	479	418	–
Heptaethylene glycol monotetradecyl ether	$C_{14}E_7$	40036-79-1	523	452	3.19
Octaethylene glycol monotetradecyl ether	$C_{14}E_8$	27847-86-5	567	486	3.08
Triethylene glycol monohexadecyl ether	$C_{16}E_3$	4484-59-7	374	328	–
Octaethylene glycol monohexadecyl ether	$C_{16}E_8$	5698-39-5	595	515	4.09
Dicarboxylic acid esters					$\log K_{wrec}^{d}$
Dimethyl suberate	DMSU	1732-09-8	202	167	1.19
Diethyl suberate	DESU	2050-23-9	230	194	1.75
Diethyl sebacate	DES	110-40-7	258	213	2.65
Dibutyl suberate	DBSU	16090-77-0	286	241	3.73
Dibutyl sebacate	DBS	109-43-3	314	269	4.46

[a]Molecular weight ($g\,mol^{-1}$)
[b]Equivalent molar volume ($cm^3\,mol^{-1}$)
[c]Partition coefficient barley wax/water K_{ww} (Schreiber et al. 1996b; Burghardt et al. 1998)
[d]Partition coefficient *Stephanotis* wax/aqueous solution containing $100\,g\,l^{-1}$ propane-1,2-diol K_{wrec} (Simanova et al. 2005)

Appendix C

Index of Plants

Botanical name	Common name	Organ	Family
Abies alba Mill.	Silver fir	Needle	Pinaceae
Abies koreana E.H. Wilson	Korean fir	Needle	Pinaceae
Allium cepa L.	Onion	Bulb scale	Liliaceae
Camellia sinensis (L.) Kuntze	Tea-plant	Leaf	Theaceae
Capsicum annuum L.	Bell pepper	Fruit	Solanaceae
Citrus aurantium L.	Seville orange	Leaf	Rutaceae
Citrus limon (L) Burm. F.	Lemon	Leaf	Rutaceae
Clivia miniata (Lindl.) Bosse	Kaffir lily	Leaf	Amaryllidacea
Cucumis sativa L.	Cucumber	Fruit	Cucurbidaceae
Euonymus japonica Thunb.	Evergreen euonymus	Leaf	Celastraceae
Ficus benjamina L.	Benjamin's-tree	Leaf	Moraceae
Ficus elastica var. *decora* Roxb.	Indian rubber plant	Leaf	Moraceae
Forsythia × *intermedia* Zabel	Golden bells	Leaf	Oleaceae
Gingko biloba L	Maidenhair tree	Leaf	Gingkoaceae
Hedera helix L.	Ivy	Leaf	Araliaceae
Hordeum vulgare	Barley	Leaf	Poacea
Juglans regia L.	English walnut	Leaf	Juglandaceae
Ligustrum vulgare L.	Prim, privet	Leaf	Oleaceae
Liriodendron tulipifera L.	Tulip-tree	Leaf	Magnoliaceae
Lycopersicon esculentum Mill.	Tomato	Fruit	Solanaceae

(*continued*)

Index of Plants (*continued*)

Botanical name	Common name	Organ	Family
Maianthemum bifolium (L.) F.W. Schmidt	May lily	Leaf	Convallariaceae
Malus domestica Borkh.	Apple	Leaf and fruit	Rosaceae
Monstera deliciosa Liebm.	Breadfruit-vine	Leaf	Araceae
Nerium oleander L.	Oleander	Leaf	Apocynaceae
Olea europaea L.	Olive	Leaf	Oleaceae
Philodendron selloum K. Koch	Philodendron	Leaf	Araceae
Picea abies (L.) Karst.	White spruce	Needle	Pinaceae
Prunus domestica L.	European plum	Leaf	Rosaceae
Prunus laurocerasus L.	Cherry laurel	Leaf	Rosaceae
Populus canescens (Aiton.) Sm.	Grey poplar	Leaf	Salicaceae
Pyrus communis L.	Pear	Leaf	Rosaceae
Schefflera actinophpylla (Endl.) Jaeger	Australian umbrella-tree	Leaf	Araliaceae
Solanum melongena L.	Eggplant	Fruit	Solanaceae
Stephanotis floribunda (R.Br.) Brongn.	Madagascar jasmine	Leaf	Asclepiadaceae
Syringa vulgaris L.	Lilac	Leaf	Oleaceae
Taxus baccata L.	English yew	Needle	Taxaceae
Vanilla planifolia Andr.	Vanilla	Leaf	Orchideaceae
Vinca major L.	Big periwinkle	Leaf	Apocynaceae

References

Abraham MH, McGowan JC (1987) The use of characteristic volumes to measure cavity terms in reversed phase liquid chromatography. Chromatographia, Springer, Heidelberg

Adamson AW (1976) Physical chemistry of surfaces. Wiley, New York

Albert A, Serjeant EP (1971) The determination of ionization constant. Chapman & Hall, London

Amelunxen F, Morgenroth K, Picksack T (1967) Untersuchungen an der Epidermis mit dem Steroscan-Elektronenmikroskop. Pflanzenphysiol 57:79–95

Andersen SO (1979) Biochemistry of insect cuticle. Annu Rev Entomol 24:29–59

Archer RJ, La Mer VK (1955) The rate of evaporation of water through fatty acid monolayers. J Phys Chem 59:200–208

Baker EA (1982) Chemistry and morphology of plant epicuticular waxes. In: Cutler DE, Alvin KL, Price CE (eds) The plant cuticle. Academic Press, London, pp 139–165

Baker EA, Procopiou J (1975) The cuticles of citrus species. Composition of the intramolecular lipids of leaves and fruits. J Sci Food Agric 26:1347–1352

Baker EA, Procopiou J, Hunt GM (1975) The cuticles of citrus species. Composition of leaf and fruit waxes. J Sci Food Agric 26:1093–1101

Baker EA, Bukovac MJ, Hunt GM (1982) Composition of tomato fruit cuticle as related to fruit growth and development. In: Cutler DF, Alvin KL, Price CE (eds) The plant cuticle. Academic Press, London, pp 33–44

Banderup J, Immergut EH (1966) Polymer handbook. Wiley, New York

Barrie JA (1968) Water in polymers. In: Crank J, Park GS (eds) Diffusion in polymers. Academic Press, London

Barrow GM (1961) Physical chemistry. McGraw-Hill, New York

Barthlott W (1990) Scanning electron microscopy of the epidermal surface of plants. In: Claugher D (ed) Scanning EM in taxonomy and functional morphology. Systematics Association's Special, Vol. 41. Clarendon, Oxford, pp 69–94

Barthlott W, Frölich D (1983) Mikromorphologie und Orientierungsmuster epicuticularer Wachs-Kristalloide: Ein neues systematisches Merkmal bei Monokotylen. Plant Syst Evol 142:171–185

Bauer L (1953) Zur Frage der Stoffbewegung in der Pflanze mit besonderer Berücksichtigung der Wanderung von Fluorochromen. Planta 42:347–451

Bauer H, Schönherr J (1992) Determination of mobilities of organic compounds in plant cuticles and correlation with molar volumes. Pestic Sci 35:1–11

Baur P (1997) Lognormal distribution of water permeability and organic solute mobility in plant cuticles. Plant Cell Environ 20:167–177

Baur P (1998) Mechanistic aspects of foliar penetration of agrochemicals and the effect of adjuvants. Recent Res Dev Agric Food Chem 2:809–837

Baur P (1999) Surfactant effects on cuticular penetration of neutral polar compounds: dependence on humidity and temperature. J Agric Food Chem 47:753–761

Baur P, Grayson T, Schönherr J (1996a) Concentration dependent mobility of chlorfenvinphos in isolated plant cuticles. Pestic Sci 47:171–180

Baur P, Marzouk H, Schönherr J, Bauer H (1996b) Mobilities of organic compounds in plant cuticles as affected by structure and molar volumes of chemicals and plant species. Planta 199:404–412

Baur P, Buchholz A, Schönherr J (1997a) Diffusion in plant cuticles as affected by temperature and size of organic solutes: similarity and diversity among species. Plant Cell Environ 20:982–994

Baur P, Marzouk H, Schönherr J, Grayson BT (1997b) Partition coefficients of active ingredients between plant cuticle and adjuvants as related to rates of foliar uptake. J Agric Food Chem 45:3659–3665

Baur P, Schönherr J, Grayson BT (1999) Polydisperse ethoxylated fatty alcohol surfactants as accelerators of cuticular penetration. 2: separation of effects on driving force and mobility and reversibility of surfactant action. Pestic Sci 55:831–842

Becker M, Kerstiens G, Schönherr J (1986) Water permeability of plant cuticles: permeance, diffusion and partition coefficients. Trees 1:54–60

Beyer M, Lau S, Knoche M (2005) Studies on water transport through the sweet cherry fruit surface: IX Comparing permeability in water uptake and transpiration. Planta 220:474–485

Bianchi G (1995) Plant waxes. In: Hamilton RJ (ed) Waxes: chemistry, molecular biology and functions. Oily Press, Dundee, pp 175–222

Bird SM, Gray JF (2003) Signals from the cuticle affect epidermal cell differentiation. New Phytol 157:9–23

Buchholz A (1998) Größenselektivität und Temperaturabhängigkeit der Diffusion organischer Verbindungen in der pflanzenlichen Kutikula. Doctoral Dissertation, Kiel University, Germany

Buchholz A (2006) Characterization of the diffusion of non-electrolytes across plant cuticles: properties of the lipophilic pathway. J Exp Bot 57:2501–2513

Buchholz A, Schönherr J (2000) Thermodynamic analysis of diffusion of non-electrolytes across plant cuticles: the effects of waxes and plasticisers on the lipophilic pathway. Planta 212:103–111

Buchholz A, Baur P, Schönherr J (1998) Differences among plant species in cuticular permeabilities and solute mobilities are not caused by differential size selectivities. Planta 206:322–328

Budavary S (ed) (1989) Merck Index, 11th edn. Merck, Rahway NJ USA

Burghardt M, Riederer M (2003) Ecophysiological relevance of cuticular transpiration of deciduous and evergreen plants in relation to stomatal closure and leaf water potential. J Exp Bot 54:1941–1949

Burghardt M, Schreiber L, Riederer M (1998) Enhancement of the diffusion of active ingredients in barley leaf cuticular wax by monodisperse alcohol ethoxylates. J Agric Food Chem 46:1593–1602

Burghardt M, Friedmann A, Schreiber L, Riederer M (2005) Comparison of sorption and diffusion by pyridate and its polar metabolite in isolated cuticular wax of *Chenopodium album* and *Hordeum vulgare*. J Agric Food Chem 53:7150–7155

Burghardt M, Friedmann A, Schreiber L, Riederer M (2006) Modelling the effects of alcohol ethoxylates on diffusion of pesticides in the cuticular wax of *Chenopodium album* leaves. Pestic Manage Sci 62:137–147

Chamel A, Pineri M, Escoubes M (1991) Quantitative determination of water sorption by plant cuticles. Plant Cell Environ 14:87–95

Crank J (1975) The Mathematics of diffusion. Clarendon, Oxford

Crank J, Park GS (1968) Diffusion in polymers. Academic Press, London

Cussler EL (1984) Diffusion, mass transfer in fluid systems. Cambridge University Press, Cambridge

Dominguez E, Heredia A (1999) Water hydration in cutinized cell walls: a physico-chemical analysis. Biochim Biophys Acta 1426:168–176

Doty PM, Aiken WH, Mark H (1946) Temperature dependence of water vapor permeability. Ind Eng Chem 38:788–791

Dunn III WH, Grigoras S, Johansson E (1986) Principal component analysis of partition coefficient data: an approach to a new method of partition coefficient estimation. In: Dunn III WJ, Block JH, Pearlman RS (eds) Partition coefficient: determination and estimation. Pergamon, Oxford, pp 21–36

Durbin RP (1960) Osmotic flow of water across permeable cellulose membranes. J Gen Physiol 44:315–326

Eckl K, Gruler H (1980) Phase transitions in plant cuticles. Planta 150:102–113

Edelmann HG, Neinhuis C, Bargel H (2005) Influence of hydration and temperature on the rheological properties of plant cuticles and their impact on plant organ integrity. J Plant Growth Regul 24 (2):116–126

Ensikat HJ, Neinhuis C, Barthlott W. (2000) Direct access to plant epicuticular wax crystals by a new mechanical isolation method. Int J Plant Sci 161 (1):143–148

Falbe J, Regitz M (eds) (1995) Römpp Chemie Lexikon. Georg Thieme, Stuttgart

Fang X, Qiu F, Yan B, Wang H, Mort AJ, Stark RE (2001) NMR studies of molecular structure in fruit cuticle lyesters. Phytochemistry 57:1035–1042

Fernández V, Ebert G (2003) Regreening of Fe-deficient plants: a review on the physiological background and the chemistry of the applied solutions. Curr Top Plant Biol 4:80–92

Fernández V, Ebert G (2005) Foliar iron fertilisation — a critical review. J Plant Nutr 28:2113–2124

Fowler JD (1999) A discussion of dependence of cuticular mobility on molecular volume. In: Tann RS, Nalejava JD, Viets A (eds) 19th symposium on pesticide formulations and application systems: global pest control formulations for the next millennium. ASTM STP 1373. American Society for Testing Materials, West Conshohocken, PA, USA

Fox RC (1958) The relationship of wax crystal structure to water vapor transmission rate of wax films. TAPPI 41:283–289

Franke W (1960) Über die Beziehungen der Ektodermen zur Stoffaufnahme durch die Blätter. II Beobachtungen an Helxine soleirolli Req. Planta 55:533–541

Franke W (1967) Mechanism of foliar penetration of solutions. Annu Rev Plant Physiol 18:281–301

Fujita B, Iwasa J, Hansch C (1964) A new substituent constant, π, derived from partition coefficients. J Am Chem Soc 86:5175–5180

Gächter R, Müller H (eds) (1990) Plastics additives handbook. Hanser, München

Geyer U, Schönherr J (1990) The effect of environment on permeability and composition of Citrus leaf cuticles: 1 Water permeability of isolated cuticular membranes. Planta 180:147–153

Glasstone S, Laidler K, Eyring H (1941) The theory of rate processes. McGraw-Hill, New York

Glover BJ, Martin C (2000) Specification of epidermal cell morphology. Adv Bot Res 31:193–217

Graca J, Schreiber L, Rodrigues J, Pereira H (2002) Glycerol and glyceryl esters of omega-hydroxyacids in cutins. Phytochemistry 61 (2):205–215

Greenspan (1977) Humidity fixed points of binary saturated aqueous solutions. J Res Natl Bur Stand (US) A Phys Chem 81A (1):89–96

Gregg SJ, Sing KSW (1982) Adsorption, surface area and porosity. Academic Press, London

Grncarevic M, Radler F (1967) The effect of wax components on cuticular transpiration-model experiments. Planta 75:23–27

Guhling O, Hobl B, Yeats T, Jetter R (2006). Cloning and characterization of a lupeol synthase involved in the synthesis of epicuticular wax crystals on stem and hypocotyl surfaces of Ricinus communis. Arch Biochem Biophys 448 (1–2):60–72

Haas K, Rentschler I (1984) Discrimination between epicuticular and intracuticular wax in blackberry leaves: ultrastructural and chemical evidence. Plant Sci Lett 36:143–147

Haas K, Schönherr J (1979) Composition of soluble cuticular lipids and water permeability of cuticular membranes from *Citrus* leaves. Planta 146:399–403

Hansch C, Leo A (1979) Substituent constants for correlation analysis in chemistry and biology. Wiley, New York

Hartley GS, Graham-Bryce IJ (1980) Physical principles of pesticide behaviour. Vols. 1 and 2. Academic Press, London

Hauke V, Schreiber L (1998) Ontogenetic and seasonal development of wax composition and cuticular transpiration of ivy (Hedera helix L.) sun and shade leaves. Planta 207 (1):67–75

Hauser PM, McLaren AD (1948) Permeation through and sorption of water vapour by high polymers. Ind Eng Chem 40:112–117

Helfferich F (1962) Ion exchange. McGraw Hill, New York

Holloway PJ (1970) Surface factors affecting the wetting of leaves. Pestic Sci 1:156–163

Holloway PJ (1974) Intracuticular lipids of spinach leaves. Phytochemistry 13:2201–2207

Holloway PJ (1982a) Structure and histochemistry of plant cuticular membranes. In: Cutler DF, Alvin KL, Price CE (eds) The plant cuticle. Academic Press, London, pp 1–32

Holloway PJ (1982b) The chemical constitution of plant cutins. In: Cutler DF, Alvin KL, Price CE (eds) The plant cuticle. Academic Press, London, pp 45–85

Holloway PJ, Brown GA, Wattendorff J (1981) Ultrahistochemical detection of epoxides in plant cuticular membranes. J Exp Bot 32:1051–1066

Howe PG, Kitchner JA (1955) fundamental properties of cross-linked polymethacrylic acid ion exchanger resins. J Chem Soc 1955:2143–2151

Hunt GM, Baker EA (1980) Phenolic constituents of tomato fruit cuticles. Phytochemistry 19:1415–1419

Israelachvili JN (1991) Intermolecular and surface forces. Academic Press, London

Jeffree CE (2006) The fine structure of the plant cuticle. Annual Plant Reviews, Vol. 23. Blackwell, Oxford, pp 11–125

Jenks MA, Rashotte AM, Tuttle HA, Feldmann KA (1996) Mutants in *Arabidopsis thaliana* altered in epicuticular wax and leaf morphology. Plant Physiol 110:377–385

Jetter R, Schaffer S, Riederer M (2000) Leaf cuticular waxes are arranged in chemically and mechanically distinct layers: evidence from Prunus laurocerasus L. Plant Cell Environ 23:619–628

Jetter R, Kunst L, Samuels L (2006) Composition of plant cuticular waxes. Annual Plant Reviews, Vol. 23, Blackwell, Oxford, pp 145–181

Kamp H (1930) Untersuchungen über Kutikularbau und kutikuläre Transpiration von Blättern. Jahrb wiss Bot 72:403–465

Kedem O, Katchalsky A (1961) A physical interpretation of the phenomenological coefficients of membrane permeability. J Gen Physiol 45:143–179

Kerler F, Schönherr J (1988a) Accumulation of lipophilic chemicals in plant cuticles: prediction from octanol/water partition coefficient. Arch Environ Contamin Toxicol 17:1–6

Kerler F, Schönherr J (1988b) Permeation of lipophilic chemicals across plant cuticles: prediction from partition coefficients and molar volumes. Arch Environ Contamin Toxicol 17:7–12

Kerstiens G (1996a) Diffusion of water vapour and gases across cuticles and through stomatal pores presumed to be closed. In: Kerstiens G (ed.) Plant cuticles- an integrated functional approach. BIOS Scientific, Oxford, pp 121–134

Kerstiens G (1996b) Cuticular water permeability and its physiological significance. J Exp Bot 47:1813–1832

Kirkwood RC (1993) Use and mode of action of adjuvants for herbicides: a review of some current work. Pestic Sci 38:93–102

Kirkwood RC (1999) Recent developments in our understanding of the plant cuticle as a barrier to the foliar uptake of pesticides. Pestic Sci 55:69–77

Kirsch T, Kaffarnik F, Riederer M, Schreiber L (1997) Cuticular permeability of the three tree species *Prunus laurocerasus* L., *Ginkgo biloba* L. and *Juglans regia* L.: comparative investigation of transport properties of intact leaves, isolated cuticles and reconstituted cuticular waxes. J Exp Bot 48:1035–1045

Kishimoto A, Maekawa E, Fujita H (1960) Diffusion coefficients for amorphous polymer and water systems. Bull Chem Soc Jpn 33:988–992

Koch K, Neinhuis C, Ensikat H-J, Barthlott W (2004) Self assembly of epicuticular waxes on living plant surfaces imaged by atomic force microscopy (AMF). J Exp Bot 55:711–719

Kolattukudy P (2004) Cutin from Plants. In: Y Doi, A Steinbüchel (eds) Polyesters I: biological systems and biotechnological production, Wiley-VCH, Weinheim, pp 1–40

Kolthoff IM, Sandell EB, Bruckenstein S (1969) Quantitative chemical analysis. MacMillan, London

Korenman YI, Nefedova TA, Taldykina SN (1977) Extraction of nitrophenols from aqueous solutions at different temperatures. Russ J Phys Chem 51:730–732

Kunst L, Samuels AL (2003) Biosynthesis and secretion of plant cuticular wax. Prog Lipid Res 42 (1):51–80

Lakshminarayanaiah N (1969) Transport phenomena in membranes. Academic Press, New York

LaMer VK, Healy TW, Aylsmore LAG (1964) The transport of water through monolayers of long-chain paraffinic alcohols. J Colloid Sci 19:673–684

Larcher W (1995) Physiological plant ecology, 3rd edn. Springer, Berlin

Le Roux JH, Loubser NH (1980) Nuclear magnetic resonance investigation of the mobile phase in paraffinic Fischer-Tropsch waxes. SA J Sci 76:157–161

Lendzian KJ, Kerstiens G (1991) Sorption and transport of gases and vapors in plant cuticles. Rev Environ Contamin Toxicol 121:65–128

Lendzian KJ, Schönherr J (1983) In-vivo study of cutin synthesis in leaves of *Clivia miniata* Reg. Planta 158:70–75

Leo A, Hansch C, Elkins D (1971) Partition coefficients and their uses. Chem Rev 71:525–616

Longsworth LG (1953) Diffusion measurements, at 25°C, of aqueous solutions of amino acids, peptides and sugars. J Am Chem Soc 75:5705–5709

Marga F, Pesacreta TC, Hasenstein KH (2001) Biochemical analysis of elastic and rigid cuticles of *Cirsium horridulum*. Planta 213:841–848

Marzouk H, Baur P, Schönherr J (1998) Relative solubilities of bifenox and naphthylacetic acid (NAA) in plant cuticles and in selected pure or aqueous glycol adjuvants. Pestic Sci 53: 278–284

McFairlane JC, Berry WL (1974) Cation penetration through isolated cuticles. Plant Physiol 53:723–727

McGowan JC, Mellors A (1986) Molecular volumes in Chemistry and biology: applications including partitioning and toxicity. Ellis Horwood, Chichester

Mérida T, Schönherr J, Schmidt HW (1981) Fine structure of plant cuticles in relation to water permeability: the fine structure of the cuticle of *Clivia miniata* Reg. leaves. Planta 152: 259–267

Meier-Maercker U (1979) Peristomatal transpiration and stomatal movement: III. Visible effects of peristomatal transpiration on the epidermis. Zeitschr Pflanzenphysiol 91:225–238

Meyer M (1938) Die submikroskopische Struktur der kutinisierten Zellmembranen. Protoplasma 29:552–586

Nawrath C (2006) Unraveling the complex network of cuticular structure and function. Curr Opin Biol 9:1–7

Nevis AH (1958) Water transport in invertebrate peripheral nerve fibre. J Gen Physiol 41:927–958

Niederl S, Kirsch T, Riederer M, Schreiber L (1998) Co-permeability of ^{3}H-labelled water and ^{14}C-labelled organic acids across isolated plant cuticles: investigating cuticular paths of diffusion and predicting cuticular transpiration. Plant Physiol 116:117–123

Nobel PS (1983) Biophysical plant physiology. Freeman, San Francisco, USA

Orgell WH (1955) Isolation of plant cuticle with pectic enzymes. Plant Physiol 30:78–80

Park GS (1968) The glassy state and slow process anomalies. In: Crank J, Park GS (eds) Diffusion in polymers, Academic Press, London

Popp C, Burghardt M, Friedmann A, Riederer M (2005) Characterization of hydrophilic and lipophilic pathways of *Hedera helix* L. cuticular membranes: permeation of water and uncharged organic compounds. J Exp Bot 56:2797–2806

Potts RO, Francoeur ML (1991) The influence of Stratum Corneum morphology on water permeability. J Investig Dermatol 96:495–499

Potts RO, Guy RH (1992) Predicting skin permeability. Pharma Res 9:663–669

Rawlings AV (1995) Skin waxes: their composition, properties, structures and biological significance In: Hamilton RJ (ed) Waxes: chemistry, molecular biology and functions. Oily Press, Dundee, pp 131–156

Renkin EM (1954) Filtration, diffusion and molecular sieving through porous cellulose membranes. J Gen Physiol 38:225–243

Reynhardt EC, Riederer M (1991) Structures and molecular dynamics of the cuticular wax from leaves of *Citrus aurantium* L. J Phys D Appl Phys 24:478–486

Reynhardt EC, Riederer M (1994) Structures and molecular dynamics of plant waxes II. Cuticular waxes from leaves of *Fagus sylvatica* L. and *Hordeum vulgare* L. Eur Biophys J 23:59–70

Riederer M (2006) Thermodynamics of the water permeability of plant cuticles; characterization of the polar pathway. J Exp Bot 57:2937–2942

Riederer M, Markstädter C (1996) Cuticular waxes: a critical assessment of current knowledge, In: Kerstiens G (ed) Plant cuticles-An integrated functional approach. BIOS Scientific, Oxford, pp 189–200

Riederer M, Schneider G (1989) Comparative study of the composition of waxes extracted from isolated leaf cuticles and from whole leaves of Citrus: evidence for selective extraction. Physiol Plant 77:373–384

Riederer M, Schneider G (1990) The effect of the environment on the permeability and composition of citrus leaf cuticles. II. Composition of soluble cuticular lipids and correlation with transport properties. Planta 180:154–165

Riederer M, Schönherr J (1984) Accumulation and transport of (2,4-dichloro phenoxy)acetic acid in plant cuticles: I. Sorption in the cuticular membrane and its components. Ecotoxicol Environ Safe 8:236–247

Riederer M, Schönherr J (1985) Accumulation and transport of (2,4-Dichlorophenoxy)acetic acid in plant cuticles: 2. Permeability of the cuticular membrane. Ecotoxicol Environ Safe 9:196–208

Riederer M, Schönherr J (1986a) Thermodynamic analysis of non-electrolyte sorption in plant cuticles: the effect of concentration and temperature on sorption of 4-nitrophenol. Planta 169:69–80

Riederer M, Schönherr J (1986b) Covalent binding of chlorophenoxyacetic acids to plant cuticles. Arch Environ Contamin Toxicol 15:97–105

Riederer M, Schönherr J (1988) Development of plant cuticles: fine structure and cutin composition of *Clivia miniata* Reg. leaves. Planta 174:127–138

Riederer M, Schönherr J (1990) Effects of surfactants on water permeability of isolated plant cuticles and on the composition of their cuticular waxes. Pestic Sci 29:85–94

Riederer M, Schreiber L (1995) Waxes – the transport barriers of plant cuticles. In: Hamilton RJ (ed) Waxes: chemistry, molecular biology and functions. Oily Press, Dundee

Riederer M, Schreiber L (2001) Protecting against water loss: analysis of barrier properties of plant cuticles. J Exp Bot 52:2023–22032

Riederer M, Burghardt M, Mayer S, Obermeier H, Schönherr J (1995) Sorption of monodisperse alcohol ethoxylates and their effects on the mobililty of 2,4-D in isolated plant cuticles. J Agric Food Chem 43:1067–1075

Robbins E, Mauro A (1960) Experimental study of the independence of diffusion and hydrodynamic permeability coefficients in collodium membranes. J Gen Physiol 43:523–532

Roelofsen PA (1952) On the submicroscopic structure of cuticular cell walls. Acta Bot Neer 1:99–114

Rosenberg G, Brotz W (1957) Beitrag zur Messung der Wärmeausdehnung von Wachsen. Fette Seifen Anstrichm 59:589–594

Round AN, Yan B, Dang S, Estephan R, Stark RE, Batteas JD (2000) The influence of water on the nanomechanical behavior of the plant biopolyester cutin as studied by AFM and solid-state NMR. Biophys J 79:2761–2767

Rust G, Herrero F (1969) Untersuchung der Wasserdampfpermeation durch Folien. Materialprüf 11:165–169

Sabljic A, Güsten H, Schönherr J, Riederer M (1990) Modeling plant uptake of airborne organic chemicals. 1. Plant cuticle/water partitioning and molecular connectivity. Environ Sci Technol 24:1321–1326

Sackmann E, Träuble H, Galla HJ, Overath P (1973) Lateral diffusion, protein mobility, and phase transitions in *Escherichia coli* membranes. A spin label study. Biochemistry 12:5360–5369

Sangster J (1997) Octanol-water partition coefficients: fundamentals and physical chemistry. Wiley, Chichester

Santos S, Schreiber L, Graça J (2007) Cuticular waxes from ivy leaves (*Hedera helix* L.): analysis of high-molecular weight esters. Phytochem Anal 18:60–69

Santrucek J, Simanova E, Karbulkova J, Simkova M, Schreiber L (2004) A new technique for measurement of water permeability of stomatous cuticular membranes isolated from Hedera helix leaves. J Exp Bot 55:1411–1422

Schatzberg P (1965) Diffusion of water in hydrocarbon liquids. J Polym Sci C 10:87–92

Schick MJ (1987) Nonionic surfactants: physical chemistry. Marcel Dekker, New York

Schlegel TK, Schönherr J (2002) Selective permeability of cuticles over stomata and trichomes to calcium chloride. Acta Hort 549:91–96

Schlegel TK, Schönherr J, Schreiber L (2005) Size selectivity of aqueous pores in stomatous cuticles of *Vicia faba* leaves. Planta 221:648–655

Schlegel TK, Schönherr J, Schreiber L (2006) Rates of foliar penetration of chelated Fe(III): role of light, stomata, species and leaf age. J Agric Food Chem 54:6809–6813

Schlotter NE, Furlan PY (1992) A review of small molecule diffusion in polyolefins. Polymers 33:3323–3342

Schmidt HW, Schönherr J (1982) Development of plant cuticles: occurence and role of non-ester bonds in cutin of *Clivia miniata* Reg. leaves. Planta 156:380–384

Schmidt HW, Mérida T, Schönherr J (1981) Water permeability and fine structure of cuticular membranes isolated enzymatically from leaves of *Clivia miniata* Reg. Z Pflanzenphysiol 105:41–51

Schönherr J (1969) Foliar penetration and translocation of succinic acid 2,2-dimethylhydrozide (SADH). MS Thesis, East Lansing, Michigan State University, USA

Schönherr J (1976a) Water permeability of isolated cuticular membranes: the effect of pH and cations on diffusion, hydrodynamic permeability and size of polar pores in the cutin matrix. Planta 128:113–126

Schönherr J (1976b) Water permeability of isolated cuticular membranes: the effect of cuticular waxes on diffusion of water. Planta 131:159–164

Schönherr J (1993a) Effects of monodisperse alcoholethoxylates on mobility of 2,4-D in isolated cuticles. Pestic Sci 38:155–164

Schönherr J (1993b) Effects of alcohols, glycols and monodisperse ethoxylated alcohols on mobility of 2,4-D in isolated plant cuticles. Pestic Sci 39:213–223

Schönherr J (2000) Calcium chloride penetrates pear leaf cuticles via aqueous pores. Planta 212:112–118

Schönherr J (2001) Cuticular penetration of calcium salts: effects of humidity, anion and adjuvants. J Plant Nutr Soil Sci 164:225–231

Schönherr J (2002) A mechanistic analysis of penetration of glyphosate salts across astomatous cuticular membranes. Pestic Manage Sci 58:343–351

Schönherr J (2006) Characterization of aqueous pores in plant cuticles and permeation of ionic solutes. J Exp Bot 57:2471–2491

Schönherr J, Bauer H (1992) Analysis of effects of surfactants on permeability of plant cuticles. In: Foy CL (ed) Adjuvants for agrochemicals. CRC, Tokyo, pp 17–35

Schönherr J, Baur P (1994) Modelling penetration of plant cuticles by crop protection agents and effects of adjuvants on their rates of penetration. Pestic Sci 42:185–208

Schönherr J, Baur P (1996) The effects of temperature, surfactants and other adjuvants on rates of uptake of organic compounds. In: Kerstiens G (ed) Plant cuticles- an integrated functional approach. BIOS Scientific, Oxford, pp 134–154

Schönherr J, Bukovac MJ (1970a) Preferential polar pathways in cuticles and their relationship to ectodesmata. Planta 92:189–201
Schönherr J, Bukovac MJ (1970b) The nature of precipitates formed in the outer cell wall following fixation of leaf tissue with Gilson solution. Planta 92:202–207
Schönherr J, Bukovac MJ (1972a) Penetration of stomata by liquids. Dependence on surface tension, wettability and stomatal morphology. Plant Physiol 49:813–819
Schönherr J, Bukovac MJ (1972b) Dissociation constants of succinic acid-2,2-dimethyl hydrazide. Agric Food Chem 20:1263–1265
Schönherr J, Bukovac MJ (1973) Ion exchange properties of isolated tomato fruit cuticular membrane: exchange capacity, nature of fixed charges and cation selectivity. Planta 109:73–93
Schönherr J, Bukovac MJ (1978) Foliar penetration of succinic acid-2,2-dimethyl hydrazide: mechanism and rate limiting step. Physiol Plant 42:243–251
Schönherr J, Huber R (1977) Plant cuticles are polyelelctrolytes with isoelectric points around three. Plant Physiol 59:145–150
Schönherr J, Lendzian K (1981) A simple and inexpensive method of measuring water permeability of isolated plant cuticular membranes. Z Pflanzenphysiol 102:321–327
Schönherr J, Mérida T (1981) Water permeability of plant cuticular membranes: the effects of humidity and temperature on permeability of non-isolated cuticles of onion bulb scales. Plant Cell Environ 4:349–354
Schönherr J, Riederer M (1986) Plant cuticles sorb lipophilic compounds during enzymatic isolation. Plant Cell Environ 9:459–466
Schönherr J, Riederer M (1988) Desorption of chemicals from plant cuticles: evidence for asymmetry. Arch Environ Contamin Toxicol 17:13–19
Schönherr J, Riederer M (1989) Foliar penetration and accumulation of organic chemicals in plant cuticles. Rev Environ Contamin Toxicol 108:1–70
Schönherr J, Schmidt HW (1979) Water permeability of plant cuticles: dependence of permeability coefficients of cuticular transpiration on vapor pressure saturation deficit. Planta 144:391–400
Schönherr J, Schreiber L (2004a) Interactions of calcium ions with weakly acidic active ingredients slow cuticular penetration: a case study with glyphosate. J Agric Food Chem 52:6546–6551
Schönherr J, Schreiber L (2004b) Size selectivity of aqueous pores in astomatous cuticular membranes isolated from *Populus canescens* (Aiton) Sm. leaves. Planta 219:405–411
Schönherr J, Ziegler H (1980) Water permeability of *Betula* periderm. Planta 147:345–354
Schönherr J, Eckl K, Gruler H (1979) Water permeability of plant cuticles: the effect of temperature on diffusion of water. Planta 147:21–26
Schönherr J, Schreiber L, Buchholz A (2001) Effects of temperature and accelerator concentration on mobility of non-electrolytes in plant cuticles. Pest Manage Sci 57:17–24
Schönherr J, Fernández V, Schreiber L (2005) Rates of cuticular penetration of chelated Fe III: role of humidity, concentration, adjuvants, temperature and type of chelate. J Agric Food Chem 53:4484–4492
Schreiber L (1995) A mechanistic approach towards surfactant/wax interactions: effects of octaethyleneglycol monododecylether on sorption and diffusion of organic chemicals in reconstituted cuticular wax of barley leaves. Pestic Sci 45:1–11
Schreiber L (2001) Effect of temperature on cuticular transpiration of isolated cuticular membranes and leaf discs. J Exp Bot 362:1893–1900
Schreiber L (2005) Polar paths of diffusion across plant cuticles: new evidence for an old hypothesis. Ann Bot 95:1069–1073
Schreiber L, Riederer M (1996a) Ecophysiology of cuticular transpiration: comparative investigation of cuticular water permeability of plant species from different habitats. Oecologia 107:426–432
Schreiber L, Riederer M (1996b) Determination of diffusion coefficients of octadecanoic acid in isolated cuticular waxes and their relationship to cuticular water permeabilities. Plant Cell Environ 19:1075–1082
Schreiber L, Schönherr J (1990) Phase transitions and thermal expansion coefficients of plant cuticles. The effects of temperature on structure and function. Planta 182:186–193

Schreiber L, Schönherr J (1992a) Analysis of foliar uptake of pesticides in barley leaves: role of epicuticular waxes and compartmentation. Pestic Sci 36:213–221

Schreiber L, Schönherr J (1992b) Uptake of organic chemicals in conifer needles: surface adsorption and permeability of cuticles. Environ Sci Technol 26:153–159

Schreiber L, Schönherr J (1992c) Leaf surface microflora may significantly affect studies on foliar uptake of chemicals. Bot Acta 105:345–347

Schreiber L, Schönherr J (1993a) Uptake of two chlorinated chemicals in conifer needles: reversibility and compartmental analysis. New Phytol 123:547–554

Schreiber L, Schönherr J (1993b) Mobilities of organic compounds in reconstituted cuticular wax of barley leaves: determination of diffusion coefficients. Pestic Sci 38:353–361

Schreiber L, Breiner H-W, Riederer M, Düggelin M, Guggenheim R (1994) The casparian strip of Clivia miniata Reg. roots: isolation, fine structure and chemical nature. Bot Acta 107:353–361

Schreiber L, Bach S, Kirsch T, Knoll D, Schalz K, Riederer M (1995) A simple photometric device analysing cuticular transport physiology: surfactant effect on permeability of isolated cuticular membranes of *Prunus laurocerasus* L. J Exp Bot 46:1915–1921

Schreiber L, Kirsch T, Riederer M (1996a) Transport properties of cuticular waxes of *Fagus sylvatica* L. and *Picea abies* (L.) Karst.: estimation of size selectivity and tortuosity from diffusion coefficients of aliphatic molecules. Planta 198:104–109

Schreiber L, Riederer M, Schorn K (1996b) Mobilities of organic compounds in reconstituted cuticular wax of barley leaves: effects of monodisperse alcoholethoxylates on diffusion of pentachlorophenol and tetracosanoic acid. Pestic Sci 48:117–124

Schreiber L, Schorn K, Heimburg T (1997) ^{2}H NMR study of cuticular wax isolated from *Hordeum* vulgare L. leaves: identification of amorphous and crystalline wax phases. Eur Biophys J 26:371–380

Schreiber L, Skrabs M, Hartmann K, Diamantopoulos P, Simanova E, Santrucek J (2001) Effect of humidity on cuticular water permeability of isolated cuticular membranes and leaf disks. Planta 214:274–282

Schreiber L, Elshatshat S, Koch K, Lin J, Santrucek J (2006) AgCl precipitates in cuticular membranes reduce rates of cuticular transpiration. Planta 223:283–290

Schwindt DA, Wilhelm KP, Maibach HI (1998) Water diffusion characteristics of human stratum corneum at different anatomical sites *in vivo*. J Investig Dermatol 111:385–389

Scott EW, Abbott JC, Trosset S (1995) Properties of paper: an introduction. TAPPI, Atlanta GA, USA

Shi T, Schönherr J, Schreiber L (2005a) Accelerators increase permeability of cuticles for the lipophilic solutes metribuzin and iprovalicarb but not for hydrophilic methyl glucose. J Agric Food Chem 53:2609–2615

Shi T, Simanova E, Schönherr J, Schreiber L (2005b) Effects of accelerators on mobility of ^{14}C-2,4-dichlorophenoxy butyric acid in plant cuticles depends on type and concentration of accelerator. J Agric Food Chem 53:2207–2212

Silva Fernandes AMS, Baker EA, Martin JT (1964) Studies on the plant cuticle. VI. The isolation and fractionation of cuticular waxes. Ann Appl Biol 53:43–58

Simanova E, Shi T, Schönherr J, Schreiber L (2005) Sorption in reconstituted waxes of homologous series of alcohol ethoxylates and n-alkyl esters and their effects on the mobility of 2,4-dichlorophenoxybutyric acid. Pest Manage Sci 61:383–389

Sitte P, Rennier R (1963) Untersuchungen an cuticularen Zellwandschichten. Planta 60:19–40

Slavik B (1974) Methods of studying plant water relations. Ecological Studies 9. Springer, Berlin

Solomon AK (1968) Characterization of biological membranes by equivalent pores. J Gen Physiol 51:335S–346S

Stannett V, Williams JL (1965) The permeability of poly(ethyl methacrylate) to gases and water vapor. J Polym Sci C 10:45–59

Stein WD (1986) Transport and diffusion across cell membranes. Academic Press, London

Strugger S (1939) Luminiszenzmikroskopische Analyse des Transpirationsstromes in Parenchymen. Biologisches Zentralblatt 50:409–442

Tadros TF (1987) Interactions at interfaces and effects on transfer and performance. Asp Appl Biol 14:1–22
Tamura H, Knoche M, Hayashi Y, Bukovac MJ (2001a) Selective solubilization of tomato fruit epicuticular wax constituents by Triton X-100 surfactant. J Pestic Sci 26:16–20
Tamura H, Knoche M, Bukovac MJ (2001b) Evidence for surfactant solubilization of plant epicuticular wax. J Agric Food Chem 49:1809–1816
Ticknor LB (1958) On the permeation of cellulose membranes by diffusion. J Phys Chem 62:1483–1485
Tomlin CDS (ed) (1997) The pesticide manual. The British Crop Protection Council, Farnham, UK
Tuckey HB (1970) The leaching of substances from plaants. Annu Rev Plant Physiol 21:305–324
Vieth WR (1991) Diffusion in and through polymers: principles and applications. Hanser, München
Villena JF, Dominguez E, Stewart D, Heredia A (1999) Characterization and biosynthesis of non-degradable polymers in plant cuticles. Planta 208:181–187
Walton TJ (1990) Waxes, cutin and suberin. Methods Plant Biochem 4:105–158
Wattendorff J, Holloway PJ (1980) Studies on the ultrastructure and histochemistry of plant cuticles: the cuticular membrane of *Agave americana* L. in situ. Ann Bot 46:13–28
Wattendorff J, Holloway PJ (1984) Periclinal penetration of potassium permanganate into mature cuticular membranes of Agave and Clivia leaves: a new implication for plant cuticle development. Planta 161:1–11
Wellons JD, Stannett V (1966) Permeation, sorption and diffusion of water in ethyl cellulose. J Polym Sci A 4:593–602
Wolosin JM, Ginsburg H (1975) The permeation of organic acids through lecithin bilayers resemblance to diffusion in polymers. Biochim Biophys Acta 389:20–33
Yamada Y, Wittwer SH, Bukovac MJ (1964) Penetration of ions through isolated cuticles. Plant Physiol 39:28–32
Yasuda H, Stannett V (1962) Permeation, solution, and diffusion of water in some high polymers. J Polym Sci 57:907–923

Index